Eukaryotic DNA Replication

The Practical Approach Series

SERIES EDITOR

B. D. HAMES
Department of Biochemistry and Molecular Biology
University of Leeds, Leeds LS2 9JT, UK

See also the Practical Approach web site at **http://www.oup.co.uk/PAS**

★ indicates new and forthcoming titles

Affinity Chromatography
Affinity Separations
Anaerobic Microbiology
Animal Cell Culture (2nd edition)
Animal Virus Pathogenesis
Antibodies I and II
Antibody Engineering
★ Antisense Technology
Applied Microbial Physiology
Basic Cell Culture
Behavioural Neuroscience
Bioenergetics
Biological Data Analysis
Biomechanics - Materials
Biomechanics—Structures and Systems
Biosensors
Carbohydrate Analysis (2nd edition)
Cell-Cell Interactions
The Cell Cycle
Cell Growth and Apoptosis
★ Cell Separation
Cellular Calcium
Cellular Interactions in Development
Cellular Neurobiology
★ Chromatin
★ Chromosome Structural Analysis
Clinical Immunology
Complement
★ Crystallization of Nucleic Acids and Proteins (2nd edition)
Cytokines (2nd edition)
The Cytoskeleton
Diagnostic Molecular Pathology I and II
DNA and Protein Sequence Analysis
DNA Cloning 1: Core Techniques (2nd edition)
DNA Cloning 2: Expression Systems (2nd edition)
DNA Cloning 3: Complex Genomes (2nd edition)
DNA Cloning 4: Mammalian Systems (2nd edition)

★ Drosophila (2nd edition)
Electron Microscopy in Biology
Electron Microscopy in Molecular Biology
Electrophysiology
Enzyme Assays
Epithelial Cell Culture
Essential Developmental Biology
Essential Molecular Biology I and I
★ Eukaryotic DNA Replication
Experimental Neuroanatomy
Extracellular Matrix
Flow Cytometry (2nd edition)
Free Radicals
Gas Chromatography
Gel Electrophoresis of Nucleic Acids (2nd edition)
★ Gel Electrophoresis of Proteins (3rd edition)
Gene Probes 1 and 2
Gene Targeting
Gene Transcription
★ Genome Mapping
Glycobiology
★ Growth Factors and Receptors
Haemopoiesis
Histocompatibility Testing
HIV Volumes 1 and 2
★ HPLC of Macromolecules (2nd edition)
Human Cytogenetics I and II (2nd edition)
Human Genetic Disease Analysis
★ Immobilized Biomolecules in Analysis
Immunochemistry 1
Immunochemistry 2
Immunocytochemistry
★ *In Situ* Hybridization (2nd edition)
Iodinated Density Gradient Media
Ion Channels
★ Light Microscopy (2nd edition)
Lipid Modification of
Lipoprotein Analysis
Liposomes
Mammalian Cell Biotechnology
Medical Parasitology
Medical Virology
MHC Volumes 1 and 2
★ Molecular Genetic Analysis of Populations (2nd edition)
Molecular Genetics of Yeast
Molecular Imaging in Neuroscience
Molecular Neurobiology
Molecular Plant Pathology I and II
Molecular Virology
Monitoring Neuronal Activity
Mutagenicity Testing
★ Mutation Detection
Neural Cell Culture
Neural Transplantation
Neurochemistry (2nd edition)
Neuronal Cell Lines
NMR of Biological Macromolecules
Non-isotopic Methods in Molecular Biology

Nucleic Acid Hybridisation
Oligonucleotides and Analogues
Oligonucleotide Synthesis
PCR 1
PCR 2
★ PCR 3: PCR In Situ Hybridization
Peptide Antigens
Photosynthesis: Energy Transduction
Plant Cell Biology
Plant Cell Culture (2nd edition)
Plant Molecular Biology
Plasmids (2nd edition)
Platelets
Postimplantation Mammalian Embryos
Preparative Centrifugation
Protein Blotting
★ Protein Expression Vol 1
★ Protein Expression Vol 2
Protein Engineering
Protein Function (2nd edition)
Protein Phosphorylation
Protein Purification Applications
Protein Purification Methods
Protein Sequencing
Protein Structure (2nd edition)
Protein Structure Prediction
Protein Targeting
Proteolytic Enzymes
Pulsed Field Gel Electrophoresis
RNA Processing I and II
★ RNA-Protein Interactions
Signalling by Inositides
Subcellular Fractionation
Signal Transduction
★ Transcription Factors (2nd edition)
Tumour Immunobiology

Last 10 published

Chromatin 1.4.98
Drosophila 2/e 19.3.98
Molecular Genetic Analysis of Populations 2/e 19.3.98
Mutation Detection 15.2.98
Antisense Technology 18.12.97
PCR 3: ISH .12.98
Genome Mapping 21.8.98
MHC Volume 2 21.8.97
MHC Volume 1 21.8.97
Signalling by Inositides

In prod

HPLC of Mac 2/e
RNA-Protein Interactions
Growth Factors and Receptors
Cell Sepn
Light Mic
ISH2/e
GEP3/e
Immobilized Biomolecules
Eukaryotic DNA Replication
Chromosone Structural Analysis
Protein Expression Vol 1

Eukaryotic DNA Replication

A Practical Approach

Edited by

SUE COTTERILL
Marie Curie Research Institute,
Oxted, Surrey

OXFORD
UNIVERSITY PRESS
1999

This book has been printed digitally in order to ensure its continuing availability

OXFORD
UNIVERSITY PRESS

Great Clarendon Street, Oxford OX2 6DP

Oxford University Press is a department of the University of Oxford.
It furthers the University's objective of excellence in research, scholarship,
and education by publishing worldwide in

Oxford New York

Auckland Bangkok Buenos Aires Cape Town Chennai
Dar es Salaam Delhi Hong Kong Istanbul Karachi Kolkata
Kuala Lumpur Madrid Melbourne Mexico City Mumbai Nairobi
São Paulo Shanghai Singapore Taipei Tokyo Toronto

with an associated company in Berlin

Oxford is a registered trade mark of Oxford University Press
in the UK and in certain other countries

Published in the United States
by Oxford University Press Inc., New York

Reprinted 2002

A catalogue record for this book is available from the British Library

Library of Congress Cataloging in Publication Data
(Data available)

ISBN 0-19-963681-8 (Hbk)

ISBN 0-19-963680-X (Pbk)

Preface

The aim of this book is to allow easy access to the techniques that are currently in use to study eukaryotic DNA replication. The subject area that falls under this heading is quite diverse, and its boundaries with related fields (e.g. the cell cycle) are also rather blurred. This span is reflected in the scope of the chapters chosen for this book which cover not only ways of working with specific replication proteins and examples of the various model systems used to study eukaryotic DNA replication but also more general techniques. By necessity many of the examples that are used in the text are quite specific but attempts have been made to put these into their relevant context, and also to include more general advice that should make the methods usable at several levels.

The field of eukaryotic DNA replication has moved fast in the last ten years and the pace of progress shows no signs of decreasing at this point. However the techniques that are covered here are all fundamental to the study of the subject area. They should therefore remain useful for some time to come and provide a good starting point for the study of any aspect of eukaryotic DNA replication.

Oxted S.C.
May 1998

Contents

Contributors

MARTHA P. ARROYO
Department of Pathology, Stanford University School of Medicine, Stanford, CA 94305–5324, USA.

GLENN A. BAUER
Department of Biology, Saint Michael's College, Colchester, VT 05439, USA.

DANIEL W. BEAN
Department of Biology, University of North Carolina, Chapel Hill, NC 27599, USA.

ANJA K. BIELINSKY
Brown University, Division of Biology and Medicine, Providence, RI 02912, USA.

JAMES A. BOROWIEC
Department of Biochemistry, and Kaplan Comprehensive Cancer Center, New York University Medical Center, 550 First Avenue, New York, NY 10016, USA.

PETER A. BULLOCK
Department of Biochemistry, Tufts University School of Medicine, Boston, MA 02111, USA.

KRISTA L. CONGER
Department of Pathology, Stanford University School of Medicine, Stanford, CA 94305–5324, USA.

WILLIAM C. COPELAND
Laboratory of Molecular Genetics, National Institute of Environmental Health Sciences, PO Box 12233, Research Triangle Park, North Carolina 27709, USA.

JOHN DIFFLEY
ICRF Clare Hall Laboratories, South Mimms, Potters Bar, Hertfordshire EN6 3LD, UK.

SHANE DONOVAN
Department of Pediatrics, 513 Parnassus Avenue, HSE302, University of California, San Francisco, CA 94143–0519, USA.

SIMON DOWELL
Glaxo-Wellcome Medicines Research Centre, Gunnels Wood Road, Stevenage, Hertfordshire SG1 2NY, UK.

ELENA FERRARI
University Zürich-Irchel, Department of Veterinary Biochemistry, Winterthurerstrasse 190, CH-8057 Zürich, Switzerland.

MARCO FOIANI
Dipartimento di Genetica e di Biologia dei Microrganismi, Universita' degli Studi di Milano, Via Celoria 26, 20133 Milano, Italy.

SUSAN A. GERBI
Brown University, Division of Biology and Medicine, Providence, RI 02912, USA.

THOMAS G. GILLETTE
Department of Biochemistry, and Kaplan Comprehensive Cancer Center, New York University Medical Center, 550 First Avenue, New York, NY 10016, USA.

LEA A. HARRINGTON
Ontario Cancer Institute/Amgen Institute, Department of Medical Biophysics, University of Toronto, 620 University Avenue, Toronto, Ontario M5G 2CI, Canada.

ULRICH HÜBSCHER
University Zürich-Irchel, Department of Veterinary Biochemistry, Winterthurerstrasse 190, CH-8057 Zürich, Switzerland.

CRISTINA IFTODE
Department of Biochemistry, and Kaplan Comprehensive Cancer Center, New York University Medical Center, 550 First Avenue, New York, NY 10016, USA. Current address: Department of Molecular Biology, Princeton University, Princeton, NJ 08544, USA.

ZOPHONÍAS O. JÓNSSON
University Zürich-Irchel, Department of Veterinary Biochemistry, Winterthurerstrasse 190, CH-8057 Zürich, Switzerland.

CHUN LIANG
Cold Spring Harbor Laboratory, Cold Spring Harbor, NY 11724, USA.

GIORDANO LIBERI
Dipartimento di Genetica e di Biologia dei Microrganismi, Universita' degli Studi di Milano, Via Celoria 26, 20133 Milano, Italy.

VICTORIA V. LUNYAK
Brown University, Division of Biology and Medicine, Providence, RI 02912, USA.

STEVEN W. MATSON
Department of Biology, University of North Carolina, Chapel Hill, NC 27599, USA.

THOMAS MELENDY
Department of Microbiology, School of Medicine and Biomedical Sciences, State University of New York at Buffalo, Buffalo, NY 14214; and Department of Cellular and Molecular Biology, Roswell Park Cancer Institute, Buffalo, NY 14263, USA.

ROMINA MOSSI
University Zürich-Irchel, Department of Veterinary Biochemistry, Winterthurerstrasse 190, CH-8057 Zürich, Switzerland.

JOHN NEWPORT
Department of Biology 0347, University of California, San Diego, La Jolla, CA 92093–0347, USA.

SIMONETTA PIATTI
Dipartimento di Genetica e di Biologia dei Microrganismi, Universita' degli Studi di Milano, Via Celoria 26, 20133 Milano, Italy.

PAOLO PLEVANI
Dipartimento di Genetica e di Biologia dei Microrganismi, Universita' degli Studi di Milano, Via Celoria 26, 20133 Milano, Italy.

ADELE ROWLEY
Glaxo-Wellcome Medicines Research Centre, Gunnels Wood Road, Stevenage, Hertfordshire SG1 2NY, UK.

NATALIA V. SMELKOVA
Department of Biochemistry, and Kaplan Comprehensive Cancer Center, New York University Medical Center, 550 First Avenue, New York, NY 10016, USA. Current address: Department of Molecular Biology, 1275 York Avenue, Memorial Sloan-Kettering Cancer Center, New York, NY 10021, USA.

MANUEL STUCKI
University Zürich-Irchel, Department of Veterinary Biochemistry, Winterthurerstrasse 190, CH-8057 Zürich, Switzerland.

FYODOR D. URNOV
Brown University, Division of Biology and Medicine, Providence, RI 02912, USA.

JOHANNES WALTER
Department of Biology 0347, University of California, San Diego, La Jolla, CA 92093–0347, USA.

TERESA S.-F. WANG
Department of Pathology, Stanford University School of Medicine, Stanford, CA 94305–5324, USA.

Abbreviations

A_{260}	absorbance at 260 nm
Abf	ARS binding factor
ACS	ARS consensus sequence
ARS	autonomously replicating sequence
BiodUTP	biotinylated dUTP
BND	benzoylated naphthoylated
bp	base pair(s)
BrdU	bromo-dUTP
BSA	bovine serum albumin
CDC	cell division cycle
CHO	Chinese hamster ovary
CPK	creatine phosphate kinase
CSF	cytostatic factor
DAPI	4,6-diamino-2-phenylindole
DEAE	diethyl aminoethyl
DEPC	diethyl pyrocarbonate
DHFR	dihydrofolate reductase
DTT	dithiothreitol
EDTA	ethylenediaminetetraacetic acid
EGTA	[ethylenebis (oxyethylenenitrilo)]-tetraacetic acid
FACS	fluorescence activated cell sorter
FEN1	flap endonuclease
FEN1/RTH1	5′-3′ nuclease
FISH	fluorescence *in situ* hybridization
FITC	fluorescein isothiocyanate
HA	haemagglutinin
hcg	human chorionic gonadotropin
Hepes	*N*-2-hydroxyethylpiperazine-*N*′-2-ethanesulfonic acid
HMK	heart muscle kinase
hTR	human telomerase RNA
iDNA	initiator DNA
IgG	immunoglobulin G
ITAS	internal telomerase amplification standard
KPA	kinase protection assay
LS	low speed
MEGA-9	*n*-nonanoyl-*N*-methyl-glucomide
MBP	maltose binding protein
MOPS	morpholinepropane sulfonic acid
MPF	maturation promoting factor
N/A	Neutral/Alkaline

NET	NaCl, EDTA, Tris
N/N	Neutral/Neutral
NP-40	Nonidet P-40
nt	nucleotide(s)
OBR	origin of bidirectional replication
OD	optical density
ORC	origin recognition complex
ORI	origin
PBS	phosphate-buffered saline
PCNA	proliferating cell nuclear antigen
PCR	polymerase chain reaction
PI	protease inhibitors
PMSF	phenylmethylsulfonyl fluoride
pol	DNA polymerase
RF-C	replication factor C
RFI	replicative form I
RI	replication intermediates
RIP	replication initiation point
RNase	ribonuclease
RNasin	ribonuclease inhibitor
RP-A	replication protein A
RT	room temperature
RTH1	rad 2 homologue 1 encoded protein
SDS	sodium dodecyl sulfate
SEA	standard elongation assay
SPA	scintillation proximity assay
SPB	spindle pole body
SSB	single-stranded DNA binding protein
ssDNA	single-stranded DNA
SV40	simian virus 40
TAE	Tris, acetate, EDTA
T-ag	SV40 T-antigen
TBE	Tris, borate, EDTA
TCA	trichloroacetic acid
TE	10 mM Tris–HCl pH 8, 1 mM Na_2EDTA
TEN	Tris, EDTA, NaCl
TLCK	L-1-chloro-3-(4-tosylamido)-4-phenyl-2-butanone
TRAP	telomerase repeat amplification protocol
TRF	telomeric repeat binding factor
tRNA	transfer ribonucleic acid

1

Methods to map origins of replication in eukaryotes

SUSAN A. GERBI, ANJA K. BIELINSKY, CHUN LIANG, VICTORIA V. LUNYAK, and FYODOR D. URNOV

1. Introduction

1.1 Overview

According to the replicon model of Jacob, Brenner, and Cuzin (1), initiation of DNA replication is achieved via the binding of a *trans*-acting protein (the initiator) to a *cis*-acting sequence (the replicator). The initiator recruits the replication machinery to the replicator, and DNA synthesis ensues. Studies in prokaryotes, fungi, and viruses of metazoans have provided ample experimental support for this hypothesis, and in such model systems as *E. coli*, *S. cerevisiae*, and SV40, a considerable amount of experimental data exist about the nature of replicators and the initiator protein(s) (see refs 2–4 for recent reviews). A useful and interesting property of replicators in these organisms is that they retain function when placed on an episome. The autonomously replicating sequence (ARS) assay that exploits this property allowed researchers to use the power of 'reverse genetics' to identify and study the replicators in these systems (for example, ref. 5). Unfortunately, the ARS assay has not been as useful for studying replicators in metazoa, where the efficiency of episome replication seems more related to the length rather than the sequence of the insert (6–8). There are additional complexities: sequences as much as 50 kb away from the origin can influence its activity (9), and proper chromatin organization of the DNA is important for replication initiation (10).

At present, the soundest approach for beginning a study of metazoan replicators is to map the physical location (origin = ORI) at which DNA synthesis initiates in a given locus. Once this has been achieved to a reasonable level of resolution, other methods, including sequence analysis, investigation of chromatin structure, and analysis of protein:DNA interactions *in vitro* and *in vivo* can be used to study the mechanisms whereby this ORI is regulated. Finally, functional assays done in the context of the chromosome may be used to test conclusions of the approaches above.

This review describes detailed procedures for achieving the first goal in this

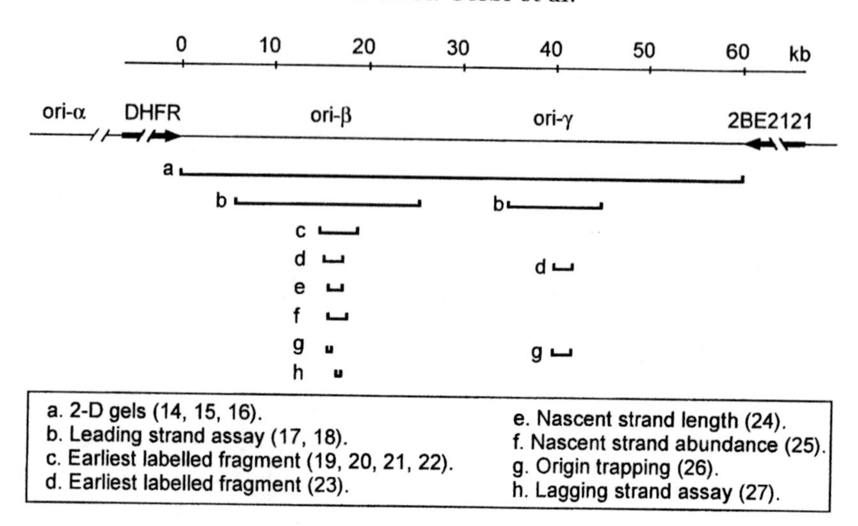

Figure 1. DNA replication ORI mapping in the DHFR locus. Modified from ref. 28. The position of the ORI as deduced by various investigators with the methods they used is designated by a bracket. For multiple applications of the same basic method, just the final conclusion is shown, unless two groups come to different conclusions, in which case both are shown. In addition, it is clear that the different methods give different results, and even some of the results shown above are subject to other interpretations (J. Hamlin, personal communication). The reader should be forewarned that, in the case of the DHFR locus, the situation remains unresolved.

approach: mapping the ORI. A comprehensive and thoughtful description of the methods for ORI mapping is available in the literature (11), as are two reviews discussing the recent progress in the field (12, 13). Since space limitations preclude us from providing recipes for every method available, we will first briefly consider several of the older techniques and then compare them to the newer methods discussed here in detail. As will be seen, these methods differ, and this can be visualized in *Figure 1*, which depicts data obtained from applying a large number of methods for ORI mapping in the dihydrofolate reductase (DHFR) gene locus in Chinese hamster ovary cells.

1.2 Methods to map origins as regions of initiation of DNA synthesis

Three methods are based on the simple notion that DNA at the ORI itself will be the first to replicate when the cell traverses the G1/S boundary:

(a) In the 'earliest labelled fragment' method (19–22), cells are arrested at the G1/S border by the use of a DNA synthesis inhibitor and then released from the block into medium containing a labelled DNA synthesis precursor. DNA is isolated after a brief period of time and the location of the ORI can be determined by hybridizing the labelled DNA to a panel of slot-blotted plasmid clones that span the area of interest. Refinement of this method by 'in-gel renaturation' (applicable to amplified systems) allowed the mapping of the ORI in the amplified DHFR domain in Chinese hamster ovary cells to a 1.8 kb region (23).

(b) In the 'replication ORI trap' method (29), random cross-links are introduced into the DNA of asynchronously dividing cells by treating them with a psoralen derivative followed by UV irradiation. After cross-linking, the cells are placed into medium with BUdR and a radioactive DNA synthesis precursor. DNA synthesis continues until the replication machinery stalls at a cross-link. Short BUdR-containing DNA—presumably, the nascent strands that were initiated at the ORI and whose synthesis was aborted once the replication machinery reached the first cross-link in the DNA—is isolated and used as a probe for mapping on genomic DNA clones. A caveat to this method is the possibility that the replication machinery might resume synthesis on the other side of a psoralen cross-link, thereby creating a population of short nascent DNA molecules that were not initiated at the ORI (30). In the DHFR locus, however, the 0.45 kb ORI fragment 'trapped' by this method in synchronized cells (26) mapped to the same stretch of DNA as the ORI mapped by the 'earliest labelled fragment' approach (23).

(c) The experimental observation that a certain percentage of the total DNA isolated from growing cells pulse labelled with BUdR is found in the 'heavy–heavy' fraction of CsCl gradients (31) led to the hypothesis that small nascent DNA strands may occasionally be extruded soon after their synthesis is initiated at the ORI. The ORI mapping method based on this phenomenon (32) involves administering a brief pulse of [^{3}H]thymidine *in vivo* to cells just released from a G1/S block, isolation of nuclei, and a brief period of subsequent DNA synthesis in the presence of Hg-dCTP to label the nascent DNA. Total DNA is purified, and then the nascent DNA chains are extruded under carefully controlled conditions. The extruded nascent DNA is purified by sulfhydryl Sepharose column chromatography and size fractionated to obtain small DNA (containing the ORI) that can be used as probes for hybridization to cloned genomic DNA (30).

1.3 Methods to map origins based on the asymmety of bidirectional replication

Two other methods exploit the asymmetry of eukaryotic chromosomal bidirectional replication as manifested in the existence of leading and lagging strands, and the continuous and discontinuous mode of DNA synthesis associated with them.

(a) The leading strand assay (= 'imbalanced DNA synthesis') (17, 18) is based on the preferential inhibition of lagging strand DNA synthesis by emetine, a protein synthesis inhibitor. An asynchronous population of cells is maintained in emetine and BUdR containing medium, nascent DNA is isolated, and hybridized to strand-specific cloned DNA fragments. The ORI is mapped by finding the location at which the hybridization signal shifts from the top to the bottom strand.

(b) The lagging strand assay (= 'Okazaki fragment distribution') (27) maps the ORI by finding the 'the origin of bidirectional replication' (OBR), i.e. the position along a given DNA strand where the transition between continuous and discontinuous DNA synthesis occurs and the first Okazaki fragment is laid down. Cells are synchronized at the G1/S boundary, permeabilized, released from the metabolic block, and the nascent DNA is labelled *in situ* with [^{32}P]dATP and BUdR for a very short length of time. The Okazaki fragment fraction of nascent DNA is purified by size fractionation on alkaline gels and immunoprecipitated with anti-BUdR antibodies. The location of the OBR is determined by hybridizing the Okazaki fragment fraction to slot-blotted strand-specific DNA and finding the location where the hybridization signal shifts from the bottom to the top strand of DNA. In the amplified DHFR domain in CHO cells, application of this method mapped the OBR to a 0.45 kb fragment (27).

While the methods described above have allowed for the mapping of ORIs in specific model systems, their general usefulness is limited by two factors. First, most of them depend upon the ability to synchronize cells at the G1/S boundary and can, therefore, be only used in model systems where synchrony can be achieved. Secondly, it remains unknown whether protein or DNA synthesis inhibitors alter the normal pattern of DNA replication initiation events.

The three sets of methods that we focus on in this review all share the important distinction of not requiring cell synchronization or treatment with metabolic inhibitors. The only requirement is that clones spanning the area of interest be available, along with some information about the DNA sequence. As far as the scale of mapping is concerned, these three methods form a natural 'magnifying glass → light microscope → electron microscope' progression that, when applied in the order in which they are described in this chapter, should allow an investigator to map an ORI (if it exists as a discrete entity) to a region of a few kilobases (2D gels), then map it to less than 1 kb (PCR analysis of nascent strands), and finally, map it to nucleotide resolution (replication initiation point mapping).

2. Two-dimensional (2D) and three-dimensional (3D) gels

2.1 Background

Two-dimensional (2D) gel electrophoresis methods allow for ORI mapping by exploiting the idiosyncratic migration of replication intermediates (RI) on agarose gels. The Neutral/Neutral (N/N) 2D gel method of Brewer and Fangman (33) allows analysis of a large stretch of DNA by Southern blot hybridization to find which restriction fragments contain a replication bubble (and hence the ORI), and which contain replication forks. In the first dimension, genomic DNA from an asynchronous cell population is digested with restriction enzymes and is run on a low percentage agarose gel at low voltage to resolve DNA molecules primarily by mass: restriction fragments of unreplicated DNA and smaller RIs will migrate faster than restriction fragments containing larger forks and bubbles. In the second dimension, RI DNA is separated from bulk DNA by running it on a higher percentage agarose gel at high voltage to separate DNA molecules on the basis of shape as well as mass. After running the second dimension and blotting the DNA onto a filter, a Southern blot performed with a particular restriction fragment as a probe will yield a 'bubble arc' pattern (see *Figure 2A*) if this fragment contains an origin of replication, or a 'fork arc' pattern if this stretch of DNA is replicated by forks passing through the region.

A complementary or alternative approach to N/N 2D gels is the Neutral/Alkaline (N/A) 2D gel method developed by Huberman *et al.* (34). This method allows one to determine the direction of replication fork movement through a particular region (in contrast, the original N/N 2D gel method cannot distinguish the direction of replication fork movement, although it can be deduced by a variant of this method utilizing restriction enzyme digestion of DNA in the gel after the first dimension has been run) (35, 36). In N/A 2D gels, ORI mapping is based on the idea that an ORI will be flanked by DNA in which replication forks are moving in opposite directions away from the ORI. In this approach, genomic DNA from an asynchronous population of cells is run on a first-dimension agarose gel under the same conditions as for the N/N 2D gel method, while the second-dimension gel is run under alkaline conditions to denature the DNA (see *Figure 2B*). The nascent strands from replication forks will form a diagonal line in the gel because of the differences in their lengths, depending on how far each replication fork has moved through the restriction fragment being tested. Nascent DNA from a fork entering the end of the fragment proximal to the ORI will be shorter than nascent DNA from a fork leaving the fragment at the end distal to the ORI. If short pieces of DNA from different areas of the restriction fragment are used as probes, the direction of fork movement can be deduced. Thus, when the probe is complementary to the end where forks enter the fragment, small as well as large nascent strands will be detected, and a complete nascent strand diagonal will be seen. If the probe is from the opposite end of the restriction fragment, then hybridization will only be found at the top of the diagonal (representing very large nascent strands). N/N and N/A 2D gels have been widely used in the field for ORI mapping in several systems; in the puff II/9A locus of the fungus fly, *Sciara*, their combined application allowed us to map the ORI to a 1 kb region (37). It remains unclear why 2D gels defined a very

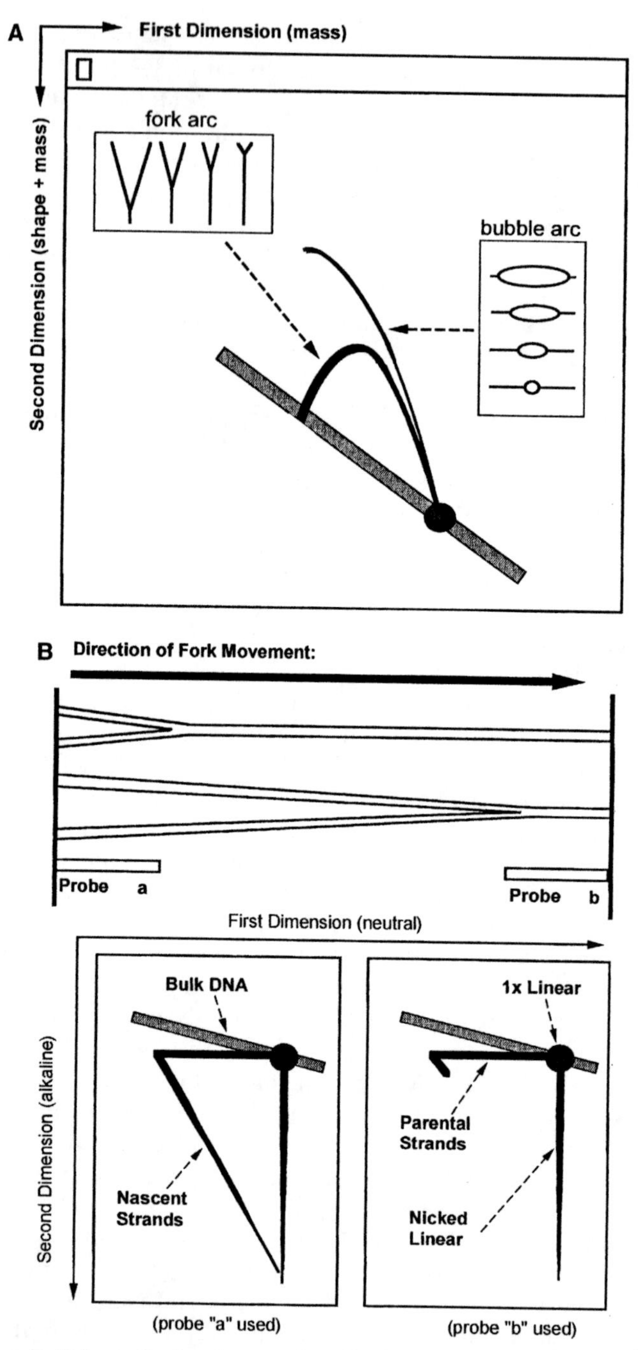

Figure 2. 2D gels. Schematic drawings to describe Neutral/Neutral 2D gels (panel A) and Neutral/Alkaline 2D gels (panel B). Modified from ref. 37.

broad zone of initiation in the DHFR gene locus (14–16), while the application of several other methods to this region indicated the presence of one to two much more localized ORIs (see *Figure 1*).

To verify and extend the conclusions obtained regarding ORI location by 2D gel methods, we developed a three-dimensional (3D) gel procedure (38) that is a composite of both 2D gel methods described above. Once again, genomic DNA from unsynchronized cells is run on a N/N 2D gel. Next, forks and bubbles that have been separated from each other by this 2D gel are resolved into parental and nascent strands by cutting the 2D gel into vertical slices, and each slice is placed in a slot for the third dimension that is run under alkaline conditions. Since the DNA-containing gel slices are rotated 90° prior to being run in this third dimension, nascent DNA fragments derived from forks and bubbles can be analysed separately. As a consequence, when used for a Southern blot probed with a panel of fragments spanning the region of interest, the 3D gel method can map the ORI by locating the fragment that generates the highest bubble/fork nascent DNA ratio. In addition, it can be used to address some long-standing unresolved issues, such as how many initiation events occur on a single DNA molecule (in *Sciara* DNA puff II/9A, the answer is only one initiation event per molecule) (38), and also whether the composite fork arc plus bubble arc pattern usually observed in N/N 2D gels of metazoan DNA is a consequence of replication bubbles breaking and generating forks (in *Sciara*, the answer is no) (38).

2.2 Preparation of replicating DNA

2.2.1 DNA preparation for 2D and 3D gels

2D and 3D gels require that the DNA is sufficiently purified so that it can be digested with restriction enzymes to completion within a few hours. In addition, the fragile replication intermediates (RI) must be preserved. Therefore, special precautions must be taken to treat the DNA gently (e.g. pipette DNA solutions slowly with wide bore Pipetman tips).

If the object of study is a plasmid, most common DNA isolation protocols including phenol:chloroform extraction should suffice. However, the most reliable method to isolate replicating chromosomal DNA is CsCl centrifugation (34), although an alternative protocol has been reported (39). For chromosomal DNA from organisms with large genomes, it may be necessary to enrich the replicating DNA by nuclear matrix isolation (40).

The following DNA isolation protocol has been developed for the budding yeast *Saccharomyces cerevisiae*. For other organisms, one can either homogenize whole cells in the TEN buffer of the yeast protocol (Day 4), or isolate nuclei and resuspend them in TEN, and then proceed as in the yeast protocol. The key is to stop cell metabolism, to keep samples cold, and to inhibit nucleases.

Protocol 1. Isolation of replicating DNA from yeast

Equipment and reagents

- Sorvall refrigerated centrifuge with GSA and SS-34 rotors with appropriate tubes
- Beckman ultracentrifuge with VTi65 or VTi75 rotor and appropriate tubes
- Glass beads: 425–600 μm, acid washed (Sigma)
- YPD broth: 1% yeast extract, 2% peptone, 2% dextrose (supplement or replace YPD with appropriate medium for selection of markers such as *ura*)
- EDTA/glycerol: 2:1 (v/v) mix of 0.2 M EDTA pH 8 and 100% glycerol
- NIB (nuclear isolation buffer): 17% (v/v) glycerol, 50 mM MOPS, 150 mM K acetate, 2 mM $MgCl_2$, 500 μM spermidine, 150 μM spermine, pH 7.2
- Hoechst dye: 5 mg/ml in dH_2O (H 33258 from Calbiochemical)
- TEN buffer (Tris, EDTA, NaCl): 50 mM Tris pH 8, 50 mM EDTA, 100 mM NaCl
- Dialysis buffer I: 200 mM NaCl, 10 mM Tris pH 8, 1 mM EDTA
- Dialysis buffer II: 50 mM K acetate pH 8, 10 mM Tris pH 8, 1 mM EDTA (dialysis buffer II is appropriate for subsequent restriction digestion with *Nco*I; if other enzymes are to be used instead, then the composition of dialysis buffer II should be adjusted to have the same salt concentration as the restriction buffer)

A. *Day 1: starter culture*

1. Inoculate a single colony of yeast into 20 ml YPD or appropriate selection medium. (A total of 12 different cultures for 12 different 2D gels can be grown and processed at one time, although six to eight cultures are easier to handle.)
2. Grow the yeast cultures overnight at 30 °C (or 23 °C for *ts* strains) on a shaker.
3. Ensure that the cultures are not contaminated by examining a sample under a microscope.
4. For each culture, prepare:
   (a) 0.5 litres YPD broth (or selection medium to maintain plasmids).
   (b) 150 ml EDTA/glycerol.

B. *Day 2: cell growth*

1. Inoculate an overnight culture into 0.5 litres YPD in a 2 litre flask to an $OD_{600} = 0.05–0.2$. Grow the yeast overnight to $OD_{600} = 1–1.2$ ($2–4 \times 10^7$ cells/ml).
2. Repeat part A, step 3.
3. Prepare nuclear isolation buffer (NIB) for day 3.
4. Prepare TEN buffer and 30% Sarkosyl for day 4.

C. *Day 3: cell harvest*

1. Cool a GSA rotor at 4 °C in a Sorvall refrigerated centrifuge.
2. For each 0.5 litre culture, put 50 ml EDTA/glycerol into each of three

250 ml centrifuge bottles for a GSA rotor. Freeze this solution in a –70°C freezer or in a dry ice/ethanol bath.

3. Add 1/100 vol. 10% sodium azide to the yeast culture and swirl.
4. Pour 166 ml of the yeast culture into each of the three GSA bottles containing the frozen EDTA/glycerol.
5. Shake bottles at room temperature until thawed. Spin 5 min at 6000 r.p.m. in a GSA rotor at 4°C.
6. Resuspend cell pellets from the three GSA bottles in a total of 30 ml chilled dH_2O and pour the pooled suspension into a 50 ml Falcon tube.
7. Pellet the yeast cells in a centrifuge by spinning at 3000 r.p.m. at 4°C for 2 min.
8. Resuspend the pellet in 5 ml cold NIB buffer and store at –70°C.

D. *Day 4: isolation of nuclei and DNA*

1. Cool a SS-34 rotor in a centrifuge at 4°C.
2. Take two, four, or six frozen tubes (each tube has all the cells from one culture) out from the –70°C freezer and swirl in room temperature water until nearly thawed. Place the tubes on ice.
3. To each tube add 7 ml glass beads (425–600 μm, acid washed).
4. Vortex the tubes in a cold room, using two strong vortex machines, with one tube in each hand, vortexing at top speed. Alternate 30 sec vortex and 30 sec on ice for a total of 10–15 times for each tube to obtain ~ 70% cell lysis (check under a phase microscope).
5. Pipette the supernatant from each culture into a cold 15 ml Corex centrifuge tube. Wash the glass beads with 7 ml ice-cold NIB. Pool the wash and the supernatant.
6. Spin at 10 000 r.p.m. at 4°C for 10 min in a Sorvall SS-34 rotor.
7. Decant the supernatant. Resuspend each pellet in 9 ml ice-cold TEN buffer using a glass pipette to stir up the pellet and then vortex briefly.
8. Add 0.5 ml 30% Sarkosyl to each tube. Invert a few times to mix gently.
9. Add 150 μl 20 mg/ml Proteinase K to each tube. Invert a few times to mix gently.
10. Incubate at 37°C for 1 h, mixing occasionally.
11. Spin tubes at 5000 r.p.m. at 4°C for 5 min a Sorvall SS-34 rotor. The supernatant will be somewhat turbid.
12. Transfer the supernatant to a 15 ml Falcon tube containing 10 g CsCl.
13. Gently dissolve the CsCl in the supernatant by slow rotation and inversion of the tube.

Protocol 1. *Continued*

14. Add 300 μl Hoechst dye solution.
15. Gently transfer (e.g. pour through a sterile Pasteur pipette or a 3 ml syringe barrel with a 16 gauge needle attached) each sample into a 12.5 ml ultracentrifuge tube for a Ti50 rotor. Top off with CsCl solution made in TEN of the same density as the DNA sample. Seal the tubes and spin at 45 000 r.p.m. for 38 h at 12 °C in a Beckman ultracentrifuge.

E. *Day 5: DNA isolation (continued)*

1. Chill dialysis buffers I and II in a cold room: 2 litres each in 4 litre beakers. Stop the spin (brake until it reaches 800 r.p.m. and then coast to a full stop).
2. Take centrifuge tubes to a dark-room (a dim cold room will also do), and prepare the following equipment:
   - stand with clamp
   - long wave UV lamp and goggles to protect eyes from UV light
   - 22 gauge and 16 gauge needles and 3 ml syringes
   - 15 ml Falcon tubes in a rack (for collecting DNA), and a 600 ml beaker (for waste liquid from the centrifuge tubes)
   - Scotch tape, Pipetman, and tips

   Clamp the centrifuge tube on a ring stand and observe it with long wave UV (*wear goggles!*). The expected pattern of bands is, from top to bottom: a fuzzy band of mitochondrial DNA, bulk genomic DNA, a minor band of ribosomal DNA, and a non-DNA band (the latter is visible without UV light; avoid it since it inhibits restriction enzymes).
3. Tape the outside of the tube where the bulk genomic DNA band appears. Pierce the top of tube with a 22 G needle, and pierce through the tape with a 16 G needle attached to a 3 ml syringe to collect the bulk genomic DNA (the rDNA band is optional unless rDNA is under study). It is essential that the collection of DNA be done slowly to minimize DNA breakage by shearing. Remove the needle from the syringe and deliver the DNA solution into a 15 ml Falcon tube.
4. Extract the DNA solution five times with an equal volume of isopropanol saturated with water and CsCl. Remove the isopropanol phase (top); monitor with long wave UV. The isopropanol phase of the last extraction should show little Hoechst dye; remove the last bit of isopropanol with a Pipetman. Discard the isopropanol phases.

5. Place the extracted aqueous samples in medium dialysis bags. Dialyse at 4 °C as follows:
   (a) 2 h to overnight in dialysis buffer I.
   (b) 6 h to overnight in dialysis buffer II.

   Alternatively, spin down the DNA as in *Protocol 13*, steps 16–18.

F. *Day 6: restriction digestion*

1. Transfer the DNA to 1.5 ml Eppendorf tubes (two tubes/culture are usually needed).
2. The DNA is now ready for restriction digestion. DNA (~ 50–100 μg) from each 500 ml yeast culture is sufficient for one to two gels. Estimate the volume of DNA solution and add Mg acetate (from a 1 M stock) to a final concentration of 13 mM, BSA to 100 μg/ml, DTT (from a 0.1 M stock) to 1 mM, and restriction enzyme such as *Nco*I (usually 10 μl of 10 U/μl for 30–50 μg DNA). If half of the DNA from each 0.5 litre culture is used for one gel, the remaining undigested DNA can be stored at –20 °C.
3. Incubate at 37 °C for 1 h to digest the DNA to completion.
4. Continue the digestion for an additional hour while you run 1 μl of the reaction mixture on an agarose gel to check for the completeness of digestion.
5. Once the digest is complete (i.e. the DNA appears as a smear on the gel) stop the reaction by adding EDTA to 5 mM and putting the sample on ice.

The digested DNA can be run on 2D gels after ethanol precipitation, or it can be enriched for RI DNA by BND cellulose column chromatography (see *Protocol 2*). The latter yields a cleaner pattern on 2D gels for yeast genomic DNA, and is almost always necessary when working with DNA from higher eukaryotes.

2.2.2 Enrichment of replication intermediates

RI DNA can constitute about 5–10% of the total DNA, and can be enriched by BND cellulose chromatography. Both double-stranded and single-stranded DNA are bound to BND cellulose under low salt (300 mM NaCl). Double-stranded DNA can be eluted with high salt (1 M NaCl), and subsequently single-stranded DNA can be eluted with high salt and caffeine (1 M NaCl plus 1.8% caffeine). Alternatively, one can also apply the DNA sample to BND cellulose in a high salt butter, such that only RI containing single-stranded portions of DNA will bind. In general, the less concentrated the sample and the bigger the column, the better the efficiency of purifying RI DNA.

Protocol 2. BND cellulose enrichment of replication intermediates

Equipment and reagents

- Beckman ultracentrifuge with SW41 rotor and appropriate tubes
- Benzoylated naphthoylated DEAE (BND) cellulose (Sigma)
- NET buffer (NaCl, EDTA, Tris): 1 M NaCl, 1 mM EDTA, 10 mM Tris pH 8
- 1.8% caffeine in NET buffer

All steps are done at room temperature. The procedure will take a few hours.

A. *Preparation of BND cellulose*

1. Weigh out 4 g BND cellulose in a 50 ml Falcon tube. Boil the BND cellulose in dH_2O for 5 min. Let it cool to room temperature, break up the particles with a rubber policeman, and spin at 2000 r.p.m. for 2 min in a centrifuge. Decant the supernatant.
2. Suspend and wash the BND cellulose once with 20 ml dH_2O.
3. Wash twice with NET buffer.
4. Store at 4°C in 20 ml NET buffer.

B. *Enrichment of replication intermediates (RI) on BND cellulose*

1. Add the BND cellulose suspension to a Bio-Rad Poly-Prep disposable column (Cat. No. 731–1550) or an Isolab QS-Q Quick-Sep column, making a 1 ml bed volume if you plan to isolate up to 20 μg RI DNA (0.5 ml bed for 5 μg RI DNA). RI DNA represents about 10% of the total DNA from asynchronized yeast cells (so load about 200 μg total nuclear yeast DNA). Wash the column extensively with NET buffer (ten volumes) until the OD_{260} is almost zero.
2. Add 5 M NaCl to the DNA solution to a final concentration of 1 M. Load the DNA solution (about 1 ml) onto the column and allow it to enter the resin by gravity; collect the flow-through.
3. Wash the column with three to five volumes NET buffer or until the OD_{260} is close to zero. Pool this salt wash fraction with the flow-through (part B, step 2) as they both contain mostly non-replicating double-stranded DNA.
4. Load one to two volumes (e.g. 1.5 ml) 1.8% caffeine in NET buffer pre-warmed to 50°C to the column. Drip off the liquid (this is the caffeine wash fraction and contains the RI DNA for further analysis).
5. Spin the caffeine wash 10 min at maximum speed in a microcentrifuge (*c.* 13000 r.p.m.) and save the supernatant. The pellet is BND cellulose particles that should be discarded.

6. Add one volume isopropanol to the double-stranded DNA (only needed for RIP mapping) and RI fractions and invert slowly to mix. Let the solution remain at 4°C for at least 30 min.
7. For larger volumes, spin the isopropanol precipitated solution in a Beckman SW41 rotor at 38 000 r.p.m. for 30 min at 4°C and decant the supernatant. For smaller volumes, spin down in an Eppendorf centrifuge at 10 000 r.p.m. for 30 min at 4°C.
8. Wash the DNA pellet with 70% ethanol. Spin for 2 min in a microcentrifuge and decant the supernatant. Briefly air dry the pellet.
9. Redissolve the DNA pellet (a few micrograms; about 5% of total DNA loaded onto the BND cellulose column) in 30–40 μl TE for 2D gels (on ice or in a cold room for ~ 30 min to several hours). If running a 2D gel, add 4–5 μl 10 × loading dye. For RIP mapping (see Section 4), the DNA should be at a final concentration of 1 μg/μl.
10. Read the absorption at 260/280 nm. If the ratio is less than 1.8, reprecipitate the DNA with 0.1 vol. 3 M Na acetate pH 5.4, and 2 vol. ethanol for at least 2 h at 4°C.

2.3 Neutral/Neutral (N/N) 2D gels and Southern hybridization

Protocol 3. Neutral/Neutral 2D gels

Reagents

- 10 mg/ml ethidium bromide stock solution
- Gel soak I: 0.15 M NaOH, 1.5 M NaCl
- 1 × SSC (standard saline citrate buffer): 0.15 M NaCl, 0.015 M Na acetate pH 7
- Wash solution: 40 mM Na phosphate buffer pH 7.2, 1% SDS, 1 mM EDTA
- Pre-hybridization solution: 0.5 M Na phosphate buffer pH 7.2, 1% SDS, 1 mM EDTA, 1% BSA
- Stripping solution: 0.1 × SSC, 0.1% SDS
- 10 × TBE: 890 mM Tris, 890 mM boric acid, 25 mM EDTA pH 8

A. *Day 1: run gel (first dimension)*

1. Prepare a 0.4% agarose gel (350–400 ml in 1 × TBE) in a large gel tray (e.g. W × L = 20 cm × 24 cm) in the cold room.[a,b]
2. Pour in 1 × TBE to just even with the gel surface. Do not flood the gel.
3. Load the DNA samples. Run at 0.6–0.7 V/cm. Flood the gel with 1 × TBE after the dye has run into the gel.[c] Run for 36–48 h (blue dye should have run ~ 10–12 cm). Do not run the first dimension with ethidium bromide.

B. *Day 3: run gel (second dimension)*

1. Boil 350–400 ml 1% agarose in 1 × TBE.[d]

Protocol 3. *Continued*

2. For each second-dimension gel, prepare 2 litres of 1 × TBE containing 0.3 μg/ml ethidium bromide (from stock solution).
3. After the first dimension, cut out the gel lanes with a razor, with the aid of a ruler. Gel lanes should be slightly wider than the wells so that the razor does not touch the DNA. The top 2 cm from the well can be discarded. Take 10 cm long slices (from 2–12 cm starting at the well) for the second dimension. This will include DNA from ~ 2 kb and up. Pick up each gel lane with a flexible, thin plastic ruler, rotate it by 90°, and put it into the gel tray for the second dimension (*Figure 3*).
4. Once the melted agarose gel has cooled to 50–60 °C, add ethidium bromide stock solution to a final concentration of 0.3 μg/ml.
5. In a cold room, pour the second-dimension gel containing 1% agarose plus 0.3 μg/ml ethidium bromide around the first-dimension gel slices. Let it set for ~ 20 min. Create wells for size markers with a hot spatula or a glass pipette.
6. Stain the size markers of the first dimension, destain, and photograph (UV Polaroid), or keep them at 4 °C (either in 1 × TBE or wrapped on a glass plate) and photograph later, together with the second dimension.
7. Pour in 2 litres 1 × TBE containing 0.3 μg/ml ethidium bromide for each gel box. Load DNA markers.
8. Run gel at 2–3 V/cm. For 35 cm long gel boxes, 80–100 V is sufficient. Run for 14–18 h. Monitor the run with long wave UV. The goal is to have the shortest DNA reach the lower right-hand corner of the gel (*Figure 4*).

C. *Day 4: Southern transfer*

1. Carefully take the gel out of the gel box (slice the gel into top and bottom halves for easier handling), and place it into a large container. Destain the gel in water for ~ 30 min. Photograph the gel under short wave UV.
2. Denature the gel in gel soak I for 30 min to 1 h. Do not depurinate the gel in HCl before DNA denaturation to avoid cutting the small nascent DNA strands into pieces that are too small to bind to the filter. Do not neutralize the gel after denaturation in gel soak I, as per the ICN Biochemicals instructions for improved Southern blot transfer.
3. Transfer the gel to a Biotran(+) membrane (ICN Biochemicals) with ~ 800 ml of 20 × SSC, 18–24 h. Put a glass plate as an even weight on top of a stack of paper towels over the membrane and gel, but do not put extra weight on top of the glass plate as it may compress the gel too much.

D. *Day 5: Southern hybridization*

1. Remove the membrane from the gel and fix the DNA onto the membrane by either UV cross-linking or baking at 80 °C for 20 min.
2. Rinse the membrane with dH_2O and then 2 × SSC. Air dry until damp or dry.
3. Make the random primed probe. For every two 12 × 20 cm membranes containing two 2D gels, use ~ 25 ng template (for the budding yeast, use the 5 kb *Nco*I fragment containing ARS1), and 5 μl [α-^{32}P]dATP (3000 mCi/mmol) for a 20–25 μl reaction volume. Incubate in the random primer reaction mixture (e.g. Boehringer Mannheim) at 37 °C for 30 min to 3 h. Stop the reaction with 2 μl 0.5 M EDTA and add TE to 100 μl. Pass through Sephadex G50 to remove unincorporated nucleotides.
4. Incubate the membrane in pre-hybridization solution (~ 10 ml for every 100 cm^2 membrane) at 65 °C for 5 min to a few hours.
5. Pour off the pre-hybridization solution. Add hybridization solution (~ 5 ml/100 cm^2 membrane) which is pre-hybridization solution containing probe and 100 μg/ml single-stranded salmon sperm DNA or calf thymus DNA (boil this together with the probe for 10 min). Incubate with shaking at 65 °C overnight.

E. *Day 6: wash Southern blots*

1. Pour hybridization solution into a Falcon tube (keep it for later reuse if needed). Rinse the hybridized membrane twice with room temperature wash solution. Add wash solution that was pre-heated to 65 °C and shake membrane in it at 65 °C for 20 min. Repeat wash two more times with a brief rinse in between. Longer washes (30–60 min) may be needed for blots of DNA from higher eukaryotes.
2. Air dry the hybridized membrane briefly until damp but not dry. Wrap the membrane in Saran Wrap. Expose it to PhosphorImager plates for 18–24 h, or longer if needed. X-ray films (two intensifying screens) require 3–14 days of exposure.

F. *Stripping and rehybridization*

1. Strip off the previous probe from the hybridized membrane by incubation in a boiling hot solution of 0.1 × SSC, 0.1% SDS with shaking for ~ 5 min. Repeat two more times. Monitor that the probe has been removed completely by exposing the filter to a PhosphorImager screen.
2. Hybridize membrane with another probe. You can rehybridize a membrane at least seven times. On the last hybridization, use the

Protocol 3. *Continued*

same probe as for the first hybridization to ensure that the pattern of hybridization is the same (i.e. that no significant amount of DNA has been removed from the membrane by the repeated stripping).

[a] The percentage of the first-dimension gel should vary according to the size of restriction fragments of interest: 0.6% for restriction fragments 1–2 kb, 0.5% for fragments 2–3 kb, 0.4% for fragments 3–7 kb, 0.3% for fragments 7–10 kb. Ordinary high melting point agarose (e.g. from Gibco BRL Life Technologies) should suffice if the percentage of the first dimension is not lower than 0.4%. Pulsed-field gel grade agarose (Boehringer Mannheim, Cat. No. 1240 609) has higher gel strength and is therefore easier to handle for first-dimension gels.
[b] The wells should be more square than rectangular and big enough for a 50 μl sample. A large gap between the wells is desirable (alternatively, load every other well to avoid cross-contamination of samples when cutting out the first-dimension gel lanes).
[c] If you are short of time, you can run the gel at 2–3 V/cm until the dye enters the gel, flood with 1 × TBE, and then decrease the voltage to 0.6–0.7 V/cm. Do not leave the gel for longer than a few hours without flooding it with 1 × TBE, or else the gel will dry out.
[d] The percentage of the second-dimension gels should also vary according to the size range of interest: 1.5–1.8% for 1–2 kb, 1.2–1.5% for 2–3 kb, 1% for 3–7 kb, and 0.7% for 7–10 kb.

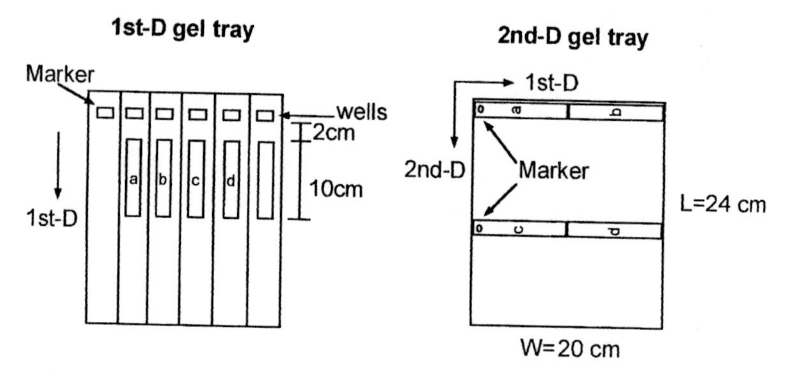

Figure 3. Neutral/Neutral 2D gel set-ups.

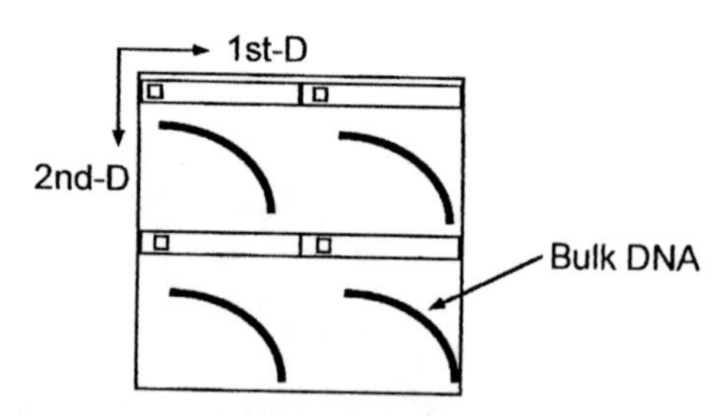

Figure 4. Bulk DNA pattern on Neutral/Neutral 2D gel.

2.4 Neutral/Alkaline (N/A) 2D gels

Protocol 4. Neutral/Alkaline 2D gels

Reagents

- See *Protocol 3*
- 1.2% agarose gel with 40 mM NaOH, 2 mM EDTA
- Running buffer: 40 mM NaOH, 2 mM EDTA
- Neutralizing buffer: 0.5 M Tris–HCl pH 7.5, 1.5 M NaCl

Method

1. For the first dimension of N/A 2D gels, prepare and run the DNA exactly as described for N/N 2D gels (see *Protocol 3*, part A, steps 1–3).
2. Immediately prior to the completion of the first dimension electrophoresis run, prepare a 1.2% agarose gel in dH_2O for the second dimension. Melt the agarose in water, cool it down to 50–60 °C, add NaOH (from 4–10 M stock) to 40 mM final concentration, and EDTA (from 0.5 M stock) to 2 mM final concentration. Keep the agarose solution in a ~ 60 °C water-bath. Pre-cool the running buffer (~ 2 litres of 40 mM NaOH and 2 mM EDTA) in a cold room.
3. After running the first dimension, cut out the gel lanes with a razor as described in *Protocol 3*, part B, step 3, and place the gel slices across the gel tray for the second dimension.
4. Pour in the agarose for the second dimension and let it set in a cold room for 20–30 min. Create the wells for size markers by depression in the gel with a hot spatula or a glass pipette as in *Protocol 3.*
5. Stain the size markers of the first dimension, destain, and photograph (UV Polaroid), or keep them at 4°C (either in 1 × TBE or wrapped in Saran Wrap on a glass plate) to photograph later, together with the markers of the second dimension.
6. Pour in the running buffer (see step 2 of this protocol) for the second dimension. Load size markers onto the gel. The second dimension should be run for 15–18 h, depending on the size range of the DNA fragments and gel conditions. The tracking dye is usually too faint to be seen after several hours in the alkaline buffer, but if it is still visible, it should run ~ 6–8 cm. Circulate the buffer occasionally (e.g. every 3–5 h) by pouring all the buffer into a beaker and back into the gel box (*turn off current from power supply before doing this*).
7. After running the second dimension, cut out the marker lane(s) and neutralize the gel in 0.5 M Tris–HCl pH 7.5, 1.5 M NaCl for ~ 1 h. Stain in ethidium bromide for 0.5–1 h, destain, and photograph (short wave UV). For the main gel, do not depurinate with HCl as small nascent

Protocol 4. *Continued*

strands from small bubbles and forks might not bind efficiently to the membrane. Soak the gel in 0.5 M NaOH, 1.5 M NaCl for 30 min. Do not neutralize the gel if Biotran(+) filters (ICN Biochemicals) are to be used. Transfer DNA from the gel to the membrane as in *Protocol 3*, part C, step 3, using 20 × SSC for 12–24 h, depending on the size range of the fragments of interest.

8. Fix the DNA onto the membrane, rinse and dry the membrane, and hybridize as in *Protocol 3*, parts D–F.

2.5 3D gels

Protocol 5. 3D gels

Reagents

- See *Protocols 3, 4,* and *9*

Method

1. Run a Neutral/Neutral 2D gel as described above, with the following modifications:
   (a) Use pulsed-field gel grade agarose for the second as well as the first dimension, so that the cut out gel slices for the third dimension are easier to handle.
   (b) Start with about 20 times as much DNA as for a Neutral/Neutral 2D gel.

   It is necessary to enrich for replication intermediates by BND cellulose chromatography when doing 3D gels.
2. Any time before completion of the second-dimension electrophoresis, cut out a piece of an unexposed, developed X-ray film to the same size as the 2D gel; this will be used as a mould for cutting gel slices for the third dimension. In addition, take a picture of the first-dimension size markers after staining with ethidium bromide, so that you will know where to cut out gel slices for the third dimension.
3. About 1 h prior to the completion of the second-dimension electrophoresis, prepare the gel for the third dimension. The gel conditions for the third dimension are the same as the second dimension of Neutral/Alkaline 2D gels (see *Protocol 4*, step 2). In a cold room, cast the gel in a large gel tray using lucite spacer bars to form two long troughs across the gel tray: one trough should be at the top and the other in the middle of the gel tray (see *Figure 5*). Each trough will hold two or three gel slices for the third dimension. Save some agarose (keep in a ~ 60 °C water-bath) for sealing the troughs later.

4. When the second-dimension run is finished, take out the gel, destain it briefly in water to remove traces of ethidium bromide, place the gel on a long wave UV box, and align the X-ray film on top of the gel. Working as quickly as possible to avoid nicking the DNA, mark the diagonal of the bulk DNA and the position of the well on the X-ray film mould under UV light.
5. Continue to destain the 2D gel in water for up to 1 h. Meanwhile, use the information of the size markers of the first dimension and the position of the bulk DNA to cut out areas on the X-ray film where the gel slices will be cut out. For study of fragments of 3–10 kb, six to eight gel slices of 6 cm long and 2 mm wide are sufficient. The asterisks in *Figure 5* mark the orientation of the gel slices for comparison with the next step.
6. Put the 2D gel onto a glass plate, with rulers at all four edges to hold the gel. Align the mould on top of the gel. Use a fresh razor to cut each of the long edges and a fresh surgical blade to cut each of the short edges of the gel slices. Take out each gel slice with a surgical blade and your gloved fingers, and place it into the trough of the third-dimension gel as diagrammed (see step 8 for blotting of the residual 2D gel). Seal the troughs with melted agarose. Let it set for ~ 15 min. Use a hot spatula to make a well for loading the size markers on the left of each trough (see *Figure 5*).
7. Soak the gel in pre-cooled alkaline buffer (40 mM NaOH, 2 mM EDTA) for 20 min to denature the DNA. Load the DNA size markers in alkaline loading buffer (see *Protocol 9* for recipe), and run the gel at 1 V/cm. Subsequent procedures are the same as in the second dimension of the Neutral/Alkaline 2D gel method (*Protocol 4*, steps 6–8), with the exception that for very short nascent DNA strands (0.2–0.5 kb) the third dimension may be run for a shorter time than the second dimension of Neutral/Alkaline 2D gels.
8. As each gel slice is cut out from the second-dimension gel to use for the third dimension (step 6), replace the missing slice in the second-dimension gel with a slice of agarose (lacking DNA). The 'stuffed' 2D gel is then photographed, denatured, and blotted (see 2D gel protocols). It is easier to handle the stuffed residual 2D gel by putting it on plastic Saran Wrap to move it about.

3. PCR analysis of nascent DNA strands

3.1 Background

This very useful and popular method for mapping an ORI by using PCR to analyse nascent DNA strands isolated from an asynchronous population of

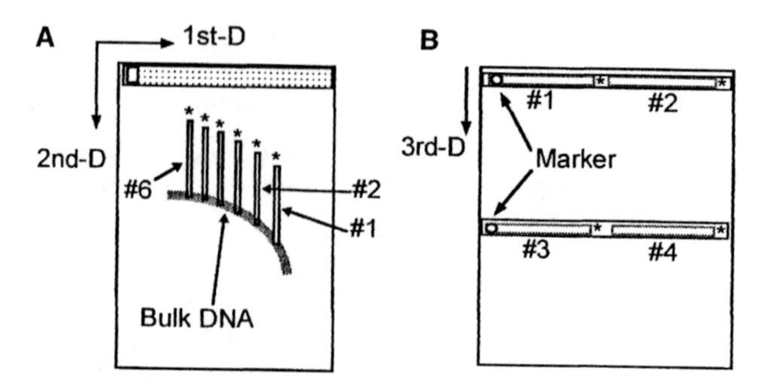

Figure 5. Schematic of 3D gel set-up. Panel (A) shows gel after running the first two dimensions. The asterisks orient the gel slices for the reader, to see how they are placed in the third-dimension gel (panel B). Modified from ref. 38.

cells was developed by Vassilev and Johnson (41). The basis of this method is that newly replicated DNA extends bidirectionally from the ORI and progressively increases in length, spanning adjacent sequences on the DNA. Total genomic DNA is isolated from cells, and the nascent DNA is purified away from unreplicated DNA; the nascent DNA is fractionated according to its size. To map an ORI using this method, a minimum of three unique DNA segments are selected that are distributed across the putative ORI region (A, B, C in *Figure 6*). The size fractionated nascent DNA is then used as a template in PCR reactions with pairs of oligonucleotide primers specific for each of the segments. Hybridization with probes for segments A, B, and C to the PCR products allows a sensitive assay of how much PCR product is present, providing a direct reflection of how much nascent DNA template was present (a moderate number of PCR cycles must be used to remain in the linear range). The goal of this analysis is to identify which stretch of DNA (e.g. segment A in *Figure 6*) has the shortest nascent strands, indicating that the ORI resides in, or very close to, the fragment of DNA used as a probe. An important control in this mapping procedure is that the smallest nascent strand sizes detected in the other segments (e.g. B and C) should increase in proportion to their distance from the deduced ORI location.

The advantage of using PCR analysis of nascent DNA to map an ORI is that the method is very sensitive and allows detection of single copy sequences; moreover, it does not require cell synchronization. This method is suitable to map an ORI whose general location has been suggested previously by other data (e.g. 2D gel mapping). This method allows one to estimate the location of a discrete ORI (should it exist) to within several hundred base pairs, and for nucleotide level mapping one has to turn to the RIP method described later in this chapter.

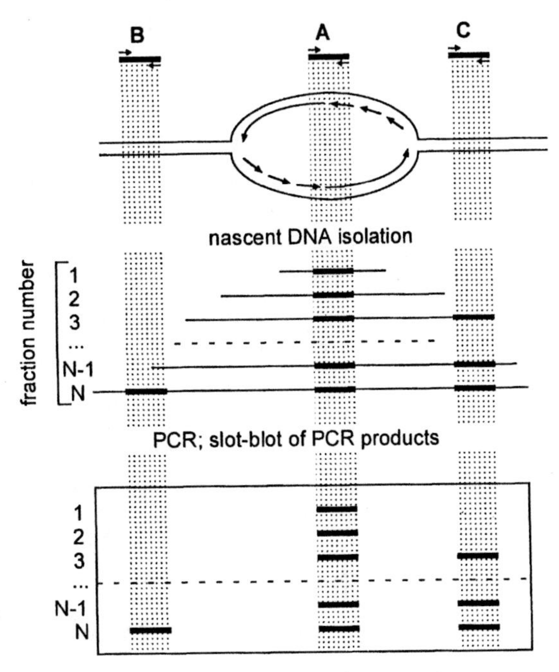

Figure 6. PCR analysis of nascent strands. Schematic drawing, modified from ref. 41. Segment A is closest to the ORI and segments B and C flank it on either side.

Two technical complications associated with the PCR method are:

(a) The nascent BUdR-containing DNA is light-sensitive and subject to breakage; this complicates intepretation of the results.

(b) The resolution reflects the accuracy of size fractionation to obtain distinct size classes of nascent DNA as template for PCR.

We have found that a good way to overcome the problems created by breakage of light-sensitive BUdR DNA is to use λ-exonuclease treatment instead of BUdR to recover nascent DNA. Details of the λ-exonuclease treatment are given in *Protocol 15*. Another advantage of using λ-exonuclease instead of BUdR is that it avoids the use of the metabolic inhibitor FUdR that might alter ORI activity. Differences in the procedure when using BUdR (flow scheme A) and λ-exonuclease (flow scheme B) are diagrammed in *Figure 7*.

3.2 DNA preparation for PCR analysis of nascent strands

Note that BUdR labelling can be omitted if using λ-exonuclease treatment (see *Protocols 14* and *15*).

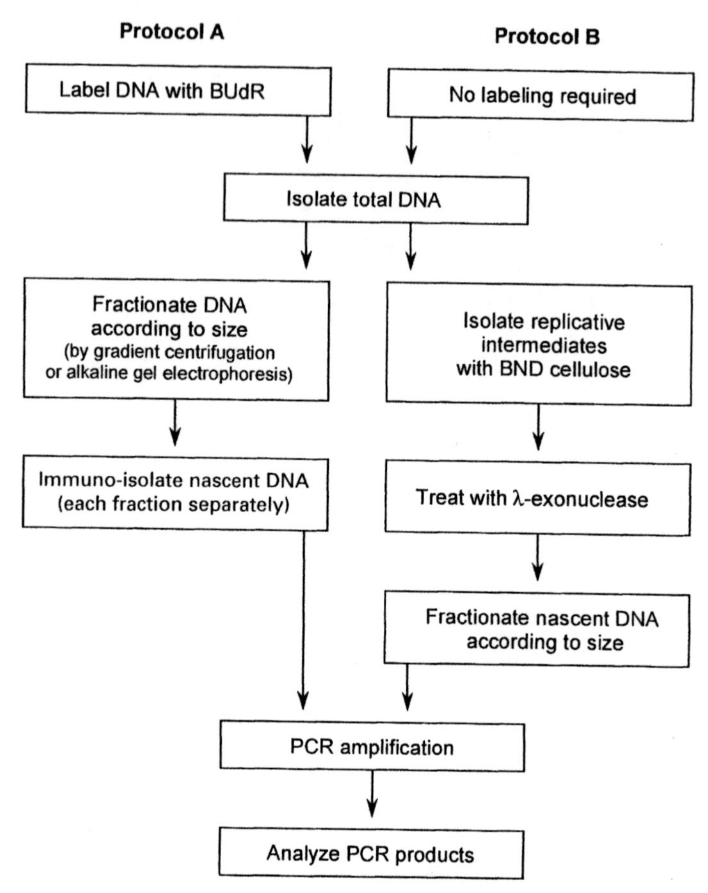

Figure 7. A flow scheme comparing protocols for BUdR and λ-exonuclease treatment to enrich nascent DNA.

Protocol 6. Labelling cells with BUdR

Reagents

- 20–100 μM BUdR, 20–100 μM FUdR in tissue culture medium
- PBS (see *Protocol 13*)

Method

1. The BUdR concentration should be customized for the particular cell type under study. For mammalian cells in culture, a pulse of 20 μM BUdR for about 10 min is satisfactory, whereas insect salivary glands in organ culture require 100 μM BUdR for 6 h.

2. The efficiency of label uptake can be improved by adding an equimolar amount of FUdR to the medium to inhibit thymidylate synthetase. In order to be able to trace the nascent DNA throughout the procedure, 5 μCi/ml [^{3}H]thymidine may be included.
3. Rinse cells or tissue twice with ice-cold PBS.[a]

[a] Once the cells have incorporated the BUdR into their DNA, in order to minimize DNA breakage all experimentation must be done in the dark or else under light passed through a yellow filter. It is important to be extremely gentle during all the DNA extraction steps and not to vortex at any of the steps.

Many DNA isolation methods can be used, but for PCR analysis DNA of nascent strands we prefer to use DNAZol (a guanidine–detergent solution) as it is rapid and less harsh than phenol:chloroform. We have applied this method to insect (*Sciara*) salivary glands and recover 15 μg DNA from 100 (amplification stage) salivary glands. The method given below is from the DNAZol manufacturer's instructions.

Protocol 7. DNA isolation from tissue samples or cultured cells

Reagents

- DNAZol (Gibco BRL Life Technologies)
- Ethanol (100%, 95%)
- 8 mM NaoH

Method

1. Add 1 ml DNAZol reagent to 25–50 mg tissue or 1–3 $\times$ 10^7 cells. Homogenize the cells or tissue samples in a hand-held glass Teflon homogenizer with a loosely fitting pestle. Homogenize with as few strokes as possible (e.g. five to ten strokes). Avoid pipetting for lysis as this will shear the DNA.
2. Centifugation (optional). Sediment the insoluble tissue fragments, RNA, and polysaccharides from the homogenate by a 10 min spin at maxium speed in a microcentrifuge. Following centrifugation, transfer the viscous supernatant containing DNA to a fresh tube.
3. Precipitate the DNA from the homogenate by the addition of 0.5 ml 100% ethanol per 1 ml DNAZol used for the isolation. Mix samples gently by inversion and store them at room temperature for 1–3 min. A cloudy precipitate of DNA should quickly become visible. Remove the DNA precipitate by spooling it onto a pipette tip and transfer it to a clean tube. Very small quantities of DNA cannot be spooled. In this case, centrifugation at 3000 *g* for 10 min in a microcentrifuge will pellet the DNA.
4. Wash the DNA twice with 1 ml 75% ethanol. For each wash, suspend

Protocol 7. *Continued*

the DNA in ethanol by inverting the tube a few times. Store the tube vertically for 1 min to allow the DNA to settle to the bottom of the tube (or briefly centrifuge the DNA); pipette off the ethanol. Air dry the DNA.

5. Dissolve the DNA at a concentration of about 0.3 μg/μl in 8 mM NaOH or alternatively in dH_2O. Either will denature the DNA at room temperature.[a]

[a]In the case of λ-exonuclease treatment, the DNA pellet should be dissolved in dH_2O or TE and not NaOH to prevent hydrolysis of the RNA primers.

3.3 Size fractionation of DNA

Newly synthesized DNA can be fractionated according to size by sedimentation through an alkaline sucrose gradient (*Protocol 8*) or by electrophoresis in an alkaline agarose gel (*Protocol 9*). The latter gives better resolution.

Protocol 8. Size separation of single-stranded DNA by alkaline gradient centrifugation

Equipment and reagents

- Beckman ultracentrifuge and SW41 rotor
- Sorvall refrigerated centrifuge and SS-34 rotor
- 0.2 M NaOH, 2 mM EDTA
- 1 M Tris pH 7.5
- 3 M Na acetate pH 5.2
- Pellet Paint (Novagen)

Method

1. Dissolve the DNA, which was isolated as described above, in 1 ml TE buffer and add NaOH to a final concentration of 0.2 M NaOH. Load the DNA solution on top of a 5–20% linear sucrose gradient containing 0.2 M NaOH and 2 mM EDTA.
2. Centrifuge for 16 h at 15°C at 50 000 *g* in a Beckman SW41 rotor.
3. Collect 0.45 ml fractions.
4. Take 50 μl aliquots from each fraction of the alkaline gradient and measure their radioactivity in a scintillation counter. Recall that the nascent DNA was labelled with ^3H.
5. Take another 50 μl aliquot from each gradient fraction and run them on an alkaline agarose gel (see *Protocol 9*) to determine their sizes.
6. Neutralize each alkaline gradient fraction with 0.1 vol. 2 M HCl. Add 1 M Tris pH 7.5, to a final concentration of 100 mM, 0.1 vol. 3 M Na acetate pH 5.2, and 2 μl of Pellet Paint co-precipitant (Novagen).

7. Add 1–2 vol. 95% ethanol, mix gently, and after incubating at –20°C for at least 1–2 h, spin down the DNA at 10000 r.p.m. in a Sorvall SS-34 rotor at 4°C for 30 min.
8. Dissolve the DNA in 20 μl 1 × TE and store at –20°C.

Protocol 9. Alkaline agarose gel electrophoresis

Reagents

- 6 × alkaline loading buffer: 300 mM NaOH, 6 mM EDTA, 18% Ficoll (type 400), 0.15% bromcresol green, 0.25% xylene cyanol FF
- 1 × alkaline buffer: 50 mM NaOH, 1 mM EDTA
- 10 × TBE: see *Protocol 3*
- DNA ladder: 100 bp or 1 kb (both from Gibco BRL Life Technologies)
- GELase (Epicentre Technologies)

Method

1. Melt agarose in dH_2O. Use 0.5% agarose for 5 kb and larger DNA, and 2.5% agarose for DNA smaller than 5 kb. If you are using analytical alkaline agarose gels to measure DNA sizes from ultracentrifugation gradient fractions (*Protocol 8*), you can run a different per cent agarose minigel for each gradient fraction. The DNA size pattern revealed in the minigels of the gradient fractions can be detected by standard Southern blot hybridization to a probe spanning the ORI region. On the other hand, if you choose to bypass the gradient, then use a preparative alkaline 1.2–1.5% low melting point agarose gel described as follows.
2. Cool the agarose solution to 60°C. Add NaOH to 50 mM and EDTA to 1 mM final concentration. Pour the gel (about 25 cm long gives good resolution) and let it harden. Add freshly made 1 × alkaline buffer and let the gel equilibrate with it overnight.
3. Load the aliquots of single-stranded DNA in alkaline loading buffer diluted to 1 × from a 6 × stock; do not overload lanes (load < 15 μg DNA/lane). Also load marker DNA (e.g. 100 bp or 1 kb DNA ladder) in the same loading buffer.
4. Alkaline agarose electrophoresis should be carried out at 50 V for 16–24 h at 4°C for a preparative alkaline gel. If low melting point agarose is used, the gel should be run in a cold room to prevent overheating.
5. Neutralize the gel with three changes of 0.5 × TBE buffer, each for 20 min.
6. Cut out the DNA marker lanes and stain them with ethidium bromide. The unstained sample DNA lanes should be cut into various size fractions (*fractions should be small and close to each other for better*

Protocol 9. *Continued*

resolution, especially if the primer sets are at short distances from one another and close to the ORI).

7. DNA is precipitated after digestion of the gel with GELase (as per the protocol from Epicentre Technologies). This fast procedure recovers DNA of any size in high yields.
8. Resuspend the pellet in dH_2O. Each size fractionated sample is sufficient for about ten PCR reactions, which will be done after selection of nascent BUdR labelled DNA (see Section 3.4).

3.4 Purification of BUdR labelled nascent DNA

Commonly, nascent DNA is purified from contaminating bulk genomic DNA by BUdR labelling and recovery of the labelled DNA by an anti-BUdR antibody, using either an immunoaffinity column (*Protocol 10*) or magnetic beads (*Protocol 11*). We prefer the latter method, as it allows for rapid and easy recovery of BUdR labelled DNA. An added convenience is that it is not necessary to uncouple the BUdR DNA:antibody complex from the magnetic beads prior to using it as a template in a PCR reaction. Use of the entire DNA:Dynabead complex for PCR avoids the use of harsh alkaline treatment and the risk that any remaining alkali would inhibit the efficiency of the PCR reaction.

Purification of BUdR labelled DNA can be replaced by λ-exonuclease treatment as described in *Protocols 14* and *15*, in which case λ-exonuclease treatment should follow the BND cellulose enrichment of RI DNA (*Protocol 2*). The λ-exonuclease procedure should be done *before* the size fractionation of DNA described in Section 3.3 (see *Figure 7*).

Protocol 10. Purification by immunoaffinity chromatography with anti-BUdR antibodies (42)

Equipment and reagents

- Sorvall refrigerated centrifuge with SS-34 rotor
- Poly-Prep 2 ml disposable chromatography column (Bio-Rad)
- Goat anti-mouse IgG coupled to CNBr-activated Sepharose 4B (Pharmacia)
- Proteinase K (200 μg/ml) in 0.5% SDS
- TBSE: 10 mM Tris–HCl pH 7.5, 150 mM NaCl, 0.1 mM EDTA
- Mouse anti-BUdR monoclonal antibody (Beckton Dickinson)
- 150 mM NaCl pH 11.5 (adjust pH with NH_4OH)
- Yeast tRNA

Method

1. Couple goat anti-mouse IgG to CNBr-activated Sepharose 4B (2.5 mg protein/ml gel) according to the manufacturer's protocol (Pharmacia).

2. Pour 0.5 ml coupled Sepharose into a Poly-Prep 2 ml disposable chromatography column.
3. Wash with at least 30 ml TBSE. Close the column with the bottom stopper, leaving 0.5 ml of buffer above the gel surface.
4. Add mouse anti-BUdR monoclonal antibody to a final concentration of 3 μg/ml.
5. Close the column with the top stopper and incubate for 2 h at room temperature with slow agitation to allow thorough mixing of the sorbent with the antibody solution.
6. Wash the column with 10 ml TBSE.
7. Denature DNA from each gradient fraction for 2 min in a boiling water-bath, and then chill on ice. Adjust the DNA sample to 0.1–10 μg DNA in 1.5–4 ml TBSE.
8. Add denatured DNA to the column. Incubate for 2 h at room temperature with slow agitation.
9. Drain out unbound DNA (assay drained solution for radioactivity).
10. Wash the column with 4 ml TBSE.
11. Elute the bound BUdR DNA:anti-BUdR mAb complex with 2 ml of 150 mM NaCl pH 11.5.
12. Neutralize the eluate to pH 8. Treat with proteinase K (200 μg/ml) in 0.5% SDS at 37 °C overnight.
13. Perform a phenol:chloroform extraction.
14. Add 20 μg yeast tRNA as carrier. Precipitate the sample with 1–2 vol. 95% ethanol at −20 °C for at least 2 h, followed by a 30 min spin in a Sorvall SS-34 rotor at 4 °C for 10 000 r.p.m.
15. Redissolve the BUdR labelled DNA in 50 μl TE.

Protocol 11. Purification of BUdR labelled nascent DNA using magnetic Dynabeads

Equipment and reagents

- Monosized (20 μm), supermagnetic, polystyrene beads with sheep anti-mouse IgG covalently bound to the surface (Dynabeads M-40 sheep anti-mouse IgG; Dynal)
- PBS plus 0.1% BSA (see *Protocol 13*)

Method

1. Heat denature the DNA in a boiling water-bath for a few minutes, cool on ice, and immediately mix with mouse monoclonal antibody

Protocol 11. *Continued*

against BUdR. Use a several-fold molar excess of antibody relative to the DNA.

2. Incubate with gentle rotation for 1 h at 4°C to form the primary antibody:DNA complex.
3. Wash the beads (suspension supplied at 4×10^8 beads/ml) according to the manufacturer's instructions (two washes in PBS buffer plus 0.1% BSA).[a]
4. Incubate the washed Dynabeads with the DNA:BUdR antibody suspension with gentle shaking for 6 h at 4°C.[b]
5. After the incubation, place the microcentrifuge tube in a magnetic test-tube holder to attract the immunocomplex to the wall of the test-tube. Carefully pipette off the solution.
6. Resuspend the beads in 50 μl of wash solution.
7. Repeat steps 5 and 6 several times to eliminate non-specific contamination.
8. Store the Dynabead complex in 50 μl of the last wash buffer at −20°C, taking special care to protect the BUdR DNA from light to minimize DNA breakage.
9. 1 μl of the entire complex can be directly placed in a PCR tube and used as template in a PCR reaction. The initial heat denaturation step of PCR makes the BUdR DNA accessible as template.

[a] Dynal recommends using 1×10^7 Dynabeads M-450 sheep anti-mouse IgG per 0.15–1.5 μg primary monoclonal antibody, but we found that it is best to determine this ratio separately for each specific case. For example, for 0.8 μg DNA from *Sciara* salivary glands we used 20 μl (250 μg/ml stock solution) of antibody against BUdR and 20 μl Dynabeads (30 mg/ml or 4×10^8 beads/ml).

[b] The incubation time and incubation temperature can be varied, with a lower concentration of Dynabeads requiring a longer incubation time and vice versa. No strong dependence of complex formation on incubation temperature was observed.

3.5 PCR amplification and Southern blot hybridization

Protocol 12. PCR reactions (43)

Equipment and reagents

- QIAquick PCR Purification kit (Qiagen)
- Nytran Plus nylon filter (Schleicher and Schuell)
- 10 × PCR buffer: 10 mM Tris–HCl pH 8.3, 50 mM KCl, 15 mM $MgCl_2$, 0.01% gelatin (w/v)
- 5 U/μl stock solution of Amplitaq DNA polymerase(Perkin Elmer)
- SSPE (1 x): 0.18 M NaCl, 10 mM Na phosphate pH 7.7, 1 mM EDTA
- Amplitaq DNA polymerase (Perkin Elmer)

Method

1. Set-up the PCR reaction mixture (all reagents were Perkin Elmer):
   - dNTPs (10 mM) — 40 μl each
   - 10 × PCR buffer — 200 μl
   - dH_2O — 160 μl
   - Amplitaq DNA polymerase — 10 μl from stock solution

   Mix well. Aliquot 26.5 μl/0.5 ml PCR tube. Store at –20 °C until ready to use. This aliquot is enough to do two 50 μl PCR reactions.
2. Primers: dilute the required primer sets to 0.5 μM of each primer per reaction, i.e. 25 pmoles primer/50 μl reaction. Use at least three primer sets encompassing the suspected ORI.
3. Template: if using plasmid DNA digested with restriction enzymes as the template (i.e. as a control to determine the PCR conditions), use 1 ng DNA per reaction. If using chromosomal DNA (e.g. from flies) as the template, use 100 ng DNA per reaction. Denature the template before adding it to the PCR reaction by heating at 94 °C for 5 min and transferring it immediately to ice. The final reaction mixture contains:
   - pre-made reaction mix — 13.25 μl
   - primers — 2 μl (from primer stock, solution of 25 pmoles/μl)
   - template — 1 μl (from stock of 100 ng/μl chromosomal DNA)
   - dH_2O to a final reaction volume of 50 μl
4. Example of PCR cycling conditions:
   (a) Denaturation: 94 °C for 20 sec.
   (b) Annealing: 50–65 °C for 90 sec.
   (c) Extension: 72 °C for 30 sec.

   Run the reaction for 25–30 cycles, but do not exceed 30 cycles.
5. Load one-third of the PCR mixture after amplification on a 4% agarose gel, and after running the gel, stain it with ethidium bromide to visualize the PCR products.
6. Purify the PCR products from the remaining two-thirds of the PCR mixture using the QIAquick PCR Purification kit.
7. Slot-blot the DNA from step 6 onto a nylon filter and hybridize it with oligonucleotides specific for each PCR product, as instructed by the manufacturer. Hybridize overnight with 1–5 × 10^6 c.p.m./ml probe (5′ end-labelled with [γ-^{32}P]ATP) (see *Protocol 17*) at 5 °C below the T_m[a] in 6 × SSPE, no SDS. The final wash can be done in 1 × SSPE, 0.5% SDS at the T_m of the hybrid.

[a] 2 °C for each AT bp and 4 °C for each GC bp.

3.6 Analysis of PCR results

If replication proceeds bidirectionally from an ORI, then the smallest size of nascent DNA that can act as template for a primer set specific for a given segment should be inversely proportional to the distance between that segment and the ORI, i.e. a segment that is far from the ORI should not have any short nascent DNA detected by the hybridization probe for that segment. In fact, however, some short nascent DNA might still be detected despite the theoretical prediction to the contrary. This may be due to several factors:

(a) The fractions of nascent DNA may overlap somewhat in terms of size.
(b) Replication might not always initiate at the same site, but rather within a broad zone.
(c) Okazaki fragments might be included in the fractions of short nascent DNA, so even a probe that detects longer nascent DNA would also detect the smaller Okazaki fragments.

None the less, the aim is to determine the shortest length of nascent DNA that can be detected by each PCR primer set. Each lane on the autoradiograms for probes A, B, and C should be scanned on a densitometer or a PhosphorImager and normalized against a control signal (for example, the same amount of plasmid DNA containing the ORI domain or bulk chromosomal DNA can be hybridized to each of the appropriate probes). An alternative approach to normalize the data is to compare the intensity of the band from fraction N (long nascent DNA; see *Figure 6*) from one panel to the next (i.e. lane to detect segment A, segment B, or segment C). The ratio obtained from such a comparison between panels should be equal to 1.0, and if it is not, a normalization factor can be derived to correct the ratio to 1.0. For each panel, the intensity of the band for each size fraction should be multiplied by the normalization factor derived for that panel to correct for the variation that may occur between different DNA segments (A, B, or C) during the amplification and hybridization steps.

Presenting the data as a ratio of the signal derived from the shortest nascent DNA detected in a given segment to the signal derived from the longest nascent DNA (fraction N in *Figure 6*) of that region will also correct for differences in specific activity of probes to different segments (such as A, B, or C) and differences in their hybridization efficiency. Since the same amount of nascent DNA is used as template for the various PCR reactions, use of a ratio as just described will also control for different efficiencies of different primer pairs in the various PCR reactions. The idea of using a ratio as a normalization procedure was developed by Yoon *et al.* (44) in their nascent strand abundance analysis, which is conceptually similar to the method described here but omits the use of PCR (and hence requires a large amount of nascent DNA as starting material).

Another way to analyse the composition of the pool of nascent DNA strands is by competitive PCR (45). This sophisticated method is based on simultaneous PCR amplification of the target nascent DNA and a competitor DNA fragment that acts as an internal standard. The competitor is a linear DNA fragment having the same sequence as target DNA except for a short unrelated insert in the middle. For each primer set, a fixed amount of the newly synthesized DNA sample is mixed with increasing amounts of the corresponding competitor and then co-amplified. The products obtained are resolved by polyacrylamide gel electrophoresis, stained with ethidium bromide, and quantified by densitometry. The ratio of signal obtained from the amplified competitor and the amplified target DNA is then plotted against the number of competitor molecules added to the reaction to solve for the initial concentration of the target. This method needs a large amount of nascent DNA to use with varying amounts of competitor DNA, and to check the parameters listed below.

Regardless of which normalization approach above is used, the following is advisable for PCR analysis of nascent strands:

(a) Several fractions of nascent DNA of different lengths should be analysed for each segment (e.g. A, B, and C). It is insufficient to analyse just short DNA fractions for each segment. As one moves away from the ORI, there will be little signal in the short DNA fractions and the size of the nascent DNA where signal first appears in these different segments can be used to confirm the position of the ORI.

(b) The resolution of this method will reflect the spacing of the segments analysed. The closer the segments are (e.g. A, B, and C) and the closer they are to the ORI, the better the resolution will be.

(c) The concentration of primers, polymerase, and nucleotides must be sufficiently large such that they are in excess and essentially constant during the initial PCR cycles, and decrease exponentially as the number of cycles increases. It is safest to use the same number of PCR cycles for all fractions and segments analysed. To ensure that the chosen number of PCR cycles is appropriate for the amount of template DNA, pilot PCR reactions should be performed on a dilution series of plasmid DNA with the ORI region (or on the competitor for competitive PCR). This will allow one to find the specific conditions that maintain a linear relationship between the various concentrations of template DNA and the intensity of the resulting PCR bands. It would be prudent to not exceed 30 cycles of amplification in order to maintain the reaction in the linear range.

4. Replication initiation point (RIP) mapping

4.1 Background

Although the progress in the development of origin of bidirectional replication (OBR) mapping techniques has allowed identification of regions containing cellular origins in eukaryotes, it is still unknown for any of those eukaryotic cell origins if DNA synthesis initiates randomly or at distinct sites within the origin region. In 1982 Hay and DePamphilis published a detailed map of initiation sites for the SV40 origin (46). Unfortunately, their elegant approach was not sensitive enough to be applied to eukaryotic origins carried on extrachromosomal plasmids or at chromosomal sites. This was the reason for us to develop a new, more sensitive assay in order to map initiation sites of DNA synthesis at the nucleotide level of resolution. This new method is called replication initiation point (RIP) mapping and does not require more than 2–5 ng of target sequence (*c.* 1000 times less than for the original Hay and DePamphilis approach). We worked out the basic methodology of RIP mapping (47) using the well understood SV40 system (48), and have successfully applied this method to yeast ARS1 (48). It should be possible to use RIP mapping to localize DNA initiation sites within origins of higher eukaryotes, especially if one studies an amplified locus like DNA puff II/9A in the fly *Sciara coprophila* or the amplified DHFR locus in CHOC 400 cells.

The principle of RIP mapping is to detect the DNA 5′ ends of nascent DNA strands by primer extension using radiolabelled oligonucleotides that anneal outside the ORI region. Clearly, the ORI should be mapped to less than a 1 kb region (see procedures above) before RIP mapping is applied. The key step that allows sucess in RIP mapping is the treatment of isolated nascent DNA with λ-exonuclease to eliminate nicked DNA that would lead to a high background in the subsequent primer extension reaction. Nascent DNA is protected by its RNA primer from the 5′ to 3′ exonuclease activity of λ-exonuclease, unlike nicked DNA ends that are digested.

Details of the RIP assay (47) are compared to the original Hay and DePamphilis approach (46) in *Figure 8*. Both methods map the junction of nascent DNA with the RNA primer (for the RIP method, this is because no available DNA polymerase nor reverse transcriptase can switch from a DNA to an RNA template).

4.2 Labelling and isolation of DNA for RIP mapping

The actual isolation procedure may vary with the kind of DNA that will be studied, e.g. viral, plasmid, or cellular chromosomal DNA. For the purification of episomal, high copy number DNA from infected or transfected mammalian cell cultures, the Hirt lysis procedure (49) is quite effective. For the isolation of chromosomal DNA from yeast, cell cultures, or tissue, procedures described earlier in this chapter are useful. In general, since the heart of this method is the presence of an RNA primer at the 5′ end of 'real' nascent DNA (as opposed to nicked DNA), special care must be taken to avoid anything that might degrade the RNA primers of nascent DNA. To ensure the highest possible efficiency for the following steps, a major first goal

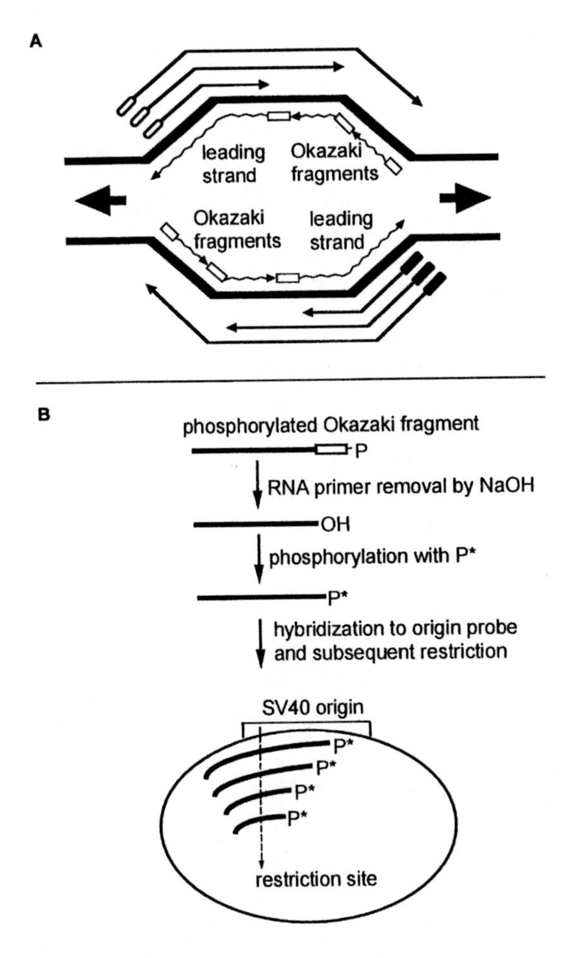

Figure 8. Comparison of replication initiation point (RIP) mapping and the 'unmasking' assay. (A) RIP mapping (47) is performed by primer extension on nascent DNA (wavy lines, open rectangles represent RNA primers) derived from each parental strand. Primer extension products are shown as arrows outside the replication bubble stopping at RNA:DNA junctions. (B) In the 'unmasking' assay of Hay and DePamphilis (46), the length of nascent DNA is detected after selective hybridization of labelled Okazaki fragments to an origin probe complementary to one or the other parental strand and subsequent restriction at a unique site.

is to isolate DNA free of proteins and RNA. That can be achieved by purifying the the DNA in a CsCl density equilibrium gradient. The following protocol describes a labelling and isolation procedure for circular molecules such as viral or plasmid DNA replicated in mammalian tissue culture cells (modified from ref. 50).

Protocol 13. Labelling and isolation of nascent (viral) DNA

Equipment and reagents

- 50 ml polypropylene tubes (Falcon 2070)
- Beckman ultracentrifuge with Ti50 and Ti60 rotors and tubes
- PBS: 137 mM NaCl, 3 mM KCl, 6.4 mM Na_2HPO_4, 1.5 mM KH_2PO_4, 0.5 mM EDTA pH 7
- Sorvall refrigerated centrifuge with SS-34 rotor
- Hirt solution: 0.6% SDS, 10 mM Tris–HCl pH 7.6, 20 mM EDTA
- TE buffer: 10 mM Tris–HCl pH 7.6, 1 mM EDTA

Method

Cell culture should be carried out in 100 mm diameter cell culture dishes. Eight of these dishes with SV40-infected monkey CV 1 cells will yield approx. 20 μg purified replicating SV40 DNA.

1. For each plate, remove culture medium from the plate 36 h after infection/transfection (time might vary for transfections). Add 1 ml culture medium containing 100 μCi [^{3}H]thymidine.
2. Incubate each plate at 37 °C for 30 min.
3. Remove culture medium from all dishes and add 10 ml ice-cold PBS to stop replication.
4. Remove PBS buffer and add 2.7 ml Hirt solution to each plate.
5. After (at least) 15 min at room temperature add 0.7 ml 5 M CsCl to each plate.
6. Swirl gently and then scrape off lysate and pool four dishes into a 50 ml polypropylene tube.
7. Leave tubes on ice for at least 12 h.
8. Transfer the contents of each tube into an autoclaved 30 ml Corex tube and spin in a Sorvall SS-34 rotor for 30 min at 14 000 r.p.m. at 4 °C.
9. Pool all supernatants into one autoclaved 100 ml graduated cylinder and add 1 g CsCl for every 1 ml solution.
10. Cover the top of the cylinder with Parafilm and dissolve the CsCl very gently (replicative intermediates are fragile) by rocking the cylinder back and forth slowly (no shaking).
11. Read refractive index (should be between 1.3994–1.4002), and adjust it if necessary with 5 M CsCl solution.
12. Let the solution sit at room temperature for 1 h; a large protein precipitate will float on top of the solution. Pipette off the precipitate with a very wide bore pipette and discard it.
13. Spin gradients in a Beckman Ti50 rotor at 45 000 r.p.m. for 18–20 h at 20 °C.

14. Fractionate each gradient into approx. 12 × 1 ml fractions. Collect corresponding fractions from duplicate gradients into the same tube.
15. Count ^{3}H c.p.m. in 10 μl aliquots of each fraction and pool ^{3}H-containing fractions (replicating DNA).
16. Add 3 vol. TE buffer, mix gently, and spin in a Beckman Ti60 rotor overnight at 50 000 r.p.m. and 15 °C to pellet the DNA.
17. Remove the supernatant very carefully, so as not to dislodge the pelleted DNA which is not tightly bound to the tube. If you make a hole in the tube slightly above the pellet, the supernatant can flow out without disturbing the sedimented DNA.
18. Transfer the DNA pellet into a sterile 1.5 ml Eppendorf tube and let it air dry.
19. Dissolve the DNA in 100 μl TE buffer with gentle shaking; do not pipette it up and down as the DNA might stick inside the pipette tip. Store at 4 °C until needed.

4.3 Enrichment of replication intermediate (RI) DNA on BND cellulose

The procedure is basically as described in *Protocol 2*, however it is important that the BND cellulose column is washed very carefully prior to loading DNA. Fine BND particles that come through the filter at the bottom of the column must be washed away. Otherwise, they might contaminate the DNA preparation and interfere with the subsequent reactions.

4.4 Phosphorylation of RI DNA by T4 polynucleotide kinase

This step serves to phosphorylate all free DNA ends (lacking an RNA primer) and thus make them recognizable as substrates for λ-exonuclease. Since a 5′ DNA hydroxyl end at a nick in double-stranded DNA is unlikely to be phosphorylated, RI DNA is heat denatured at 100 °C for 2 min and then immediately cooled to 0 °C in ice water.

Protocol 14. Phosphorylation of RI DNA

Reagents

- 50 mM ATP stock solution
- T4 polynucleotide kinase and kinase buffer
- 5% Sarkosyl
- 250 mM EDTA
- 20 mg/ml proteinase K stock solution

Method

1. The T4 polynucleotide kinase reaction is carried out in a total volume of 20 μl:
   - 10 μl heat denatured RI DNA (one-tenth of the sample from *Protocol 13*)

Protocol 14. *Continued*

- 2 μl ATP (50 μM final concentration, diluted from a 50 mM stock solution)
- 1 μl T4 polynucleotide kinase (10 U)
- 2 μl 10 × T4 polynucleotide kinase buffer
- 5 μl dH_2O (autoclaved and filter sterilized)

2. Incubate at 37 °C for 30 min.
3. To stop the reaction add:
   - 1 μl 5% Sarkosyl
   - 2 μl 250 mM EDTA
   - 2 μl proteinase K (625 μg/ml, diluted from stock solution)
4. Incubate for 1 h at 37 °C.
5. Extract the sample once with 25 μl phenol:chloroform:isoamyl alcohol (25:24:1, by vol.) and once with chloroform:isoamyl alcohol (24:1, v/v).
6. Precipitate the DNA with 0.1 vol. 3 M Na acetate pH 5.2, and 2 vol. ethanol at –20 °C overnight.
7. Spin DNA in an Eppendorf centrifuge at 10 000 r.p.m. at 4 °C for 15 min.
8. Wash the pellet once with 70% ethanol, and let it air dry.
9. Resuspend the DNA in 20 μl 10 mM Tris pH 8 and keep it on ice if proceeding immediately to the next step. Alternatively, DNA can be stored at 4 °C in 10 μl TE and later add 10 μl H_2O just before proceding to the λ-exonuclease digestion (the dilution is important as too much EDTA will inhibit λ-exonuclease). Since the sample still contains free ATP, a calculation of the recovered amount of DNA by spectrophotometry gives unreliable results. However a 1 μl aliquot can be run on a small 0.9% agarose gel to estimate the amount of DNA in the sample.

4.5 λ-exonuclease treatment of RI DNA

Protocol 15. λ-exonuclease digestion

Reagents

- 2.5 × λ-exonuclease buffer: 167.5 mM glycine–KOH pH 8.8, 6.25 mM $MgCl_2$, 125 μg/ml BSA
- λ-exonuclease (Gibco BRL Life Technologies)

Method

1. Split the 20 μl sample of phosphorylated RI DNA in half, so you can use each half for a separate experiment.

2. Digestion with λ-exonuclease is carried out in a total volume of 20 μl:
   - 10 μl RI DNA
   - 8 μl 2.5 × λ-exonuclease reaction buffer
   - 2 μl λ-exonuclease (3–3.5 U/μl)
3. Incubate at 37 °C for 12 h. λ-exonuclease is slow to digest heat denatured, single-stranded DNA. The optimal pH for λ-exonuclease is 9.4. However, degradation of RNA primers can occur at pH 9.4, since RNA can be hydrolysed in weak alkali as low as pH 9. Hence, the pH of the reaction buffer used here is titrated to pH 8.8.
4. As a control, check λ-exonuclease activity on restricted, non-replicating phosphorylated DNA from the flow-through fraction of the BND cellulose column. If the DNA is restricted to produce 5′ phosphorylated ends, it has to be heat denatured for 2 min at 100 °C prior to incubation with λ-exonuclease.
5. Treat the control as described in steps 2 and 3. As an additional control, use another sample from step 4 to treat in parallel to steps 2 and 3 but omit λ-exonuclease.
6. Run an aliquot of the two controls on a 0.9% agarose gel to check that the λ-exonuclease digestion was complete.
7. If the digestion was incomplete, incubate at 37 °C for an additional 3–4 h, after adding:
   - 2 μl 2.5 × λ-exonuclease reaction buffer
   - 1 μl λ-exonuclease
   - 2 μl dH_2O
8. Repeat step 6 (0.9% agarose gel) to check the completeness of digestion. Also check that reaction buffers and enzyme preparations do not have RNase activity by incubating 2 μg tRNA in a total volume of 10 μl as described in steps 2 and 3, and running a 2.5% agarose gel as in step 6.
9. Once the λ-exonuclease digestion is complete, proceed by heating the samples to 75°C for 10 min to inactivate the λ-exonuclease and then immediately cool on ice.
10. Extract once with chloroform:isoamyl alcohol (24:1, v/v) and store at 4°C until use in primer extension reactions.

4.6 Primer extension reaction

Choice of primers: the length of the primers may vary between 23–27 nucleotides, and their GC content should be at least 40% and preferably 50% or more. The T_m should be 68–70 °C. In rare cases, primers that fulfil these requirements might not work on replicative intermediates, although they work in the control reaction (see below). The reason for that is unclear.

Protocol 16. Primer phosphorylation with [γ-^{32}P]ATP

Equipment and reagents

- See *Protocol 13*
- Sorvall GLC-2B centrifuge
- Sephadex G25 spin column (fill QS-Q Quick-Sep column, Isolab, with pre-swollen Sephadex)

Method

1. Place a sterile Eppendorf tube on ice. The labelling reaction is carried out in a total volume of 10 μl containing:
   - 2 μl primer (200 ng)
   - 1 μl 10 × T4 polynucleotide reaction buffer
   - 1 μl T4 polynucleotide kinase (10 U/μl)
   - 5 μl dH_2O
   - 1 μl [γ-^{32}P]ATP (5000 Ci/mmol, 150 mCi/ml) (use only fresh label)
2. Incubate on ice for 1 h.
3. Prepare a Sephadex G25 spin column.
4. Spin for 3 min at 2000 r.p.m. in a Sorvall GLC-2B centrifuge.
5. Add 1 ml TE buffer and spin again; do this step twice.
6. Add 40 μl TE buffer to the labelling reaction and load it onto the column.
7. Spin for 3 min at 2000 r.p.m. in a Sorvall GLC-2B centrifuge.
8. Add 50 μl TE buffer to the column and spin again. This eluate will be pooled with the eluate of step 7 that still remains at the bottom of the tube.
9. Count 1 μl of the eluate in a scintillation counter. (Specific radioactivity should be 10^8 c.p.m./μg, assuming 100% yield.)
10. Store the labelled primer at –20°C until needed; it is best to use it within one week.

Protocol 17. Primer extension reaction

Equipment and reagents

- PCR machine
- 10 mM dNTP stock solutions (Clontech)
- 100 mM $MgSO_4$
- Vent (exo-) DNA polymerase and 10 × buffer (New England Biolabs)
- 3 M Na acetate pH 5.2
- Formamide loading buffer: 95% formamide, 0.025% bromphenol blue, 0.025% xylene cyanol, 0.5 mM EDTA, 0.025% SDS

Method

Use double-stranded, non-replicating DNA restricted at a unique site as a positive control. This will give you a single extension product of defined size, having extended up to the restriction site. This control will indicate if the primer anneals at any other places besides the intended place, in which case you will get more than one band. The amount of RI DNA used as template is calculated from the amount that was determined *before* λ-exonuclease treatment.

1. Mix on ice with a sterile Pipetman tip containing an aerosol-resistant filter, in this order:
   - 1 μl dNTP (from each dNTP stock solution)
   - 3.5 μl 100 mM $MgSO_4$
   - 3 μl 10 × Vent (exo-) DNA polymerase buffer
   - 1 μl Vent (exo-) DNA polymerase (2 U/μl)
   - 17 μl dH_2O (autoclaved and filter sterilized)
   - 2 μl template DNA (2–5 ng for highly purified DNA such as viral or plasmid DNA, or 200–500 ng for yeast chromosomal DNA)
   - 2.5 μl radiolabelled primer
2. Overlay with oil (a good precaution, even for PCR machines with heated lids).
3. Run 30 cycles with an initial denaturation step at 95 °C for 4 min:
   - 1 min at 94 °C
   - 1 min at 70 °C
   - 1.5 min at 72 °C

   (conditions for Perkin Elmer PCR System 2400).
4. Prepare as many Eppendorf tubes as you have primer extension samples and add to each:
   - 1 μl tRNA (0.5 μg/μl)
   - 2.5 μl 3 M Na acetate pH 5.2
   - 16.5 μl TE buffer
5. Add the 30 μl primer extension mixture from steps 1–3.
6. Extract with 50 μl chloroform:isoamyl alcohol (24:1, v/v).
7. Add 100 μl 95% ethanol and precipitate on dry ice for 20 min, or at least 12 h at –20 °C.
8. Spin down in an Eppendorf centrifuge at 10 000 r.p.m. for 15 min at 4 °C.
9. Decant and let the pellet air dry.
10. Dissolve the pellet in formamide loading buffer.

Protocol 17. *Continued*

11. Heat samples to 80 °C for 5 min. Cool on ice.
12. Load samples onto a 6–8% acrylamide–urea sequencing gel in 1 × TBE, pre-run until the gel has reached 40 °C.
13. Load sequencing reactions performed with the same primer (T7 Sequenase v 2.0 sequencing kit, Amersham) in adjacent gel lanes.
14. Run gel until bromphenol blue marker has just run out of the gel.
15. Dry the gel under vacuum and expose it to a PhosphorImager plate overnight or to X-ray film.

Acknowledgements

We thank Bonita Brewer, Joel Huberman, and York Marahrens for laboratory protocols on isolation of yeast replicating DNA and 2D gels; Michael Leffak and Mauro Giacca for advice on PCR analysis of nascent strands; Joyce Hamlin for stimulating discussion; Carol Newlon and Joel Huberman for fruitful discussions when we developed the RIP mapping method; Corrado Santocanale (Diffley lab) for sharing the protocols on primer labelling and extension; Clifton McPherson (Zaret lab) for sharing protocols on handling primer extension products; and Carmen González for typing this chapter. This work was supported by NIH GM35929 to S. A. G., and all authors contributed equally to this article.

References

1. Jacob, F., Brenner, S., and Cuzin, F. (1964). *Cold Spring Harbor Symp. Quant. Biol.*, **43**, 129.
2. Marczynski, G. T. and Shapiro, L. (1993). *Curr. Opin. Genet. Dev.*, **3**, 775.
3. Donovan, S. and Diffley, J. F. X. (1996). *Curr. Opin. Genet. Dev.*, **6**, 203.
4. Hassell, J. A. and Brinton, B. T. (1996). In *DNA replication in eukaryotic cells* (ed. M. L. DePamphilis), pp. 639–78. Cold Spring Harbor Laboratory Press, Plainview, NY.
5. Marahrens, Y. and Stillman, B. (1992). *Science*, **255**, 817.
6. Krysan, P. J. and Calos, M. P. (1991). *Mol. Cell. Biol.*, **11**, 1464.
7. Krysan, P. J., Smith, J. G., and Calos, M. P. (1993). *Mol. Cell. Biol.*, **13**, 2688.
8. Smith, J. G. and Calos, M. P. (1995). *Chromosoma*, **103**, 597.
9. Aladjem, M. I., Groudine, M., Brody, L. L., Dieken, E. S., Fournier, R. E. K., Wahl, G. M., *et al.* (1995). *Science*, **270**, 815.
10. Gilbert, D. M., Miyazawa, H., and DePamphilis, M. L. (1995). *Mol. Cell. Biol.*, **15**, 2942.

11. Vassilev, L. T. and DePamphilis, M. L. (1992). *Crit. Rev. Biochem. Mol. Biol.*, **27**, 445.
12. Hamlin, J. L. and Dijkwel, P. A. (1995). *Curr. Opin. Genet. Dev.*, **5**, 153.
13. DePamphilis, M. L. (1996). In *DNA replication in eukaryotic cells* (ed. M. K. DePamphilis), pp. 45–86. Cold Spring Harbor Laboratory Press, Plainview, NY.
14. Vaughn, J. P., Dijkwel, P. A., and Hamlin, J. L. (1990). *Cell*, **61**, 1075.
15. Dijkwel, P. A., Vaughn, J. P., and Hamlin, J. L. (1991). *Mol. Cell. Biol.*, **11**, 3850.
16. Dijkwel, P. A. and Hamlin, J. L. (1992). *Mol. Cell. Biol.*, **12**, 3715.
17. Handeli, S., Klar, A., Meuth, M., and Cedar, H. (1989). *Cell*, **57**, 909.
18. Burhans, W. C., Vassilev, L. T., Wu, J., Sogo, J. M., Nallaseth, F. S., and DePamphilis, M. L. (1991). *EMBO J.*, **10**, 4351.
19. Heintz, N. H. and Hamlin, J. L. (1982). *Proc. Natl. Acad. Sci. USA*, **79**, 4083.
20. Burhans, W. C., Selegue, J. E., and Heintz, N. H. (1986). *Proc. Natl. Acad. Sci. USA*, **83**, 7790.
21. Burhans, W. C., Selegue, J. E., and Heintz, N. H. (1986). *Biochemistry*, **25**, 441.
22. Heintz, N. H. and Stillman, B. (1988). *Mol. Cell. Biol.*, **8**, 1923.
23. Leu, T.-H. and Hamlin, J. L. (1989). *Mol. Cell. Biol.*, **9**, 523.
24. Vassilev, L. T., Burhans, W. C., and DePamphilis, M. L. (1990). *Mol. Cell. Biol.*, **10**, 4685.
25. Pelizon, C., Divacco, S., Falaschi, A., and Giacca, M. (1996). *Mol. Cell. Biol.*, **16**, 5358.
26. Anachkova, B. and Hamlin, J. L. (1989). *Mol. Cell. Biol.*, **9**, 532.
27. Burhans, W. C., Vassilev, L. T., Caddle, M. S., Heintz, N. H., and DePamphilis, M. L. (1990). *Cell*, **62**, 955.
28. DePamphilis, M. L. (1993). *Annu. Rev. Biochem.*, **62**, 29.
29. Russev, G. and Vassilev, L. (1982). *J. Mol. Biol.*, **161**, 77.
30. Razin, S. V., Kekelidze, M. G., Lukanidin, E. M., Scherrer, K., and Georgiev, G. P. (1986). *Nucleic Acids Res.*, **14**, 8189.
31. Zannis-Hadjopoulos, M., Persico, M., and Martin, R. G. (1981). *Cell*, **27**, 155.
32. Kaufmann, G., Zannis-Hadjopoulos, M., and Martin, R. G. (1985). *Mol. Cell. Biol.*, **5**, 721.
33. Brewer, B. J. and Fangman, W. L. (1987). *Cell*, **51**, 463.
34. Huberman, J. A., Spotila, L. D., Nawotka, K. A., El-Assouli, S. M., and Leslie, R. D. (1987). *Cell*, **51**, 473.
35. Fangman, W. L. and Brewer, B. J. (1991). *Annu. Rev. Cell Biol.*, **7**, 375.
36. Brewer, B., Lockshon, D., and Fangman, W. (1992). *Cell*, **71**, 267.
37. Liang, C., Spitzer, J. D., Smith, H. S., and Gerbi, S. A. (1993). *Genes Dev.*, **7**, 1072.
38. Liang, C. and Gerbi, S. A. (1994). *Mol. Cell. Biol.*, **14**, 1520.
39. Wu, J.-R. and Gilbert, D. M. (1995). *Nucleic Acids Res.*, **23**, 3997.
40. Dijkwel, P. A., Vaughn, J. P., and Hamlin, J. L. (1991). *Mol. Cell. Biol.*, **11**, 3850.
41. Vassilev, L. and Johnson, E. M. (1989). *Nucleic Acids Res.*, **19**, 7693.
42. Contreas, G., Giacca, M., and Falaschi, A. (1992). *Biotechniques*, **12**, 824.
43. Trivedi, A. A., Waltz, S. E., and Leffak, M. (1998). In preparation.
44. Yoon, Y., Sanchez, J. A., Brun, C., and Huberman, J. A. (1995). *Mol. Cell. Biol.*, **15**, 2482.
45. Diviacco, S., Norio, P., Zentilin, L., Menzo, S., Clementi, M., Biamonti, G., *et al.* (1992). *Gene*, **122**, 313.
46. Hay, R. and DePamphilis, M. L. (1982). *Cell*, **28**, 767.

47. Gerbi, S. A. and Bielinsky, A.-K. (1997). In *Methods: a companion to methods in enzymology 13* (ed. M. De Pamphilis), pp. 271–80. Academic Press, NY.
48. Bielinsky, A.-K. and Gerbi, S. A. (1998). *Science*, **279**, 95.
49. Hirt, B. (1967). *J. Mol. Biol.*, **26**, 365.
50. DePamphilis, M. L. (1995). In *Methods in enzymology* (ed. J. Campbell), pp. 628–70. Academic Press, NY.

2

Purification of yeast replication origin binding proteins

SHANE DONOVAN, SIMON DOWELL, JOHN DIFFLEY, and ADELE ROWLEY

1. Introduction

This chapter provides practical advice on the biochemical purification of proteins that bind to replication origins in the budding yeast *Saccharomyces cerevisiae*. Genetic and biochemical analysis in budding yeast has facilitated much of our current understanding about the initiation of replication of chromosomal DNA in eukaryotic cells, in large part because only in this organism have the DNA sequences that serve as replication origins been identified and characterized in detail. Replication origins in *S. cerevisiae* consist of discrete 100–300 bp sequences termed autonomously replicating (ARS) sequences that are both necessary and sufficient for replication of plasmid DNA. At least a subset of ARSs function as origins of replication in their native chromosomal context. The combined power of biochemical and genetic approaches available in yeast has led to the identification of proteins required for initiation of DNA replication that interact, either directly or indirectly, with ARS sequences. Furthermore, recent progress has demonstrated the periodic assembly of complexes involving these proteins at replication origins during the cell cycle and that this process is regulated by cyclin-dependent kinases (CDKs). The goal of this chapter is to provide protocols and necessary information for the purification of three origin interacting factors: ORC, Abf1p, and Cdc46p, a member of the Mcm family of proteins. Recent evidence also implicates Cdc6p, Dbf4p, and Cdc45p as origin interacting factors. The characteristics of Cdc6p, Dbf4p, and Cdc45p will be briefly reviewed, however no protocols are yet available for the purification of these proteins from yeast. For more comprehensive discussion of ARS sequence characteristics, ARS binding proteins, and cell cycle regulation of initiation the reader is directed to a number of recent reviews (1–6).

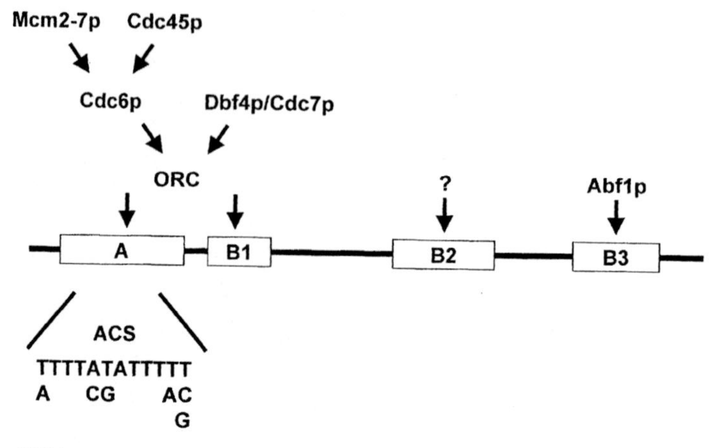

Figure 1. ARS1 structure. Functional elements of ARS1 are shown as boxes and the binding sites for ORC and Abf1p are indicated. Cdc6p and Dbf1p interact with origins indirectly via ORC. The interaction of Mcm proteins with origins is dependent upon ORC and Cdc6p. The Cdc45p interaction with origins is similar to that of Mcm proteins and is speculatively shown to be through Cdc6p, although this has not yet been established.

2. Origin sequences

A large number of studies have identified key features of budding yeast replication origins, as exemplified by ARS1, an ARS that functions both on plasmids and in its normal chromosomal location (*Figure 1*). All ARSs contain an exact or very close match to a core 11 bp consensus sequence called the ARS consensus sequence (ACS) that is absolutely required for the initiation of replication. Detailed linker scanning mutagenesis of ARS1 has revealed that the ACS in ARS1 is contained within a slightly larger functional domain termed domain A. This analysis has also demonstrated that domain A is flanked by the larger domain B which is essential but contains three important but individually non-essential elements: B1, B2, and B3. Mutations in B1 have the greatest effect on replication initiation while those in B2 are more deleterious than B3 mutations (7). While all ARSs so far characterized contain an equivalent of the B domain, its composition varies in some details, for example not all ARSs have a B3 equivalent. Subsequent detailed mutagenesis of a second ARS, ARS307, has shown that this origin also has B1 and B2 subdomains and that these, despite little sequence conservation, are functionally interchangeable with B1 and B2 of ARS1 (8, 9).

3. Sequence-specific DNA binding proteins

3.1 ORC

The ACS of all ARSs is recognized *in vivo* and *in vitro* by the six subunit origin recognition complex (ORC) (10, 11). ACS recognition by ORC is essential for the initiation of DNA replication *in vivo* and for DNA binding *in vitro*. Essential genes (*ORC1–6*) encoding each of the six ORC subunits (Orc1p, 120 kDa; Orc2p, 72 kDa; Orc3p, 62 kDa; Orc4p, 56 kDa; Orc5p, 53 kDa; Orc6p, 50 kDa) have been cloned and mutations in *ORC2* and *ORC5* have been shown to cause elevated plasmid loss rates and a significant reduction in origin firing (12–18). In *orc2* and *orc5* mutants this plasmid maintenance defect can be suppressed by the incorporation of multiple ARSs, consistent with a role for ORC at replication origins (16). In addition to its role in DNA replication, ORC is also involved in transcriptional silencing (12, 15–21).

In addition to an intact ACS, ORC binding to the ACS also requires ATP (10). Recent studies have demonstrated that both Orc1p and Orc5p bind ATP and that Orc1p can hydrolyse ATP. ATP binding to Orc1p is responsible for the ATP dependence of sequence-specific ORC–ACS interactions (22). While the ACS and ATP are essential for ORC binding to ARS sequences the B1 element of ARS1 also plays an important role (23, 24). The ORC footprint on all ARSs encompasses not only the ACS but also ACS flanking sequences in domain B. B2 and B3 mutations have no effect on ORC binding but B1 mutations decrease ORC binding both *in vitro,* under certain footprinting conditions, and *in vivo* (6, 24).

Genomic footprinting analysis, in which cells are permeabilized and immediately treated with nuclease to examine protein:DNA interactions in chromatin, has demonstrated that budding yeast replication origins exist in two distinct states during the cell cycle (25). The post-replicative state persists from early S phase to late mitosis and detailed comparisons of genomic and *in vitro* footprints at ARS1 have revealed that this state can be accounted for by the binding of ORC and Abf1p (25) (Section 3.2). Consistent with this, in *orc2–1* temperature-sensitive mutants the post-replicative state is thermolabile (26). The pre-replicative state is present from late mitosis until the G1/S transition and is characterized by an additional expanded region of protection across domain B and the suppression of a predominant DNase I hypersensitive site in B1. Recent evidence confirms that ORC is associated with replication origins throughout the cell cycle, consistent with the idea that ORC is present both in post-replicative and pre-replicative complexes (27–30). ORC may thus serve as a base, defining the sites on DNA to which additional factors are recruited to construct pre-replicative complexes. Protocols for genomic footprinting have recently been described (31).

Despite the differences in origin structure that have made it difficult to identify and study origins in other eukaryotes, ORC itself has been highly conserved during evolution and ORC subunits have recently been identified in *Schizosaccharomyces pombe*, *Kluyveromyces lactis*, and in higher eukaryotes such as *Xenopus*, *Drosophila*, human, and mouse (1). Protein complexes similar to yeast ORC have been purified from *Drosophila* (32) and *Xenopus*. In the latter there is also good evidence that the well established *in vitro* replication system is dependent upon ORC binding to chromatin (33–35).

ORC does not appear to be an abundant protein in budding yeast and we have previously estimated that the Orc2p subunit is present at little more than one copy for each of the approximately 300 origins per haploid cell (23). ORC has been purified from budding yeast by two groups and our protocol is presented below. It is also likely that an ARS121 binding activity termed core binding factor (CBF) by Estes *et al.* (36) is related to ORC although this factor has not been purified to homogeneity.

3.1.1 Purification of ORC

The protocol described here for the purification of ORC from a whole cell extract of budding yeast (*Protocol 1*) is a previously described (23, 25) modification of a nuclear extract purification first described by Bell and Stillman (10). Recombinant yeast ORC has also been purified from insect cells infected with baculovirus expressing each of the six ORC subunits and this protein complex appears functionally indistinguishable from that purified from yeast (17).

Yeast proteases are notoriously problematic and therefore one of the first considerations, as in any protein purification from yeast, is the choice of a strain with multiple protease deficiencies. A comprehensive discussion of the protease problem and solutions can be found in Jones, 1991 (37). We have successfully purified ORC from strains BJ926 (MATa/α, *pep4–3/pep4–3, prc1–126/prc1–126, prb1–1122/prb1–1122*) and PY26 (MAT**a**, *pep4–3, prc1–407, prb1–1122, ura3–52, trp1-Δ1, leu2–3,112, NUC1::LEU2*) but any strain with appropriate protease mutations should be suitable.

ORC DNA binding activity is monitored by DNase I footprinting analysis using end-labelled ARS1 probes prepared using a simple PCR procedure (*Protocols 2* and *3*) (*Figure 2*). We have routinely used an ARS1 footprinting template but any ORC binding site could be used, ORC footprints on ARS307, H4 ARS, ARS121, and at silencers have been described in the literature (10, 14). A gel retardation assay for ORC binding has also been described and could be used to monitor activity during purification (24). The gel retardation assay requires the addition of competitor DNA to binding reactions. In the DNase I footprinting assay we have previously determined that ORC:DNA interactions differ in the presence or absence of competitor DNA. In the presence of competitor ORC binding is more dependent upon the presence of the B1 element and ORC:DNA interactions around B3 are inhibited (23). The footprint observed under these conditions most closely resembles that observed in chromatin (25). Nevertheless, ORC footprints with purified protein are more readily detected in the absence of competitor so we recommend that during purification competitor DNA is not used in footprinting reactions beyond the dsDNA cellulose column. Antibodies to ORC subunits have been raised by several groups, allowing individual ORC subunits to be detected by Western blotting throughout the preparation. As in any purification however the presence of particular subunits does not

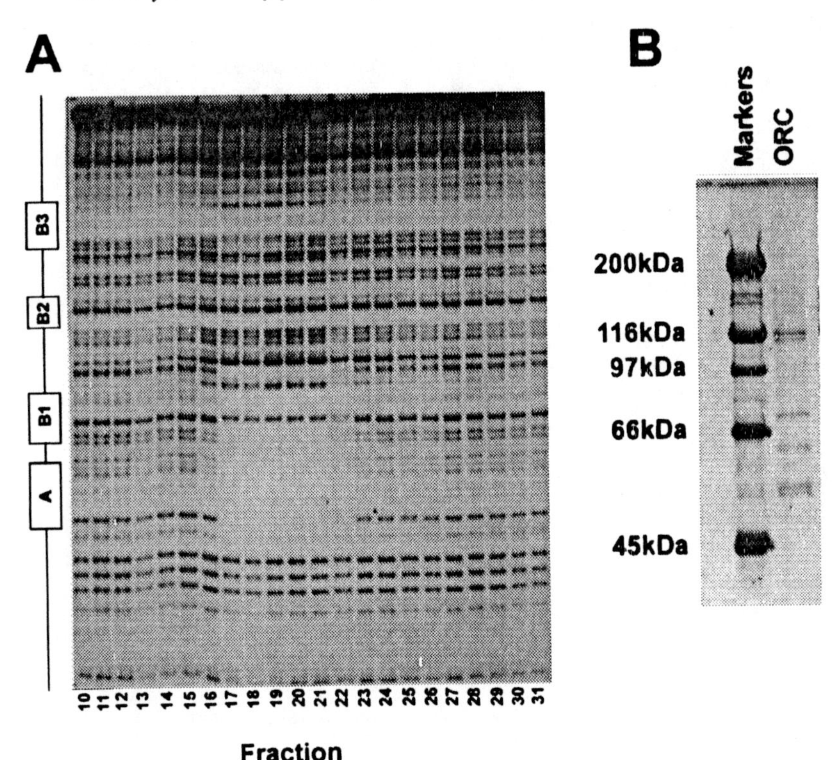

Figure 2. ORC purification. (A) DNase I footprinting analysis of 2 μl of the final FPLC Mono S column fractions in *Protocol 1*. The functional elements on the ARS1 template are indicated on the left. Note the DNase I hypersensitive sites generated in the B domain, notably the site in the middle of B1 that is seen before protection over A. (B) 7.5 μl of the final 200 μl FPLC Mono S column fractions run on a 10% SDS–polyacrylamide gel stained with Coomassie blue.

necessarily indicate that a particular fraction is active and is recommended only in addition to a DNA binding assay.

Protocol 1. ORC purification

Equipment and reagents

- Cell disruptor, e.g. Beadbeater (Stratech Scientific/Biospec Products)
- Acid washed 0.5 mm glass beads (Sigma)
- Double-stranded oligonucleotide affinity column prepared using synthetic ACS oligonucleotides ACS oligonucleotides (5'-GATCTAAACATAAAATCTGTAAAA-3' and 5'-GATCTTTTACAGATTTTATGTTTA-3') (10) as described (38)
- FPLC Mono S column (Pharmacia)
- *Saccharomyces cerevisiae* strain, e.g. BJ926 (YGSC)
- YEPD plus adenine medium: 1% yeast extract, 2% peptone, 2% glucose, 20 mg/litre adenine sulfate
- Buffer 1: 100 mM Tris–acetate pH 7.9, 50 mM potassim acetate, 10 mM $MgSO_4$, 2 mM EDTA, 20% glycerol, 2 mM DTT, 0.5 mM PMSF, 2 μM pepstatin A, 0.6 μM leupeptin, 1 mM benzamidine

Protocol 1. *Continued*

- Buffer H: 50 mM Hepes pH 7.5, 1 mM EDTA, 1 mM EGTA, 5 mM magnesium acetate, 0.02% NP-40, 10% glycerol, 2 mM DTT, 0.5 mM PMSF, 2 μM pepstatin A, 1 mM benzamidine—H/0.1 contains 100 mM KCl, H/0.2 contains 200 mM KCl, and H/0.5 contains 500 mM KCl
- Buffer H^+: buffer H with the addition of 0.5 mM ATP, 5 mM $CaCl_2$, and 10 mM $MgCl_2$
- Buffer T: 25 mM Tris–HCl pH 8, 1 mM EDTA, 1 mM EGTA, 5 mM magnesium acetate, 0.02% NP-40, 10% glycerol, 1 mM DTT, 0.5 mM PMSF, 2 μM pepstatin A, 1 mM benzamidine—T/0.18 contains 180 mM KCl, and T/0.45 contains 450 mM KCl
- dsDNA cellulose (Sigma)
- S-Sepharose (Pharmacia)
- Q-Sepharose (Pharmacia)

Method

1. Grow cells overnight to saturation in YEPD medium at 30 °C. To obtain enough cells for ORC purification it is most convenient to grow cells in a fermenter of approximately 10–12 litre capacity with vigorous agitation and aeration. It is best to inoculate the fermenter with an actively growing culture and to optimize the inoculation time and inoculum size so that the fermenter culture just reaches saturation during the incubation period. The following morning read the A_{600} of the culture and add a further 20 g/litre dry glucose. Incubate further, monitor the A_{600} over the next hour, and harvest the cells when the A_{600} has at least doubled. The cells should not be allowed to reach saturation for a second time. Harvest the cells by centrifugation and determine the combined wet weight of the cell pellets.
2. Completely resuspend 250 g (wet weight) cells in a minimal volume (approx. 250 ml) of buffer 1. Lyse the cells by mechanical agitation with acid washed glass beads in, for example, a Beadbeater. Monitor the level of cell breakage by microscopic examination and continue until greater than 90% breakage is achieved. Prevent the cell lysate heating up during cell disruption by intermittent chilling on ice.
3. Measure the volume of lysate and slowly add 4 M $(NH_4)_2SO_4$ to a final concentration of 0.9 M. Stir gently at 4 °C for 30 min and then pellet unbroken cells, cell debris, and unwanted protein aggregates at 4 °C for 30 min (e.g. Sorval GSA rotor at 13 000 r.p.m.).
4. Slowly adjust (over approx. 15 min) the $(NH_4)_2SO_4$ concentration by adding a further 0.35 g/ml (to 75% saturation) with finely ground solid $(NH_4)_2SO_4$. Stir for a further 45 min at 4 °C and centrifuge at 4 °C for 1 h as above (step 3) to harvest protein precipitate.
5. Resuspend in approx. 200 ml buffer H/0.1. Dialyse against several changes of H/0.1 overnight and if necessary dilute protein solution with H buffer lacking salt to lower conductivity to H/0.1 equivalent.
6. Apply protein solution to 300 ml fast flow S-Sepharose column pre-equilibrated in H/0.1. Wash with two to three volumes (600–900 ml) H/0.2 and step elute with three volumes (900 ml) H/0.5 in 15 ml fractions. Perform protein assay on fractions and pool all protein-containing fractions. It is not necessary to assay activity at this step.

7. Dialyse overnight against T/0.18. Adjust conductivity with T buffer lacking salt if necessary.
8. Apply protein solution to 100 ml Q-Sepahrose column pre-equilibrated in T/0.18. Wash with two to three volumes (200–300 ml) T/0.18 and step elute with 300 ml T/0.45 collecting 5 ml fractions. Pool all protein-containing containing fractions.
9. Dialyse pooled fractions overnight against H/0.1, and if necessary dilute protein solution with H buffer containing no KCl to lower conductivity to H/0.1 equivalent.
10. Apply material to 30 ml dsDNA cellulose column pre-equilibrated in H/0.1. Wash with two to three volumes H/0.1 and elute 5 ml fractions in a linear gradient of H buffer containing 0.1–0.6 M KCl. Assay protein-containing fractions by DNase I footprint analysis (*Protocols 2* and *3*). Fractions can be quick-frozen in liquid N_2 and stored at –70°C at this point.
11. Pool all active fractions and dilute protein solution with H buffer containing no KCl to lower conductivity to H/0.1 equivalent. Adjust to 10 mM $MgCl_2$, 5 mM $CaCl_2$, and add ATP to 0.5 mM (H^+). Apply material to 10 ml ACS oligonucleotide affinity column pre-equilibrated in H^+/0.1. Wash with 30 ml H^+/0.1 and elute in a linear gradient from H^+/0.1 to H^+/0.6 collecting 2 ml fractions. Measure conductivity of fractions and assay across gradient fractions by DNase I footprint analysis (*Protocols 2* and *3*). Fractions can be frozen at –70°C at this point.
12. Pool all active fractions and dilute protein solution with H buffer containing no KCl to lower conductivity to H/0.1 equivalent. For the final preparative step H buffer glycerol concentration is lowered to 5%. Apply material to 1 ml FPLC Mono S column pre-equilibrated in H/0.1. Wash with 3 ml H/0.1 at 0.5 ml/min, elute with 5 ml H/0.1 to H/0.6 linear gradient at 0.5 ml/min collecting 200 μl fractions. Measure conductivity and assay by DNase I footprinting (*Protocols 2* and *3*) across 0.2–0.5 M KCl range.

Protocol 2. DNase I footprinting: preparation of labelled footprinting template

Reagents

- 10 × kinase buffer: 700 mM Tris pH 7.5, 100 mM $MgCl_2$, 50 mM DTT
- 10 × PCR buffer: 500 mM KCl, 100 mM Tris–HCl pH 8.3, 25 mM $MgCl_2$, 0.1% gelatin
- dNTP mix: 2.5 mM each dATP, dCTP, dGTP, dTTP
- 10 U/μl polynucleotide kinase (NEB)
- pUC forward (5′-GTAAAACGACGGCCAGT-3′) and reverse (5′-CAGGAAACAGCTATGAC-3′) sequencing primers
- ARS1-containing template, e.g. pARS1.4.1 containing ARS1 domains A and B cloned into the multiple cloning site of pUC19 (39)
- 10 μCi/μl [γ-^{32}P]ATP (Amersham)

Protocol 2. *Continued*

Method

1. 5′ end-label forward pUC forward sequencing primer for 1 h on ice in the following reaction mix: 1.5 μl freshly prepared 10 × kinase buffer, 7.5 μl [γ-^{32}P]ATP, 35 pmol primer, 1 μl T4 polynucleotide kinase, H_2O to 15 μl. Heat to 95 °C for 15 min to inactivate kinase.
2. Amplify ARS1-containing product from plasmid DNA using end-labelled forward and unlabelled reverse sequencing primers in the following reaction: 15 μl labelled oligo (step 1), 50 ng ARS1-containing template, 8 μl 10 × PCR buffer, 35 pmol unlabelled reverse sequencing primer, 5 μl dNTP mix, 5 U *Taq* polymerase, H_2O to 80 μl. Suitable conditions using this template primer combination are: 94°C/60 sec (denaturation), 37 °C/40 sec (annealing), 72 °C/60 sec (extension) for 25 cycles. A final PCR cycle incorporating a 10 min extension at 72°C should be added to ensure that all products are fully extended.
3. Separate labelled PCR product from unincorporated [γ-^{32}P]ATP and unincorporated primers, for example on a 1 ml Sephadex G25 column.
4. Evaluate the quality and quantity of PCR product by running 1% of total product on a 6% polyacrylamide minigel with appropriate size standards of known concentration. Stain the gel with ethidium bromide. Expose the gel briefly (15 min) to X-ray film to confirm that a single labelled product is present. Obtain c.p.m./μl by scintillation counting.

Protocol 3. DNase I footprinting reaction

Reagents

- End-labelled ARS1 template (*Protocol 2*)
- DNase I (Sigma): dilute to 3 Kunitz U/μl in 50% glycerol
- Stop buffer: 100 mM NaCl, 87.5 mM EDTA pH 8, 80 mM Tris–HCl pH 8.5, 36 μg/ml *Escherichia coli* tRNA (Sigma)
- 10 × nucleotide mix: 2 mM CTP, 2 mM GTP, 2 mM UTP, 40 mM ATP, 1.25 mM dCTP, 1.25 mM dGTP, 1.25 mM TTP, 0.25 mM dATP
- Loading buffer: 1 mg/ml each bromphenol blue and xylene cyanol FF in formamide containing 200 μl/10 ml 0.5 M EDTA pH 8
- 2.5 × footprinting buffer: 100 mM Hepes–KOH pH 7.5, 25 mM magnesium acetate, 7.5 mM EGTA, 1.25 mM EGTA, 5% polyvinyl alcohol, 12.5% glycerol
- H/0.35: H buffer (*Protocol 1*) containing 350 mM KCl

Method

1. Prepare sufficient reaction mix for the number reactions required. Each reaction requires 10 μl 2.5 × footprinting buffer, 2 μl 0.5 M creatine phosphate pH 6.5, 1 μl creatine phosphokinase 0.06 U/μl, 2.5 μl 10 × nucleotide mix, 2 ng labelled template (approx. 50 000 c.p.m.) (*Protocol 2*).

2. Quick-thaw protein fractions and add up to 8.5 μl fraction to 16.5 μl reaction mix in 1.5 ml Eppendorf tube on ice. When using less than 8.5 μl fraction make up to 25 μl with H buffer containing 350 mM KCl.
3. Transfer tube to 30°C at T = 0. Incubate for exactly 10 min. Add 2 μl 3 U/μl DNase I and incubate for a further 1 min. Add 140 μl stop buffer, mix thoroughly, and place on ice.
4. When all reactions are complete extract with 166 μl phenol:chloroform: isoamyl alcohol (25:24:1). Spin at maximum speed for 30 min in micro-centrifuge. Precipitate aqueous phase with 2.5 vol. cold EtOH.
5. Carefully remove all EtOH, dry pellets, and resuspend in 5 μl loading buffer. Heat to 68°C for 5 min and place on ice.
6. Electrophorese on 5% sequencing gel. Fix gel and expose to autoradiographic film for approx. 18 h.

3.2 Abf1p

ARS binding factor 1 (Abf1p) was the first ARS binding factor identified and recognizes *in vitro* a consensus sequence contained within B3 of ARS1 and in several other ARSs (39–41). *In vitro* footprints with purified Abf1p are indistinguishable from the footprint observed over B3 in chromatin throughout the cell cycle (11, 25) demonstrating that Abf1p binds to ARSs *in vivo* as suggested by the plasmid stability defects observed in *abf1* temperature-sensitive mutants (42). *ABF1* is an essential gene (43–45) but the exact role of Abf1p in the initiation DNA replication is not clear and Abf1p binding sites are also found in promotors and in transcriptional silencers where Abf1p can function as either a negative or positive regulator of transcription (46). Indeed, Abf1p binding sites in replication origins can be replaced without loss of function by binding sites for other unrelated transcription factors (7). Abf1p binding sites are not found in all ARSs but since it is clear that Abf1p can function at a distance this may indicate that most if not all ARSs are in functional proximity of an Abf1p binding site (47).

3.2.1 Abf1p purification

Abf1p is an abundant protein that can be purified relatively easily from a budding yeast whole cell extract using the following protocols (*Protocols 5–7*). Nevertheless, we have found that overexpression from an ectopic plasmid in yeast can lead to a 25-fold increase in yield per cell volume (*Figure 3*). Since overexpression of Abf1p is toxic to yeast cells, this is best achieved using a regulated promoter such as that of *GAL1*. A simple but effective way to induce expression of *ABF1* from this promoter is outlined in *Protocol 4*. The protocol given is for a small scale preparation but can be easily scaled-up, especially if a fermenter is used to grow a larger volume of cells. Induction of Abf1p can be monitored using an Abf1p gel retardation assay (*Protocol 6*)

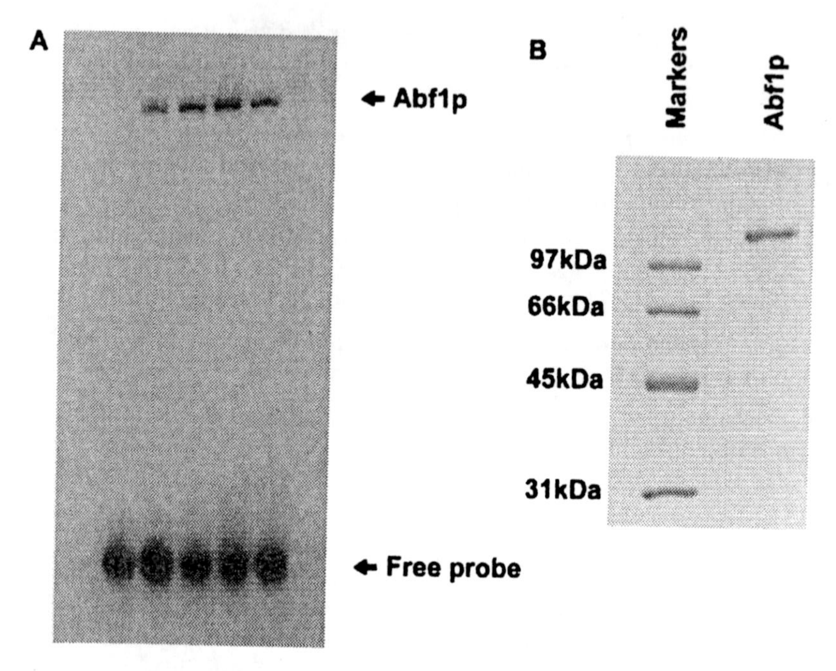

Figure 3. Abf1p purification. (A) Induction time course of Abf1p overexpression monitored by a gel retardation assay using a double-stranded Abf1p binding site radiolabelled probe. (B) 10 μl of the final 1 ml Abf1p preparation run on a 10% SDS–polyacrylamide gel and stained with Coomassie blue.

and this assay is also used to detect Abf1p throughout the purification procedure (*Protocol 7*). The purification protocol makes use of four chromatographic separation steps including a DNA affinity purification step using an Abf1p oligo affinity resin prepared as in *Protocol 5*.

Protocol 4. Abf1p overexpression

Reagents

- *Saccharomyces cerevisiae* strain, e.g. PY26, carrying an Abf1p overexpression plasmid
- Synthetic complete medium with appropriate selection for Abf1p overexpression plasmid

Method

1. Grow 100 ml yeast culture of with appropriate selection in synthetic medium containing 2% glucose, to a density of approx. 1×10^7 cells/ml at 30 °C.

2. Add these cells to a litre of synthetic medium containing 4% galactose, instead of glucose, and grow for a further 16–24 h.
3. Overexpression of Abf1p will be clear from the phenotypes of misshapen cells typically with long projections.

Protocol 5. Abf1p affinity resin preparation

Equipment and reagents

- Sun lamp (Vector Laboratories)
- Photoprobe (Vector Laboratories)
- Column storage buffer: 1 mM EDTA, 2.5 M NaCl, 0.02% NaN_3, 50 mM Tris–HCl pH 7.5
- Abf1p binding site oligonucleotides, e.g. based on the S33 Abf1p binding site (ligated as described (38)
- Streptavidin agarose (Sigma)

Method

1. Resuspend 1 mg ligated oligo in 100 μl 100 mM EDTA.
2. Add 500 μl distilled water to one vial of photoprobe in the dark. Do not expose the stock to light. Add 50 μl of this photoprobe stock to the DNA.
3. Split the 150 μl into three colourless Eppendorfs tubes and place them with the lids open in a tightly packed ice bucket. Irradiate each tube under the sun lamp at a distance of 10 cm from lamp to tube rim for 15 min.
4. To each tube add 200 μl 100 mM Tris pH 9.5 to stop the reaction.
5. Add 250 μl water saturated butanol and mix vigorously. Discard the upper phase. Repeat.
6. Add 25 μl 3 M sodium acetate pH 5.5, mix, then add 550 μl EtOH. Precipitate on ice for 15 min, then spin at maximum speed for 15 min in a microcentrifuge. The precipitate should be brownish/orange if biotinylation has been successful (it changes colour from bright orange as the reaction occurs).
7. Wash the pellets with 70% ethanol, then resuspend each DNA pellet in 100 μl column storage buffer, and pool them.
8. Resuspend 0.5 ml streptavidin agarose (Sigma) in column storage buffer. Spin and remove the supernatant. Add the 300 μl of re-suspended DNA. Incubate on a rotator at 4°C overnight. If the oligos are radioactively labelled it is possible to measure the amount of oligonucleotide incorporated onto the agarose by the difference between counts in 1 μl immediately after adding the biotinylated oligonucleotides to the streptavidin agarose and after incubation overnight. Typically, 70–80% incorporation is achieved, i.e. about 2 mg DNA/ml streptavidin agarose.
9. Prepare as a column.

Protocol 6. Abf1p DNA binding assay

Reagents

- Freshly prepared 10 × binding buffer: 25 μl 1 M Hepes/Na^+ pH 7.5, 5 μl 1 M $MgCl_2$, 2 μl 1 M DTT, 20 μl 10 mg/ml BSA acetylated, 48 μl H_2O
- Boiled, sonicated, phenol extracted salmon sperm DNA
- ^{32}P-labelled double-stranded Abf1p binding site oligonucleotides
- Sucrose loading dye: 0.25% bromphenol blue, 40% sucrose in water (49)

Method

1. Assemble a master mix as follows (per reaction): 1 μl 10 × binding buffer, 10 μg salmon sperm DNA, H_2O to 10 μl.
2. Add protein sample to master mix and pre-incubate for 10 min at room temperature.
3. Add 100 000 c.p.m. labelled oligonucleotide. Incubate for 20 min.
4. To each sample add 2 μl sucrose loading dye, mix, and run on a 10% Tris–glycine gel.
5. Dry the gel on a gel dryer and expose to film. Abf1p activity will retard the probe and be seen as a band towards the top of the gel (*Figure 3*).

Protocol 7. Abf1p purification

Equipment and reagents

- Cell disruptor, e.g. Beadbeater (Stratech Scientific/Biospec Products)
- Acid washed 0.5 mm glass beads (Sigma)
- 1 ml FPLC Mono Q column (Pharmacia)
- 2 × buffer X: 100 mM Pipes, 20 mM sodium metabisulfite, 2 mM EDTA, 40% glycerol (v/v)
- Approx. 35 g cell pellet (*Protocol 4*)
- 2 × buffer Y: 50 mM Tris–HCl pH 8, 20 mM sodium metabisulfite, 2 mM EDTA, 40% glycerol (v/v)
- High salt buffer: 2.5 M NaCl, 50 mM Tris–HCl pH 7.5, 1 mM EDTA
- Heparin agarose (Sigma)
- Calf thymus ssDNA cellulose (Sigma)
- Oligonucleotide affinity agarose (*Protocol 6*)

Method

1. Prepare 1 × buffer X, adding fresh DTT to 10 mM and PMSF to 1 mM, plus the appropriate salt. The high concentration of DTT helps to maintain the zinc finger within the DNA binding domain of Abf1p. Take conductivity measurements of buffer X with a range of salt concentrations and make a standard curve. This will be necessary to calculate the salt concentration of fractions eluting from the column.
2. Resuspend approx. 35 g cell pellet in 3 vol. of 1 × buffer X, and break the cells in a Beadbeater with an equal volume of acid washed glass beads in ten 30 sec bursts. Slowly add $(NH_4)_2SO_4$ from a 4 M stock to

a final concentration of 0.2 M. Leave on ice for 30 min, then centrifuge at 10 000 *g* for 20 min to remove cell debris.

3. Adjust the conductivity to 160 mM $(NH_4)_2SO_4$ using buffer X.
4. Prepare a 50–100 ml column of heparin agarose and wash with high salt buffer. Equilibrate in buffer X containing 160 mM $(NH_4)_2SO_4$ until the flow-through has the identical pH and conductivity as the starting buffer.
5. Apply the extract to the column. Collect and save the flow-through; quick-freeze in a dry ice/ethanol bath and store at –70 °C.
6. Wash the column with 300 ml buffer X containing 160 mM $(NH_4)_2SO_4$. Collect fractions of approx. 3 ml. Apply five column volumes of a linear gradient from 160–500 mM $(NH_4)_2SO_4$ in buffer X collecting 3 ml fractions.
7. Measure the conductivity of the fractions and assay the protein content using, e.g. Bio-Rad Bradford assay. Use the DNA binding assay (*Protocol 5*) to identify the active fractions including samples of the starting extract, the column flow-through, and wash as controls. Abf1p activity elutes at about 300 mM $(NH_4)_2SO_4$. Combine the main active fractions. If a lot of Abf1p activity resides in the flow-through it may be necessary to re-run the column.
8. Dialyse the pooled fractions overnight at 4°C against buffer X containing 50 mM NaCl. Measure the volume of the dialysate.
9. Prepare a 25 ml column of calf thymus ssDNA cellulose. Wash the column with high salt buffer and equilibrate with 1 × buffer X containing 50 mM NaCl.
10. Load the dialysed extract, collect the flow-through as before, and wash the column in 50 ml 1 × buffer X containing 50 mM NaCl.
11. Elute the column with a 125 ml total linear gradient from 50–350 mM NaCl in 1 × buffer X. Collect fractions of approx 2.5 ml. Assay the fractions as before.
12. Dialyse the pooled fractions overnight at 4°C against buffer X containing 50 mM NaCl. As before, measure the volume and load the dialysed extract onto the prepared oligonucleotide affinity column, pre-equilibrated with buffer X/50 mM NaCl.
13. Elute in a 50 ml linear gradient from 50 mM to 1 M NaCl in buffer X and collect approx. 25 × 2 ml fractions. Assay fractions as before, but note, at this stage the high salt content in the higher fraction numbers can reduce the efficiency of the DNA binding assay. Dialysis of fractions in a microdialyser against 50 mM NaCl/buffer X can optimize this reaction to enable identification of the true activity peak.

Protocol 7. *Continued*

14. Prepare two buffers: a 'no salt' buffer (1 × buffer X, 10 mM DTT, 1 mM PMSF), and a 'high salt' buffer (1 × buffer X, 10 mM DTT, 1 mM PMSF, 1 M NaCl). Pool the active fractions from the oligonucleotide affinity column and dialyse against 1 × buffer Y containing 10 mM DTT and 1 mM PMSF.
15. Prepare a (1 ml) FPLC Mono Q column and equilibrate it using 5 ml no salt buffer, then 10 ml high salt buffer, then a further 5 ml no salt buffer. Load the protein extract at 0.2 ml/min in no salt buffer, then wash in 3 ml no salt buffer.
16. Execute an elution programme as follows: 3 ml 80 mM NaCl wash phase, 5 ml linear gradient 80–135 mM NaCl, 3 ml high salt (1 M NaCl) step, and finally a 3 ml no salt wash. During this time, maintain a flow rate of approx. 0.2 ml/min, and collect fractions of approx. 0.25 ml.
17. Assay fractions as before. Fractions can now be run on a 10% SDS–polyacrylamide gel and stained with Coomassie blue. Abf1p should appear as a clean purified species eluting at around 250 mM NaCl and running at approx. 135 kDa on an SDS–polyacrylamide gel (*Figure 3*).

4. Origin interacting proteins

4.1 Mcm proteins

The Mcm2–7p family of proteins have been conserved in evolution from yeast to humans. These proteins play a critical role in initiating DNA replication and in limiting DNA replication to once per cell cycle. There are six members of the *MCM* family of genes in budding yeast: *MCM2*, *MCM3*, *CDC46*, *CDC47*, *CDC54*, and a homologue of the fission yeast *mis5*$^{+}$ gene encoding proteins of 101, 107, 87, 95, 105, 113 kDa respectively. *MCM2*, *MCM3*, *CDC46*, *CDC47*, *CDC54* are all individually essential for viability.

The budding yeast *MCM* genes were originally identified in two different genetic screens. In the first approach, which led to the isolation of the *MCM2* and *MCM3* genes, mutants that were defective in the maintenance of a plasmid carrying a single origin of replication were isolated. The inability to maintain these plasmids depended on the ARS but not the centromere sequence used strongly suggesting that the strains were defective in DNA replication rather than chromosome segregation (50, 51). This was supported by the finding that the *mcm2-1* and *mcm3-1* mutants are defective in initiation at chromosomal origins of replication using two-dimensional gel electrophoresis to examine DNA replication intermediates (52). The second approach relied on identifying cold-sensitive mutants that where defective in cell cycle progression. Mutations in two genes, *CDC45* (Section 4.2) and

CDC54, were identified in this screen. Suppressors of the cold-sensitive phenotype of *cdc45* and *cdc54* mutants were identified that exhibited temperature-sensitive phenotypes on their own. This led to the identification of the *CDC46* and *CDC47* genes (53, 54). Of these, *CDC46*, *CDC47*, and *CDC54* are members of the *MCM2–7* family and *CDC46* is allelic to *MCM5*, another mutant isolated in the minichromosome maintenance screen (55).

mcm mutants show defects in firing of all origins tested. Together with ample evidence showing genetic interactions among *MCM* genes these results strongly suggests that the six Mcm proteins act together to initiate DNA replication from each origin. Experiments in *Xenopus* and yeast extracts demonstrate that Mcm proteins associate with chromatin in G1 in an ORC- and Cdc6p-dependent reaction, strongly suggesting that Mcm proteins are part of the pre-replicative complex (56, 57). There is also compelling evidence in the *Xenopus* system that Mcm proteins are required for the licensing reaction (58–60). Recent chromatin immunoprecipitation experiments have shown that in G1 the Mcm proteins bind in the vicinity of replication origins but also may remain associated with DNA, but no longer specifically with replicator sequences, in S phase (27, 28). This latter observation has been taken to suggest that Mcm proteins may be a component of the replicative helicase since members of this family all contain a 240 amino acid domain that contains the motifs characteristic of DNA-dependent ATPases. Weak helicase activity has recently been associated with a complex of Mcm4p, Mcm6p, and Mcm7p in HeLa cell extracts (61).

4.1.1 Cdc46p purification

At present, no enzymatic activity has been ascribed to the Mcm proteins in yeast. In order to begin to characterize the Mcm proteins, we have developed procedures to purify them from budding yeast extracts using an epitope tagging strategy. In this strategy, homologous recombination is used to construct a protease-deficient yeast strain containing a single epitope tagged *MCM* gene. This strategy is outlined in *Figure 4A* and utilizes the plasmid pMHT, a vector derived from the yeast integrating vector pUC119/URA3A. pMHT contains a short tag consisting of nine histidine residues and an epitope from the c-myc gene which is recognized by the 9E10 monoclonal antibody (*Figure 4B*). Briefly, a C terminal fragment from the gene to be tagged is amplified by PCR so that the termination codon is removed and either a *Sal*I or *Xho*I site is introduced at the end of the coding sequence ensuring that this sequence is in the same reading frame as the epitope tag. Either a *Bam*HI or *Bgl*II site can be introduced into the oligonucleotide at the 5′ end of the PCR product to allow simple directional cloning of the PCR product between the *Bam*HI and *Xho*I sites of pMHT. Integration is then directed to the *MCM* gene by linearizing the plasmid with a restriction enzyme that cuts within the coding sequence. Transformants are screened by either PCR or Southern

A

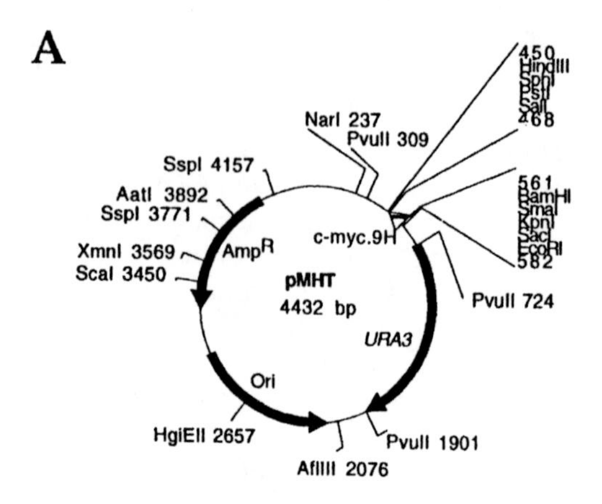

B

```
BamHI       XhoI                     HindIII
GGA TCC AAA CTC GAG ATG GAA CAA AAG CTT ATT TCT GAA GA
GLY SER LYS LEU GLU MET GLU GLN LYS LEU ILE SER GLU GL
                                   9E10 epitope

GAC TTG AGA TCC CAC CAC CAC CAC CAC CAC CAC CAC CAC CA
ASP LEU ARG SER HIS HIS HIS HIS HIS HIS HIS HIS HIS GL
                               9 His

BglII       SalI        PstI    SphI    HinDIII
GAT CTT GAG TCG ACC TGC AGG CAT GCA AGC TTG GCG TAA
ASP LEU GLU SER THR CYS ARG HIS ALA SER LEU ALA OCH
```

Figure 4. Mcm protein tagging strategy. (A) Map of pMHT plasmid used to tag Mcm proteins. (B) Sequence of insert containing c-myc and His9 tags in pMHT.

blotting to identify correctly targeted integrants. The epitope tagged protein can be visualized by immunoblotting with the 9E10 antibody.

We have used this epitope tagging strategy to purify Cdc46p (*Protocol 8*) (*Figure 5B*). Using this purification scheme Cdc46p is purified uncomplexed to other Mcm protein family members, probably because Mcm protein complexes are sensitive to the chloride buffers. A recent study in fission yeast described the use of glutamate or acetate buffers to purify intact Mcm complexes (62).

4.2 Cdc45p

CDC45 was first identified in a genetic screen that also led to the identification of Mcm family members (Section 4.1). Recently cloned, Cdc45p is not related to Mcm proteins at the amino acid sequence level but exhibits genetic and biochemical interactions consistent with a role for this protein in the initiation of DNA replication within the pre-replicative complex (63–66). Cross-linking

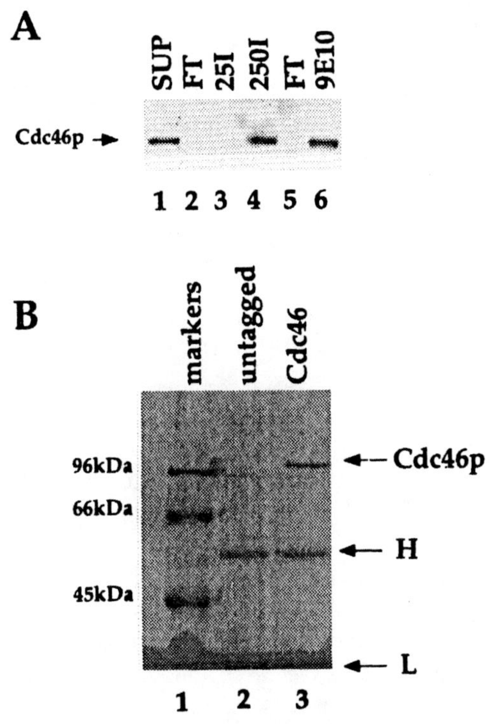

Figure 5. Mcm family purification. (A) Immunoblot analysis with 9E10 antibody of protein fractions from Cdc46p purification. The fractions in lanes 1–6 are: starting supernatant, (SUP, lane 1); flow-through from the Ni-NTA agarose column (FT, lane 2); 25 mM imidazole wash from Ni-NTA agarose column (25I, lane 3); 250 mM imidazole elution from Ni-NTA column (250I, lane 4); flow-through from the 9E10 immunoaffinity column (FT, lane 5); and bound fraction from the 9E10 immunoaffinity column (9E10). (B) SDS–PAGE of final fraction from the Cdc46 purification from tagged and untagged strains stained with Coomassie blue. The position of Cdc46p as well as the heavy (H) and light (L) chains of 9E10 are indicated.

experiments place Cdc45p close to origins in G1 and, like Mcm proteins, Cdc45p may remain associated with chromatin in S phase, but not specifically with replicator sequences suggesting that Cdc45p may move with the replication fork (28). Cdc45p has not yet been purified.

Protocol 8. Purification of tagged Mcm proteins

Reagents

- Lysis buffer: 25 mM Tris–HCl pH 7.5, 5 mM $MgCl_2$, 100 mM NaCl, 10% glycerol (v/v), 1 mM PMSF (0.1 M in isopropanol), 2 mg/ml leupeptin, 1 mg/ml pepstatin A (1 mg/ml in methanol), and 0.1% Triton X-100
- Tagged Mcm yeast strain, e.g. Cdc46p
- Ni-NTA agarose resin (Qiagen)
- Imidazole
- 9E10 immunoaffinity resin (*Protocol 9*)

Protocol 8. *Continued*

Method

1. Prepare cell extract from a 1 litre cell culture grown to 3×10^7 cell/ml by resuspending the cell pellet in three volumes of lysis buffer. To lyse the cells pour liquid N_2 into a mortar and slowly drip the cell slurry into the N_2. Crush the frozen cell slurry beads into a powder with the pestle until 70% of the cells are lysed, as assessed microscopically. Continually poured fresh liquid N_2 into the mortar during the crushing procedure to keep the cell extract frozen.
2. Rapidly thaw extracts and remove debris by centrifugation in Corex tubes at 10 000 r.p.m. in an HB4 rotor (Beckman) for 20 min.
3. Incubate the supernatant with 0.5 ml Ni-NTA agarose resin for 2 h at 4 °C with rotation. Pour the mixture into a 10 ml column and allow the liquid to flow through the resin. Pass ten column volumes of the lysis buffer through the resin. Pour a further ten column volumes of the lysis buffer with 25 mM imidazole pH 6.8 through the resin.
4. To elute the tagged proteins apply three column volumes of the lysis buffer containing 250 mM imidazole pH 6.8 to the resin. Allow 1.5 vol. of this solution to pass through the column, with collection of the eluate. Then stop the flow and leave the resin in this solution for 20 min, after which restart the flow and collect the final eluate. Save aliquots from each step in the above procedure to monitor the recoveries of the tagged proteins during the purification using an immunoblot assay with the 9E10 antibody.
5. Incubate the eluate from the Ni-NTA agarose resin with 100–200 μl 9E10 resin by batch binding for 2 h at 4 °C. Gently pellet the resin by centrifugation in a microcentrifuge at 1500 r.p.m. for 30 sec. Wash the resin four times in ten volumes of the original lysis buffer and pellet each time as above. Finally resuspend the resin in one volume of 1 × Laemmli buffer and boil for 3 min at 95 °C. Alternatively, some Mcm protein can be recovered by incubation of the resin with 1 mg/ml c-myc peptide in lysis buffer for at least 24 h at 4 °C. Save aliquots from each step in this procedure and use to monitor the recoveries of the tagged proteins during the purification step, using an immunoblot assay with the 9E10 antibody.

Protocol 9. Constructing 9E10 immunoaffinity resin

Equipment and reagents

- 20 ml Poly-Prep column (Bio-Rad)
- Affi-Gel 10 active ester agarose (Bio-Rad)
- 9E10 monoclonal antibody
- PBS: 140 mM NaCl, 2.5 mM KCl, 10 mM Na_2HPO_4, 1.4 mM KH_2PO_4 adjusted to pH 7.2

Method

1. Pre-treat Affi-Gel 10 active ester agarose according to the manufacturer's instructions. Incubate 5 mg of the 9E10 monoclonal antibody in 5 ml PBS with 1 ml Affi-Gel 10 resin with rotation at 4°C for 16 h.
2. Add 100 μl 1 M ethanolamine pH 8 and incubate for a further hour. Pour the mixture into a 20 ml Poly-Prep column and allow the resin to settle. Allow the liquid to flow through the resin, saving an aliquot to measure the efficiency of the antibody binding to the resin.
3. Wash the column with 20 column volumes of PBS at 4°C, and then with 20 column volumes of PBS containing 0.5 M NaCl. Re-equilibrate the column with ten volumes of PBS.
4. Store a suspension of the immunoaffinity resin in PBS with 0.02% NaN_3. The concentration of the resin is usually about 4 mg of 9E10 antibody per ml of Affi-Gel 10 resin.

4.3 Dbf4p

The *DBF4* gene was first identified by a conditional cell cycle mutation resulting in a terminal phenotype similar to that of other essential genes involved in DNA replication. Genetic interactions between *DBF4* and *CDC7* led to the finding that Dbf4p is the catalytic subunit of the Cdc7p protein kinase. Cdc7p, like Dbf4p is required for the G1/S transition and periodic Cdc7p kinase activity peaks at this time (67). A genetic screen to identify ARS1 interacting factors led Dowell *et al.* to discover that Dbf4p (and by inference Cdc7p) interacts with ARS sequences. This interaction is absolutely dependent upon an intact ACS and is lowered by mutations in B1, and to a lesser extent B2, suggesting that the interaction of Dbf4p with origin sequences may be through ORC (68). These findings strongly suggest that the Cdc7p/Dbf4p kinase functions to activate the initiation of DNA replication by phosphorylating an essential factor bound to origins. *cdc7* and *dbf4* mutations can be bypassed by a mutation in the *CDC46/MCM5* gene (Section 4.1) suggesting that Mcm proteins may be targets for the Cdc7p/Dbf4p protein kinase (69). This idea is reinforced by the recent identification of an allele of *DBF4* as a suppressor of the *mcm2–1* mutation. Subsequent analysis demonstrated that Mcm2p, Mcm3p, Mcm4p, and Mcm6p are Dbf4p/Cdc7p substrates *in vitro* (70). Purification procedures to isolate Dbf4p from yeast have not yet been established.

4.4 Cdc6p

The *CDC6* gene was first identified in budding yeast by the *cdc6–1* temperature-sensitive mutation (71, 72). Even at the permissive temperature *cdc6–1* mutant cells exhibit a defect in plasmid stability that can be suppressed by the incorporation of multiple ARSs, consistent with a role for the Cdc6p protein at origins of DNA replication (73). The Cdc6p gene product is an unstable nuclear phosphoprotein that has been conserved during evolution: Cdc6p homologues have been identified in *S. pombe* (Cdc18p), *Xenopus*, and humans (1). Like *CDC6*, *S. pombe* $cdc18^+$ is required for the initiation of DNA replication; deletion strains fail to initiate DNA replication and *cdc6* mutants exhibit reduced origin firing and genetic interactions with genes encoding ORC subunits (13, 74, 75). Overexpression of $cdc18^+$ in fission yeast causes cells to undergo multiple rounds of DNA replication without an intervening mitosis suggesting that, in this organism, $cdc18^+$ encodes a rate-limiting regulator of DNA replication (76, 77). Consistent with these results in yeast, *Xenopus* Cdc6p is required for DNA replication in oocyte extracts (57).

The *CDC6* gene is transcriptionally regulated during the cell cycle (78, 79) and synthesis of new Cdc6p post-mitosis is required for the formation of pre-replicative complexes described in Section 3.1. Furthermore, in a *cdc6–1* mutant pre-existing pre-replicative complexes are thermolabile, strongly suggesting that Cdc6p is a component of the pre-replicative complex and that the pre-replicative complex is required for initiation of DNA replication (26, 80). Recent experiments in budding yeast substantiate this model by demonstrating that Cdc6p protein associates with chromatin during the G1 phase of the cell cycle and that the DNA associated with Cdc6p is enriched in origin sequences (27, 29). Cdc6p association with origin squences depends upon an intact ACS suggesting that Cdc6p interacts with origins through ORC. Cdc6p binding is required in turn for the association of Mcm family proteins with origins during G1 (27, 28, 56). This is consistent with evidence in *Xenopus* egg extracts where depletion of *Xenopus* Cdc6p prevents Mcm protein binding to chromatin and depletion of ORC prevents Cdc6p binding (57).

While there is clear evidence that Cdc6p is an origin interacting protein in multiple species this has not yet been demonstrated with purified Cdc6p *in vitro* and the biochemical activities of Cdc6p are so far unidentified. The *CDC6* sequence predicts the presence of a nucleotide binding motif and one study suggests that Cdc6p may indeed exhibit ATPase activity (57, 74, 78, 81). Although published protocols for Cdc6p purification are not yet available this situation is unlikely to continue for long given the intense interest in this protein.

5. Future prospects

The last ten years has seen, through a combination of genetic and biochemical approaches, the identification of probably the majority of gene products involved in the initiation of chromosomal DNA replication in eukaryotic cells. The problem that remains is to determine the mechanisms by which these products interact with one another, with origin DNA, with replication enzymes, and with cell cycle and checkpoint machinery to ensure that replication is restricted to a once per cell cycle event and to ensure that the entire genome is replicated. Already significant progress is being made in this direction. For example, we have recently learned that protein complexes involving ORC, Cdc6p, and Mcm proteins assemble at replication origins. There are also tantalizing insights into how this process may be cell cycle regulated by cyclin-dependent kinases. We still know almost nothing however about the biochemical characteristics of any of the proteins implicated in initiation complexes and how their activities are regulated and co-ordinated. Furthermore we know very little about the way that the initiation apparatus facilitates origin unwinding and to polymerase recruitment.

It is likely that yeast will continue to play a leading role in the elucidation of these functions since only in yeast do we know the identity of the proteins that interact with origins and the structure and function of origin DNA sequences. That studies in yeast will be informative for eukaryotic organisms in general is without doubt given the conservation of proteins involved in replication initiation and their roles in various eukaryotic sytems. The near future will certainly see the purification from yeast of Cdc6p, Cdc45p, and Cdc7p/Dbf4p, and further biochemical analysis will reveal how these proteins interact with origins, ORC, and Mcm proteins. Ultimately it should be possible to faithfully reconstitute the regulated initiation process *in vitro*, a development that will facilitate the biochemical dissection of the events leading to initiation of replication in yeast and in higher eukaryotes.

References

1. Dutta, A. and Bell, S. P. (1997). *Annu. Rev. Cell Dev. Biol.*, **13**, 293.
2. Donovan, S. and Diffley, J. F. X. (1996). *Curr. Opin. Genet. Dev.*, **6**, 203.
3. Diffley, J. F. X. (1996). *Genes Dev.*, **10**, 2819.
4. Stillman, B. (1996). *Science*, **274**, 1659.
5. Nasmyth, K. (1996). *Science*, **274**, 1643.
6. Rowley, A., Dowell, S. J., and Diffley, J. F. X. (1994). *Biochim. Biophys. Acta*, **1217**, 239.
7. Marahrens, Y. and Stillman, B. (1992). *Science*, **255**, 817.
8. Theis, J. F. and Newlon, C. S. (1994). *Mol. Cell. Biol.*, **14**, 7652.
9. Rao, H., Marahrens, Y., and Stillman, B. (1994). *Mol. Cell. Biol.*, **14**, 7643.
10. Bell, S. P. and Stillman, B. (1992). *Nature*, **357**, 128.
11. Diffley, J. F. X. and Cocker, J. H. (1992). *Nature*, **357**, 169.
12. Micklem, G., Rowley, A., Harwood, J., Nasmyth, K., and Diffley, J. F. X. (1993). *Nature*, **366**, 87.
13. Li, J. J. and Herskowitz, I. (1993). *Science*, **262**, 1870.
14. Bell, S. P., Kobayashi, R., and Stillman, B. (1993). *Science*, **262**, 1844.
15. Foss, M., McNally, F. J., Laurenson, P., and Rine, J. (1993). *Science*, **262**, 1838.
16. Loo, S., Fox, C. A., Rine, J., Kobayashi, R., Stillman, B., and Bell, S. (1995). *Mol. Biol. Cell*, **6**, 741.
17. Bell, S. P., Mitchell, J., Leber, J., Kobayashi, R., and Stillman, B. (1995). *Cell*, **83**, 563.

18. Fox, C. A., Loo, S., Dillin, A., and Rine, J. (1995). *Genes Dev.*, **9**, 911.
19. Bell, S. P., Kobayashi, R., and Stillman, B. (1993). *Science*, **262**, 1844.
20. Fox, C. A., Ehrenhofer-Murray, A. E., Loo, S., and Rine, J. (1997). *Science*, **276**, 1547.
21. Triolo, T. and Sternglanz, R. (1996). *Nature*, **381**, 251.
22. Klemm, R. D., Austin, R. J., and Bell, S. P. (1997). *Cell*, **88**, 493.
23. Rowley, A., Cocker, J. H., Harwood, J., and Diffley, J. F. X. (1995). *EMBO J.*, **14**, 2631.
24. Rao, H. and Stillman, B. (1995). *Proc. Natl. Acad. Sci. USA*, **92**, 2224.
25. Diffley, J. F. X., Cocker, J. H., Dowell, S. J., and Rowley, A. (1994). *Cell*, **78**, 303.
26. Santocanale, C. and Diffley, J. F. X. (1996). *EMBO J.*, **15**, 6671.
27. Tanaka, T., Knapp, D., and Nasmyth, K. (1997). *Cell*, **90**, 649.
28. Aparicio, O. M., Weinstein, D. M., and Bell, S. P. (1997). *Cell*, **91**, 59.
29. Donovan, S., Harwood, J., Drury, L. S., and Diffley, J. F. X. (1997). *Proc. Natl. Acad. Sci. USA*, **94**, 5611.
30. Liang, C. and Stillman, B. (1997). *Genes Dev.*, **11**, 3375.
31. Santocanale, C. and Diffley, J. F. X. (1997). In *Methods in enzymology* (ed. W. C. Dunphy), Vol. 283, pp. 377–90. Academic Press, London.
32. Gossen, M., Pak, D. T., Hansen, S. K., Acharya, J. K., and Botchan, M. R. (1995). *Science*, **270**, 1674.
33. Carpenter, P. B., Mueller, P. R., and Dunphy, W. G. (1996). *Nature*, **379**, 357.
34. Rowles, A., Chong, J. P. J., Brown, L., Howell, M., Evan, G. I., and Blow, J. J. (1996). *Cell*, **87**, 287.
35. Romanowski, P., Madine, M. A., Rowles, A., Blow, J. J., and Laskey, R. A. (1996). *Curr. Biol.*, **6**, 1416.
36. Estes, H. G., Robinson, B. S., and Eisenberg, S. (1992). *Proc. Natl. Acad. Sci. USA*, **89**, 11156.
37. Jones, E. W. (1991). In *Methods in enzymology* (ed. C. Guthrie and G. R. Fink), Vol. 194, p. 428. Academic Press, London.
38. Kadonaga, J. T. and Tjian, R. (1986). *Proc. Natl. Acad. Sci. USA*, **83**, 5889.
39. Diffley, J. F. X. and Stillman, B. (1988). *Proc. Natl. Acad. Sci. USA*, **85**, 2120.
40. Buchman, A. R., Kimmerly, W. J., Rine, J., and Kornberg, R. D. (1988). *Mol. Cell. Biol.*, **8**, 210.
41. Shore, D., Stillman, D. J., Brand, A. H., and Nasmyth, K. A. (1987). *EMBO J.*, **6**, 461.
42. Rhode, P. R., Elsasser, S., and Campbell, J. L. (1992). *Mol. Cell. Biol.*, **12**, 1064.
43. Halfter, H., Kavety, B., Vandekerckhove, J., Kiefer, F., and Gallwitz, D. (1989). *EMBO J.*, **8**, 4265.
44. Diffley, J. F. X. and Stillman, B. (1989). *Science*, **246**, 1034.
45. Rhode, P. R., Sweder, K. S., Oegema, K. F., and Campbell, J. L. (1989). *Genes Dev.*, **3**, 1926.
46. Diffley, J. F. X. (1992). In *Molecular biology of Saccharomyces* (ed. L. A. Grivell), Vol. 62, p. 25. Kluwer Academic Publishers, London.
47. Walker, S. S., Francesconi, S. C., and Eisenberg, S. (1990). *Proc. Natl. Acad. Sci. USA*, **87**, 4665.
48. Dorsman, J. C., van Heeswijk, W. C., and Grivell, L. A. (1990). *Nucleic Acids Res.*, **18**, 2769.

49. Sambrook, J., Fritsch, E. F., and Maniatis, T. (ed.) (1989). *Molecular cloning: a laboratory manual*, 2nd edn, Ed. 3 vol., Cold Spring Harbour Laboratory Press, NY.
50. Maine, G. T., Sinha, P., and Tye, B.-K. (1984). *Genetics*, **106**, 365.
51. Yan, H., Gibson, S., and Tye, B.-K. (1991). *Genes Dev.*, **5**, 944.
52. Yan, H., Merchant, A. M., and Tye, B.-K. (1993). *Genes Dev.*, **7**, 2149.
53. Hennessy, K. M., Lee, A., Chen, E., and Botstein, D. (1991). *Genes Dev.*, **5**, 958.
54. Moir, D., Stewart, S. E., Osmond, B. C., and Botstein, D. (1982). *Genetics*, **100**, 547.
55. Chen, Y., Hennessy, K. M., Botstein, D., and Tye, B.-K. (1992). *Proc. Natl. Acad. Sci. USA*, **89**, 10459.
56. Donovan, S. (1997). Ph. D. Thesis. King's College, London.
57. Coleman, T. R., Carpenter, P. B., and Dunphy, W. G. (1996). *Cell*, **87**, 53.
58. Madine, M. A., Khoo, C.-Y., Mills, A. D., and Laskey, R. A. (1995). *Nature*, **375**, 421.
59. Chong, J. P. J., Mahbubani, H. M., Khoo, C.-Y., and Blow, J. J. (1995). *Nature*, **375**, 418.
60. Kubota, Y., Mimura, S., Nishimoto, S.-I., Takisawa, H., and Nojima, H. (1995). *Cell*, **81**, 601.
61. Ishimi, Y. (1997). *J. Biol. Chem.*, **272**, 24508.
62. Adachi, Y., Usukura, J., and Yanagida, M. (1997). *Genes to Cells*, **2**, 467.
63. Zou, L., Mitchell, J., and Stillman, B. (1997). *Mol. Cell. Biol.*, **17**, 553.
64. Owens, J. C., Detweiler, C. S., and Li, J. J. (1997). *Proc. Natl. Acad. Sci. USA*, **94**, 2521.
65. Dalton, S. and Hopwood, B. (1997). *Mol. Cell. Biol.*, **17**, 5867.
66. Hopwood, B. and Dalton, S. (1997). *Proc. Natl. Acad. Sci. USA*, **93**, 12309.
67. Jackson, A. L., Pahl, P. M., Harrison, K., Rosamond, J., and Sclafani, R. A. (1993). *Mol. Cell. Biol.*, **13**, 2899.
68. Dowell, S. J., Romanowski, P., and Diffley, J. F. X. (1994). *Science*, **265**, 1243.
69. Hardy, C. F., Dryga, O., Seematter, S., Pahl, P. M., and Sclafani, R. A. (1997). *Proc. Natl. Acad. Sci. USA*, **94**, 3151.
70. Lei, M., Kawasaki, Y., Young, M. R., Kihara, M., Sugino, A., and Tye, B. K. (1997). *Genes Dev.*, **11**, 3365.
71. Hartwell, L. H. (1973). *J. Bacteriol.*, **115**, 966.
72. Hartwell, L. H. (1976). *J. Mol. Biol.*, **104**, 803.
73. Hogan, E. and Koshland, D. (1992). *Proc. Natl. Acad. Sci. USA*, **89**, 3098.
74. Kelly, T. J., Martin, G. S., Forsburg, S. L., Stephen, R. J., Russo, A., and Nurse, P. (1993). *Cell*, **74**, 371.
75. Liang, C., Weinreich, M., and Stillman, B. (1995). *Cell*, **81**, 667.
76. Nishitani, H. and Nurse, P. (1995). *Cell*, **83**, 397.
77. Muzi-Falconi, M., Brown, G. W., and Kelly, T. J. (1996). *Proc. Natl. Acad. Sci. USA*, **93**, 1566.
78. Zwerschke, W., Rottjakob, H.-W., and Küntzel, H. (1994). *J. Biol. Chem.*, **269**, 23351.
79. Piatti, S., Lengauer, C., and Nasmyth, K. (1995). *EMBO J.*, **14**, 3788.
80. Cocker, J. H., Piatti, S., Santocanale, C., Nasmyth, K., and Diffley, J. F. X. (1996). *Nature*, **379**, 180.
81. Zhou, C., Huang, S.-H., and Jong, A. Y. (1989). *J. Biol. Chem.*, **264**, 9022.

3

Eukaryotic DNA polymerases

TERESA S.-F. WANG, KRISTA L. CONGER, WILLIAM C. COPELAND, and MARTHA P. ARROYO

1. Introduction

DNA polymerases are the fundamental enzymes that synthesize cellular DNA to produce two identical daughter chromosomes. Cellular DNA polymerases in conjunction with DNA repair enzymes play critical roles for the transmission and maintenance of error-free genetic information from one generation to the next. In the last decade, genetic studies in yeast, biochemical studies of *in vitro* replication of the simain virus 40 (SV40) origin-containing DNA, and isolation of genes and cDNAs of DNA polymerases from budding yeast, fission yeasts, and mammalian cells led to the identification of five distinct DNA template-dependent DNA polymerases, named polymerase α, β, γ, δ, and ε (1, 2). Their enzymatic properties, protein structures, associated activities and effectors, sequence conservation, genetic loci and structures, gene expression during cell growth and the cell cycle, and their proposed functional roles in replication and repair have been extensively described in several previous reviews (1–9). The distinguishing characteristics of each DNA polymerase, their properties, and assay conditions are summarized in *Table 1*. A noteworthy distinction among these cellular DNA polymerases is their associated enzymatic activities. DNA polymerases α and β are the two cellular DNA polymerases that do not contain an associated proof-reading 3′-5′ exonuclease, whereas polymerases γ, δ, and ε have an intrinsic proof-reading 3′-5′ exonuclease associated with the polymerase catalytic subunit. Furthermore, DNA polymerase α is uniquely associated with an DNA primase activity and therefore has the ability to initiate DNA synthesis. This chapter briefly describes the properties of each polymerase and provides a practical approach to the analysis of these eukaryotic DNA polymerases.

2. General assay protocol for DNA polymerases

DNA polymerase activity is assayed by incorporation of radiolabelled dNMP into acid insoluble DNA primer:template. A common DNA template is

Table 1. Properties of eukaryotic DNA polymerases

	α	β	γ	δ	ε
Protein structure					
Mammalian					
Catalytic (kDa)	165[a]	40	140	125	255
Associated (kDa)	70 58 49	None	35–55	48	55
S. cerevisiae					
Catalytic (kDa)	165	68	140	124	250
Associated (kDa)	86 58 49				
Auxiliary proteins	None	None	None	PCNA[b]	None
Cellular location	Nuclear	Nuclear	Mito[c]	Nuclear	Nuclear
Associated activities					
3′–5′ exonuclease	None	None	Yes	Yes	Yes
Primase	Yes	None	None	None	None
Processivity	Moderate	Low	High	High (PCNA)	High
Fidelity	Medium	Low	High	High	High
Inhibitors					
Aphidicolin	Sen[d]	Not sen	Sen	Sen	Sen
N-ethylmaleimide	Sen	Not sen	Sen	Sen	Sen
Butylphenyl dGTP or dATP	High sen	Not sen	Not sen	Mod sen	Mod sen
Dideoxynucleotide triphosphates	Not sen	Sen	Sen	Not sen	Not sen
Optimal assay conditions[e]					
pH	8	8.9	8	6.5	7.5
Primer:template	Gapped DNA (60–50 nt)	Gapped DNA (14–20 nt)	Gapped DNA or poly(rA) oligo(dT)	Poly(dA)-oligo(dT) or poly(dA-T)	Gapped DNA or poly(dA) oligo(dT)
Metal activator	Mg^{2+}	Mg^{2+}/ Mn^{2+}	Mg^{2+}/ Mn^{2+}	Mg^{2+}	Mg^{2+}

[a] Predicted molecular weight from amino acid sequence.
[b] Proliferating cell nuclear antigen.
[c] Mitochondrial.
[d] Sensitive.
[e] Mammalian polymerases.

gapped DNA, made by treating double-stranded DNA with DNase I under limiting conditions as described in *Protocol 1*. Homopolymeric primer: templates as described in *Protocol 2* can also be used. *Protocol 3* is given to demonstrate general concepts about how DNA polymerase activity is measured. It should also serve as a good starting point for assaying polymerases from all sources. However the specific assay conditions such as pH, metal ion concentration, dNTP concentrations, incubation time, and temperature to achieve optimal enzymatic activity should be empirically tested for each individual DNA polymerase.

Protocol 1. Preparation of optimally gapped DNA primer:template for DNA polymerase reaction

Equipment and reagents

- Omnimixer[a]
- Water-bath at 77 °C for quick heat denaturation and an ice-bath for rapid cooling
- Salmon sperm or calf thymus DNA (Sigma)
- 100 mg/ml BSA, crystallized and lyophilized (Sigma)
- DNase I solution: dissolve a bottle of Worthington or Boehringer best grade DNase I in 1 mM HCl at 5 mg/ml; aliquot the DNase I into 100 μl portions in screw-cap Eppendorf tubes, and store at –20 °C
- 5 × buffer A: 5 mg/ml BSA, 500 mM Tris–HCl pH 7.5, 25 mM $MgCl_2$

Method

1. Soak 1 g salmon sperm or calf thymus DNA in 100 ml water for two days at 4 °C.
2. Chop the DNA solution in an omnimixer for 2–3 min at maximum speed.[a]
3. Add to the DNA solution 10 ml 1 M Tris–HCl pH 7.5, 1 ml 1 M $MgCl_2$, and 1 ml 100 mg/ml BSA.
4. Set-up seven 0.5 ml aliquots of the DNA and pre-incubate them at 37 °C with shaking.
5. Take out one vial of the DNase I solution and dilute to 500 μg/ml with buffer A.
6. Add DNase I to the DNA-containing tubes to give final concentrations of 10, 50, 100, 200, 360, and 500 ng/nl DNase I.
7. Digest the DNA samples at 37 °C for 15 min. Then denature the DNase I in the 77 °C water-bath with vigorous shaking for 5 min, and quickly cool the digested DNA samples in order to melt the small nicked fragments of DNA.
8. Test the different levels of DNase I digested DNA sample as primer:template for your selected DNA polymerase. Select the DNA sample that give the highest level of DNA polymerization activity as the optimally gapped DNA primer:template.[b]
9. If many assays are to be done a large batch of optimally gapped template can be prepared. Care should be taken when scaling-up however as changes in the viscosity of the digestion solution could alter the amount of DNase I required to generate the optimal gap.

[a] It is also possible to use sonication for this step.
[b] Since each DNA polymerase has different optimal gap size for polymerization, if possible the optimal time of DNase I digestion and should be determined for each individual polymerase. If no polymerase samples are available, a crude idea of the optimal size required can be measured using commercially available polymerase, e.g. Klenow.

Protocol 2. Preparation of homopolymeric primer:template

Reagents

- Approx. 1 mg/ml solutions of poly(dA) and oligo(dT)[a] (commercially available from several sources, e.g. Boehringer)

Method

1. Mix together poly(dA) and oligo(dT) in 40 mM KCl at the ratio needed to give the required theoretical gap size.[b]
2. Heat to 65°C for 10 min and then allow to cool slowly to room temperature.
3. Store in aliquots at –20°C until needed.

[a] Other types of homopolymeric pairs could also be used.
[b] The average gap size of the template is dependent on the relative coverage of the poly(dA). If the number of moles nucleotides of oligo(dT) is close to that of the poly(dA) the gaps will be small. Conversely if oligo(dT) is present at much lower levels then the gaps will be relatively large. For an example of the use of this type of substrate see *Protocol 11*.

Protocol 3. Assay of DNA polymerase activity using optimally gapped DNA as primer:template[a,b]

Equipment and reagents

- Glass fibre filters such as Whatman GF/C filters
- Filtration apparatus (*Figure 1*)
- Apparatus for scintillation counting
- Cold 20% solution of trichloroacetic acid (TCA)
- Optimally gapped DNA template prepared by DNase I treated calf thymus DNA
- Stop buffer: 100 mM sodium pyrophosphate, 5 mg/ml sonicated DNA as carrier, and 0.5 mg/ml BSA
- Four dNTP solutions
- Labelled dNTP: usually either ^{32}P-labelled in the α-phosphate of dNTP, or labelled with ^{3}H or ^{14}C in the nucleotide base—the amount to be added varies depending on the enzymatic activity of the polymerase and the objective of the experiment
- Reaction mixture (5 ×): 10 mM 2-mercaptoethanol or DTT, 1 mg/ml BSA, $MgCl_2$ or $MnCl_2$ as metal activator (concentration of the metal ion depending on each individual polymerase)

Method[a]

1. For each assay set-up a 30 μl or 50 μl incubation containing 1 × reaction mixture, appropriate quantities of dNTPs with one radiolabelled dNTP, gapped DNA at concentration optimal for each individual DNA polymerase, and the source containing the polymerase.[c]
2. Incubate at 37°C for an appropriate period of time.[d] The time of incubation depends on the activity of the enzyme and the objective of your experiment. Each DNA polymerase should be tested to determine the time of incubation most suitable for the experimental purpose.

3. Stop the reactions by the addition of 0.5 ml stop buffer.
4. Precipitate the incorporated radioactive nucleotides in the terminated reaction by addition of an equal volume of cold 20% TCA.[e]
5. Collect the precipitates on a glass fibre filter. Wash extensively in 1 M HCl as described in ref. 10, and then either air dry or dry under a heat lamp.
6. Determine the amount of insoluble radioactivity on the filter by counting in a scintillation counter.[f]
7. The specific activity of the DNA polymerase is expressed as units of polymerase activity per milligram of protein assayed. One unit is defined as one nanomole of labelled dNMP (or four total dNMP) incorporated per hour at 37 °C.[g]

[a] Summarized in *Figure 1*.
[b] More detailed protocols for individual polymerases are included in later protocols in the chapter.
[c] When using a new source of polymerase several different concentrations should be tested to ensure that the assay is operating in the linear range.
[d] Often samples are taken at several different time points, also to ensure that the assay remains within the linear range.
[e] Alternatively, a defined aliquot of the reaction product can be spotted onto a DE81 paper disc washed three times with 0.15 M LiCl and ethanol.
[f] If ^{32}P is used as the labelled nucleotide the samples can be counted directly in a scintillation counter by Cerenkov counting. If a lower activity isotope is used it will be necessary to add scintillation fluid.
[g] If the assay is carried out using a crude extract nuclease activity may interfere with the amount of insoluble radioactivity observed. An indication of the extent of the nuclease problem can be obtained by including an additional tube containing a mix of the crude extract and a characterized polymerase (e.g. the Klenow fragment of *E. coli* DNA polymerase I).

3. DNA polymerase α

DNA polymerase α (pol α) is thought to be responsible for initiation of DNA synthesis due to its associated two subunits containing DNA primase activity (1, 2). The expression of the human polymerase α gene has been studied in different cell types, in transformed cells and in normal cells, in proliferative and terminally differentiated cells, and at different stages of the cell cycle (11–14). The enzymological properties, catalytic mechanisms, associated activities, effectors, genetic structure, and physiological roles of polymerase α are described in two review articles (1, 2) and summarized in *Table 1*.

3.1 Protein structure, subunits, cDNAs, and catalytic function

DNA pol α has four subunits: a catalytic subunit of 165–180 kDa, a 70 kDa subunit also known as B-subunit with no detectable catalytic activity (15, 16),

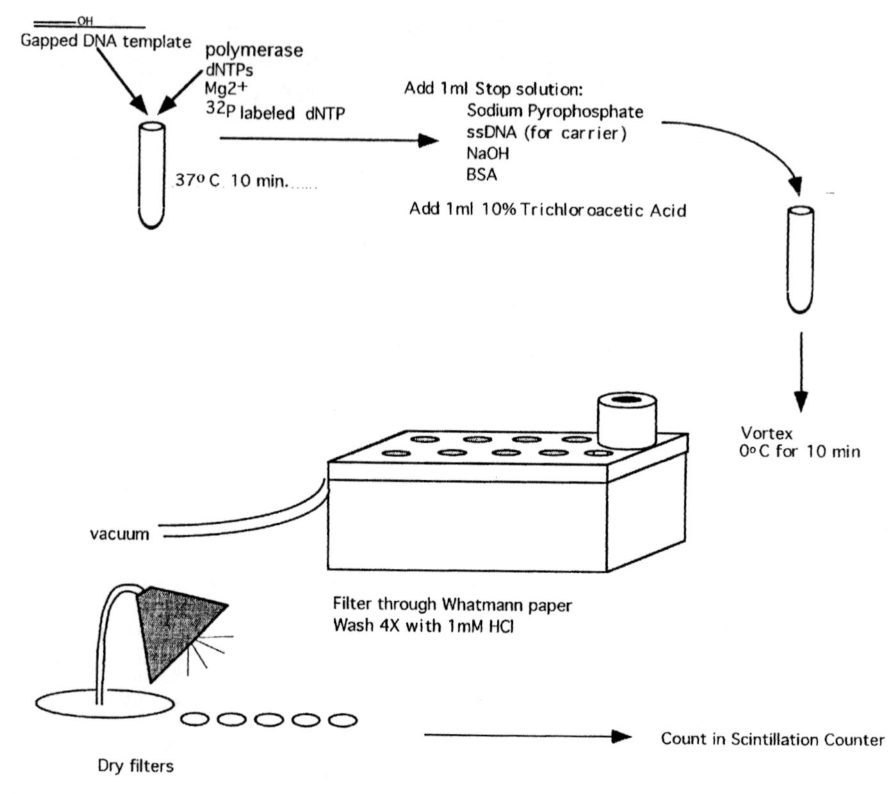

Figure 1. A diagram of how to assay DNA polymerases.

and two subunits of 49 and 58 kDa which contain the DNA primase activity (1, 17). cDNAs for all four subunits from human, mouse, *Drosophila melanogaster* embryo, and the gene from *Saccharomyces cerevisiae* (*S. cerevisiae*) have been cloned and characterized (15, 18–28). The catalytic subunit of human and *Schizosaccharomyces pombe* (*S. pombe*) polymerase α is both phosphorylated and glycosylated (13, 29) and the phosphorylation occurs in a cell cycle-dependent manner (13, 30). The human p70 subunit (B-subunit) is phosphorylated only in the G2/M phase of the cell cycle (13) and the *S. cerevisiae* B-subunit is also phosphorylated in a cell cycle-dependent manner (16, 31). Studies of the two subunits of mammalian primase indicate that the 49 kDa subunit functions in the synthesis of the RNA primer while both subunits are required for synthesizing the first dinucleotide (17). Studies of *S. cerevisae* and *Drosophila melanogaster* primase showed that the smaller subunit of the primase (48 kDa) is sufficient for both initiation (i.e. synthesis of the initial dinucleotide) and RNA primer synthesis (28, 32). The mammalian 49 kDa subunit has also been found to have a role in mediating species specificity of polyomaviral *in vitro* replication (33).

The primary sequence of polymerase α contains several regions that are highly conserved in DNA polymerases from both prokaryotic and eukaryotic organisms, bacteriophage, and animal viruses. DNA polymerases that contain these conserved regions are classified as α-like polymerases (19, 34–36). Site-directed mutagenesis studies have shown that the three most conserved regions of the α-like DNA polymerases contain the components of the active site. In addition, residues responsible for interacting with the metal activator (Mg^{2+}), primer terminus, and dNTPs have been identified. Additional studies have also identified residues that are responsible for DNA synthetic fidelity such as misinsertion, mispair primer extension, and metal-induced infidelity (2, 37–42). A model of residues in the active site of α-like DNA polymerases functioning in DNA synthesis is described in detail in refs 2 and 41.

3.2 Biological function

In both budding and fission yeast, the DNA polymerase α gene is essential for cell viability (43, 44). Due to the associated primase activity, polymerase α is thought to be the principal cellular DNA polymerase responsible for initiation. Studies of cell-free and reconstituted *in vitro* replication of SV40 origin-containing DNA and cell-free replication of bovine papillomavirus (BPV) origin-containing DNA have shown that the catalytic subunit of polymerase α (p180) physically interacts with SV40 large T-antigen and BPV E1, and this interaction is necessary for viral DNA replication *in vitro* (45–48). Furthermore, the transition of SV40 DNA replication from pre-initiation phase to initiation phase requires the physical interaction of a specific amino terminal domain of human polymerase α catalytic subunit with SV40 T-antigen in the pre-initiation complex (47). In addition, the 70 kDa subunit (B-subunit) is also able to interact with T-antigen in the cell-free SV40 replication assay (15). The catalytic subunit of polymerase α has also been shown to physically interact with herpes simplex virus UL9 protein (49). Together, these results from *in vitro* viral DNA replication suggest polymerase α plays a critical role in synthesizing the initiator DNA (iDNA) during cellular chromosome replication.

3.3 Purification of DNA polymerase α

Polymerase α has been purified from human cells (50–54), calf thymus (55, 56), mouse cells (57, 58), *Drosophila melanogaster* embryo (59, 60), and *S. cerevisiae* (61, 62). The purification protocols for polymerase α vary widely depending on the organism and cell type and often involve a mixture of standard and immunoaffinity purification (54). Readers interested in purifying the enzyme from characterized sources should consult the relevant references. The purification of polymerase α can be achieved from new sources by assaying at each stage in the protocol using the protocol outlined in *Protocol 4*. Polymerase α activity can be distinguished from others by specific antibody or by inhibitors as described in *Table 1*.

Protocol 4. Assay of DNA polymerase α activity using optimally gapped DNA template

Equipment and reagents

- See *Protocol 3*

Method

1. For each sample to be assayed set-up a 30 μl or 50 μl reaction containing: 20 mM Tris–HCl pH 8, 2 mM 2-mercaptoethanol, 200 μg/ml bovine serum albumin (BSA), 10 mM $MgCl_2$, 50 μM each of the 4 dNTPs with one radiolabelled dNTP, 800 μg/ml optimally gapped DNA.
2. Incubate and process the product as described in *Protocol 3*.

4. DNA primase

4.1 Properties

DNA primase exists as two subunits, p58 and p49, of the DNA polymerase α holoenzyme complex (1, 2). Molecular clones encoding both primase subunits have been isolated from yeast, *Drosophila*, human, and mouse (20, 25–28, 63). The yeast primase genes are single copy in haploid cells and both are essential for cell growth (26, 64). Primase initiates *de novo* synthesis with a purine, resulting in the 5′ nucleotide as a triphosphate nucleotide (51, 65). The average unit length of products is eight or nine nucleotides (51, 66, 67). After primer synthesis by the primase subunits, the DNA polymerase α/primase complex switches to allow elongation of the primer by DNA polymerase α (66, 67). Only primers ≥ 7 nucleotides long are extended by the DNA polymerase α catalytic subunit p180 (67). The initiation of primer synthesis occurs by the formation of a dinucleotide with a purine triphosphate at the 5′ end (65–67). It is thought that primase contains two separate ribonucleotide triphosphate binding sites in close proximity to each other. One of the binding sites has a specificity for binding purine triphosphates, with this site offering the 3′ OH group of the purine triphosphate to the second nucleotide which is located in the RNA polymerase domain of the primase complex. It is not yet known whether these two activities are performed by only one polypeptide of the two primase subunits or by both subunits.

4.2 Biological function

The biological function of primase has been investigated in yeast. The genes of the two primase subunits are called *PRI1* and *PRI2* (43). Thermosensitive mutants of these two genes have been generated (64, 68, 69). Genetic studies of these primase mutants have suggested that primase plays an essential role in a subset of the RAD53-dependent cell cycle progression checkpoint pathway in response to DNA damage. A specific mutant allele of *PRI1*, *pri1-M4*, affects the rate of S phase progression and partially delays the G1 to S phase transition in response to DNA damage by MMS and UV irradiation (70).

4.3 Purification

Separation and purification of the two primase subunits from the polymerase α/primase four subunit holoenzyme complex have not been successful. Isolation of cDNAs for the two primase subunits has allowed the use of expression systems for purification of the two primase subunits. During the course of purification from such systems the primase activity can be assayed using the methods described in *Protocols 2* and *3*. As an example a two-plasmid system has been developed for the co-expression of both primase subunits of the human p58 and p49 (71) in *E. coli*.

4.4 Assay protocols for DNA primase activity

Several different methods can be used to assay the primase activity. Primase activity can be measured 'directly' as the incorporation of radiolabelled rATP on a poly(dT) template in the absence of DNA polymerase, or determined by a 'coupled assay' with DNA polymerase as described (17, 51).

The coupled assay measures the incorporation of radiolabelled dAMP into acid insoluble DNA which is dependent on the presence of a primer synthesized by the primase.

Protocol 5. Direct assay for DNA primase activity

Equipment and reagents

- 15% polyacrylamide sequencing gel containing 1 × TBE and 7 M urea
- 2.5 pmol poly(dT)_{290}
- 100 mM ATP solution
- [α-^{32}P]ATP
- TBE: Tris, buffer, EDTA
- 5 × reaction buffer: 250 mM Tris–HCl pH 8, 20 mM DTT, 1 mg/ml acetylated BSA (regular BSA is often contaminated with nucleases), 10 mM $MgCl_2$, 100 mM KCl
- Sequencing loading buffer: 95% formamide, 10 mM EDTA pH 8, 0.1% xylene cyanol, 0.1% bromphenol blue

Method

1. For each assay set-up a 10 μl reaction containing 1 × reaction buffer, 2.5 pmol poly(dT)_{290}, 100 μM [α-^{32}P]ATP, and the sample to be assayed for primase activity.
2. Incubate reactions for 15 min at 30 °C.
3. Stop the reaction by the addition of a equal volume of 95% formamide sequencing loading buffer.

Protocol 5. *Continued*

4. Separate the products of the reaction on a sequencing gel and subject directly to autoradiography without drying the gel. Primer products can be quantitated by autoradiography or by PhosphorImager.
5. One unit is defined as the amount of enzyme to incorporate 1 nmol of radiolabelled AMP into riboprimer in 1 h at 30 °C.

Protocol 6. Coupled assay using Klenow polymerase[a]

Equipment and reagents

- Apparatus for carrying out TCA precipitation reactions (see *Protocol 3*)
- Poly(dT)$_{290}$
- 100 mM ATP solution
- 100 mM dATP solution
- [α-^{32}P]dATP
- 5 × reaction buffer: 250 mM Tris–HCl pH 8, 20 mM DTT, 1 mg/ml acetylated BSA (regular BSA is often contaminated with nucleases), 10 mM $MgCl_2$, 100 mM KCl
- Klenow fragment of *E. coli* polymerase I
- Stop buffer (see *Protocol 3*)

Method

1. For each assay set-up a 30 μl reaction containing 1 × reaction buffer, 3.3 μg/ml poly(dT)$_{290}$, 1.5 mM ATP,[b] 25 μM [α-^{32}P]dATP, appropriate units of Klenow depending on the purpose of the experiment, and the sample to be assayed for primase activity.
2. Incubate reactions at 30 °C for 30 min.
3. Terminate the reactions and TCA precipitate the insoluble materials as described in *Protocol 3* (step 3 onwards).

[a] This reaction could also be carried out using M13 as a template, and the rNTP and dNTP conditions as described in *Protocol 4*.
[b] It should be noted that in reactions with high ATP concentrations, the $MgCl_2$ levels should always be maintained at a ratio of twofold higher than the total nucleotide concentration.

Protocol 7. Coupled assay utilizing naturally associated polymerase activity[a]

Equipment and reagents

- Apparatus for carrying out TCA precipitation reactions (see *Protocol 3*)
- M13 ssDNA (New England Biolabs)
- 100 mM NTP solutions
- 100 mM dNTP solutions
- [α-^{32}P]dATP or dTTP
- 5 × reaction buffer for M13 assay: 250 mM Tris–HCl pH 8.6, 20 mM DTT, 1 mg/ml BSA, 12.5 mM $MgCl_2$
- Stop buffer (see *Protocol 3*)

Method

1. For each assay set-up a 30 μl reaction containing 1 × reaction buffer, 50 μM of each dNTP with 2000–6000 c.p.m./pmol of [α-^{32}P]dATP,

600 μM ATP, 100 μM each GTP, CTP, and UTP, single-stranded M13 DNA at 3 μg/ml, and polymerase–primase enzyme.

2. Incubate reactions at 30 °C for 30 min.
3. Terminate the reactions and TCA precipitate the insoluble material as described in *Protocol 3* (step 3 onwards).

[a] This reaction could also be carried out using poly(dT)$_{290}$ as a template, and the ATP and dATP conditions as described in *Protocol 3*.

5. DNA polymerase β

5.1 Structure and domain functions

DNA polymerase β is the simplest eukaryotic polymerase and therefore is the best studied DNA polymerase in eukaryotic cells. The enzyme isolated from vertebrates is a 39 kDa protein devoid of any associated enzymatic activities (1, 2, 72). A 68 kDa homologue has been reported in budding yeast (73–75). Mammalian polymerase β has been overexpressed and purified from *E. coli* (72, 76) which greatly facilitates the physical analysis of the protein structure and the understanding of the domain structure–function relationship of this enzyme (72, 77, 78). The primary sequence of polymerase β deduced from its cDNA indicates that polymerase β is not a member of the α-like cellular DNA polymerase family (35, 36). The crystal structure of rat polymerase β has been reported by two independent laboratories (79, 80). Furthermore, structures of ternary complexes of rat polymerase β with a DNA primer: template and dideoxycytidine triphosphate (ddCTP) have also been described (81, 82). The crystal structure and biochemical analysis of polymerase β reveals distinct domains required for enzymatic function. Reviews of these structure domains and functional comparison with the *E. coli* Klenow fragment and *HIV-1* reverse transcriptase are described in refs 2, 72, and 80.

5.2 Biological functions

In vitro, polymerase β prefers to fill gapped DNA with small gap sizes (approximately 14 nt gap size) and is capable of performing limited strand displacement on nicked DNA (83–85). Expression of polymerase β is not elevated when quiescent cells are induced to proliferate and polymerase β is constitutively expressed at low level even in differentiated tissues such as testis and brain (86). Thus, polymerase β is thought to be involved in short-patch excision repair (2, 72, 75, 85, 87). This notion is further supported by the finding that DNA damaging agents induce the transcription of the polymerase β message (88). In addition, polymerase β has also been implicated in mismatch repair (89) and in replicating single-stranded DNA injected into *Xenopus* oocytes (90). Studies of the yeast homologue POL4 (YCR14C) have suggested that polymerase β may also play a role in meiotic function (73, 75, 91).

5.3 Purification and assay protocol

Recombinant rat and human polymerase β can be overexpressed in *E. coli* (72, 76). Bacterial expression, purification, and the assay of the recombinant mammalian polymerase β are described in detail in ref. 72. To purify polymerase β from a novel source the activity can be followed using the assay conditions described in *Protocol 5*. Polymerase β activity can be distinguished from the other cellular polymerases by its high sensitivity to dideoxynucleotide triphosphate (ddNTPs) (see *Table 1*).

Protocol 8. Assay for polymerase β activity

Equipment and reagents

- Apparatus for carrying out TCA precipitation reactions (see *Protocol 3*)
- Optimally gapped DNA template: polymerase β prefers small gap size DNA as primer: template—the optimal size gap is 14 nt
- 5 × reaction buffer: 250 mM Tris–HCl pH 8.9, 5 mM 2-mercaptoethanol, 1 mg/ml bovine serum albumin (BSA), 100 mM $MgCl_2$ (or 25 $MnCl_2$), 500 mM KCl
- 100 mM dNTP solutions
- Labelled dNTP: usually either ^{32}P-labelled in the α-phosphate position of dNTP, or labelled with ^{3}H or ^{14}C in the nucleotide base—the specific activity of the labelled dNTP used in the reaction depends on the activity of the enzyme
- Stop buffer (see *Protocol 3*)

Method

1. For each sample set-up a 50 μl reaction containing 1 × buffer, 100 μM each dATP, dGTP, dCTP, dTTP, radioactively labelled dNTP, optimally gapped DNA template, and source of polymerase β (84, 92–94). If $MgCl_2$ is used as metal activator, greater than 10 μg of optimally gapped DNA is required for a 50 μl reaction. If $MnCl_2$ is used as metal activator, 175 ng of optimally gapped DNA is required for a 50 μl reaction.
2. Incubate at 37°C for an appropriate time.
3. Process the products as described in *Protocol 3*.

6. DNA polymerase γ

6.1 Protein and genetic structure

DNA polymerase γ is a nuclear encoded enzyme that replicates mitochondrial DNA. Polymerase γ contains an intrinsically associated exonuclease (*Table 1*). The DNA polymerase γ genes and cDNA from *S. cerevisiae*, *S. pombe*, *Drosophila melanogaster* (*D. melanogaster*), *Gallus gallus* (*G. gallus*),

Xenopus, and *Homo Sapiens* have been isolated (95–99). The predicted primary sequence of the human polymerase γ polypeptide is 1239 amino acids with a calculated molecular weight of 139.5 kDa. The human amino acid sequence is 41.6%, 43%, 48.7%, and 77.6% identical to *S. pombe*, *S. cerevisiae*, *D. melanogaster*, and the C terminal half of *G. gallus* respectively The conservation of sequence is about 10% higher in the polymerase domain compared to the exonuclease domain. These sequences all contain the three DNA polymerase motifs common to all DNA polymerases and the three conserved 3′-5′ exonuclease motifs common to exonuclease proficient DNA polymerases (100). Amino acid sequence alignment with the *S. cerevisiae* polymerase γ shows that the *S. cerevisiae* polymerase γ is unique, having an additional 250 amino acids in the C terminus (96).

6.2 Biological functions

The human mitochondrial DNA is a 16 569 bp, circular, double-stranded molecule that encodes 13 protein subunits of four oxidative phosphorylation complexes, 22 tRNAs, and two rRNAs required for their synthesis (101, 102). Mitochondrial DNA is 16 times more prone to oxidative damage and evolves 10–20 times faster than nuclear DNA. Many of the known mitochondrial diseases have been attributed to mutations in the mitochondrial genome (103). These mutations most likely occur during DNA replication. Therefore, any defect in polymerase γ activity may manifest itself in a defect in oxidative phosphorylation and consequently ATP production.

6.3 Enzymatic properties

DNA polymerase γ is very sensitive to nucleotide analogues such as AZT and ddI or ddC. Clinical evidence demonstrates that AZT induced mitochondrial myopathy slowly and cumulatively develops during AZT treatment (104). Thus, patients being treated with such antiviral drugs as AZT and ddI may develop symptoms of energy decline in certain organs mimicking mitochondrial genetic diseases.

DNA polymerase γ was originally identified in the cell as a reverse transcriptase activity. This reverse transcriptase activity was measured on homopolymeric RNA and was inactive on natural RNA. Even though DNA polymerase γ is active on a wide variety of substrates, the reverse transcriptase assay is still the assay of choice for DNA polymerase γ because of its selectivity and sensitivity.

6.4 Purification of mitochondrial polymerase

A wide variety of protocols have been used to purify polymerase γ from mitochondria of different sources. Different purification protocols used by different investigators to purify polymerase γ from different sources (105–

108). The following assay (*Protocol 9*) can be used to identify polymerase γ during the purification. Polymerase γ is sensitive to aphidicoln which is also an inhibitor for polymerases α, δ, and ε (*Table 1*). Polymerase γ is sensitive to dideoxynucleotide triphosphate (ddNTP) which is also a potent inhibitor for polymerase β (*Table 1*). Thus, there is not a known specific inhibitor that can be used to distinguish polymerase γ from other cellular polymerases.

Protocol 9. Assay procedure for mitochondrial DNA polymerase γ

Equipment and reagents

- Apparatus for carrying out TCA precipitation reactions (see *Protocol 3*)
- Poly(rA)/oligo(dT), annealed as described in *Protocol 2*
- 100 mM dNTP solutions
- [α-^{32}P]dTTP (25 μCi/ml)
- 5 × reaction buffer: 250 mM Hepes–KOH pH 8, 500 mM NaCl, 1 mg/ml acetylated BSA, 2.5 mM $MnCl_2$, 12.5 mM 2-mercaptoethanol
- Stop buffer (see *Protocol 3*)

Method

1. For each sample set-up a 30 μl reaction containing 1 × buffer, 50 μg/ml poly(rA)/oligo(dT), 10 μM [α-^{32}P]dTTP, and source of polymerase γ.
2. Incubation at 37 °C for an appropriate time.
3. Terminate the reactions and process as described in *Protocol 3*.

7. DNA polymerase δ

DNA polymerase δ (also named *POL3* in budding yeast) is an essential enzyme for replicating eukaryotic chromosomal DNA. Reconstituted DNA virus (SV40) replication studies strongly suggest that DNA polymerase δ is the principal enzyme that synthesizes the bulk of viral chromosomal DNA (2, 109, 110). The general enzymatic properties of polymerase δ are summarized in *Table 1*.

7.1 Subunit, cDNA, and catalytic function

Initial biochemical purification of polymerase δ from human, bovine, and *S. cerevisiae* have indicated that polymerase δ contains only two subunits, a 124 kDa DNA polymerase catalytic subunit with an intrinsic 3′-5′ exonuclease and a small subunit of 50 kDa (1, 2). The cDNAs and gene of the 124 kDa catalytic subunit from human, bovine, *S. cerevisiae*, and *S. pombe* have been isolated (23, 111–115). The apparent molecular mass of the polymerase δ catalytic subunit purified from different organisms seems to be identical with the molecular mass deduced from the cDNA sequence. This suggests that the polymerase δ catalytic subunit is not extensively modified post-translationally.

The deduced primary sequence indicates that polymerase δ is a member of the α-like DNA polymerases (111). The cDNAs of the human 124 kDa and 50 kDa subunits have been cloned and the recombinant proteins have been expressed in baculovirus-infected insect cells either as individual proteins or as the heterodimeric enzyme (116).

The heterodimeric enzyme is distributive during the polymerization reaction and requires an accessory protein, proliferating cell nuclear antigen (PCNA) for highly processive DNA synthesis (117). In contrast to the heterodimeric enzyme, neither the human nor the *S. pombe* polymerase δ catalytic subunit (124 kDa subunit) expressed from baculovirus-infected insect cells is responsive to the stimulation by PCNA (118, 119). The 50 kDa small subunit, although it has no known catalytic activity, is required for the functional interaction between the heterodimeric polymerase δ and PCNA for processive DNA synthesis (116). In *S. pombe*, the gene encoding the 50 kDa subunit is named *cdc1*, the gene product of which physically interacts with the polymerase δ catalytic subunit and PCNA, and also with the gene product of *cdc27* (120). The biological significance of the interaction between the *cdc27* gene product and polymerase δ is not yet known. A recent study by both biochemical and genetic approaches has indicated that *S. pombe* polymerase δ contains five subunits of apparent molecular weights of 125, 55, 54, 42, and 22 kDa (121).

7.2 Biological roles

Results of reconstituted SV40 replication suggest that once polymerase α synthesizes the initiator DNA (iDNA), an ATP-dependent structure-specific DNA binding protein, replication factor C (RF-C), binds the iDNA, and then loads the polymerase δ accessory protein PCNA onto the replication fork. PCNA then binds polymerase δ and a polymerase switching occurs. Polymerase δ displaces polymerase α-primase from the DNA template. DNA polymerase δ assumes the leading strand synthesis at the SV40 origin (122, 123). On the lagging strand, a similar polymerase switching occurs also via the RF-C binding to the 3′ end of the iDNA for subsequent loading of PCNA and polymerase δ to complete the synthesis of each Okazaki fragment. Therefore, results from the SV40 *in vitro* replication system suggest that polymerase δ has a role in synthesis of the leading as well as for completion of the lagging strand, while polymerase α is the enzyme responsible for synthesis of iDNA on both the leading and lagging strands (109, 124).

DNA polymerase δ is also implicated in base excision repair in *S. cerevisiae* (125, 126). In a genetic background containing a combination of a polymerase α (*pol1*) or a polymerase ε (*pol2*) mutation with polymerase δ (*pol3*) mutant, the cells show repair defects, suggesting that polymerase δ might play a role in repair synthesis (126). Polymerase δ has also been shown to be required for mismatch repair synthesis in *S. cerevisiae* (127).

7.3 Purification and assay protocol

Purification of mammalian polymerase δ and conditions used to assay polymerase δ with the accessory protein PCNA are described in ref. 128. Purification and assay procedures for budding yeast polymerase δ are described in refs 62 and 129. Polymerase δ can be distinguished from other DNA polymerases by its high sensitivity to an inhibitor, carbonyldiphosphonate.

Protocol 10. Assay conditions for polymerase δ

Equipment and reagents

- Apparatus for carrying out TCA precipitation reactions (see *Protocol 3*)
- Poly(dA):oligo(dT) at 10–20 poly(dA) to 1 oligo(dT) ratio in nucleotide equivalents as described in *Protocol 2*
- 100 mM dTTP
- Radiolabelled dTTP
- 5 × reaction buffer: 200 mM bis–Tris pH 6.5, 30 mM $MgCl_2$, 200 μg/ml BSA, 10% glycerol, and appropriate units of polymerase δ
- PCNA preferably from the same species as the polymerase δ
- Stop buffer (see *Protocol 3*)

Method

1. For each sample set-up a 30–50 μl reaction mixture containing 1 × reaction buffer, 40 μM radiolabelled dTTP, 15 μg/ml poly(dA): oligo(dT), 2 μg/ml PCNA, and the sample containing polymerase δ.
2. Incubate at 37 °C for an appropriate time.
3. Terminate the reactions and process the insoluble products as described in *Protocol 3*.

8. DNA polymerase ε

DNA polymerase ε is also known as *POL2* in budding yeast and Cdc20 in fission yeast. This enzyme was originally identified in budding yeast as DNA polymerase II (130, 131). It is one of the least understood cellular DNA polymerases. *POL2* gene is essential for viability in budding yeast, suggesting that polymerase ε is involved in chromosome replication. Like polymerase δ, polymerase ε also contains an intrinsic 3′-5′ exonuclease (*Table 1*). Therefore, polymerase ε purified from calf thymus and human cells was previously mistaken as another form of polymerase δ (132–135). Polymerase ε has also been isolated from human cells as a DNA repair factor and mistaken as a polymerase δ (133) and it was later found to be a structurally and catalytically distinct cellular DNA polymerase (136, 137).

8.1 Protein structure, cDNA, gene, and catalytic properties

The *POL2* gene from budding yeast encodes a protein of 255 kDa. The deduced primary sequence indicates that *POL2*, like *POL3* (polymerase δ), is a member of the α-like polymerases (138). The cDNA of polymerase ε from human and calf cells has been isolated (139). The subunit composition of polymerase ε has not yet been resolved. Polymerase ε and polymerase δ both contain an intrinsic proof-reading 3′-5′ exonuclease (*Table 1*). One enzymological characteristic that distinguishes these two exonuclease-containing polymerases is that polymerase δ absolutely requires PCNA for processive DNA synthesis (117). In contrast, polymerase ε is able to synthesize DNA processively without PCNA (4, 135–137, 140–142). Under certain reaction conditions with ATP, however, polymerase ε can form a stable complex with the accessory replication factors, RF-C and PCNA (143–148).

8.2 Biological function

An SV40 origin-containing plasmid can be replicated *in vitro* by purified protein factors without polymerase ε (124). However, genetic evidence from both budding and fission yeast shows that yeast cells arrested by polymerase ε thermosensitive mutants display an S phase arrested phenotype, suggesting a role of polymerase ε in chromosomal DNA replication (138, 149–152). Polymerase ε was originally isolated as a protein factor required for DNA repair synthesis in permeabilized human fibroblasts (133). It has also been reported that polymerase ε is involved in yeast excision repair (125). An *in vitro* reconstitution of the DNA excision repair reaction with 30 purified protein factors from mammalian cells requires polymerase ε as one of the repair factors (153). Thus, these findings implicate polymerase ε in both replication and repair.

8.3 Purification and assay protocol

A detailed description of the purification and assay procedures for the mammalian polymerase ε is reported in refs 141 and 142. Thus far, there is no known specific or preferable inhibitor to distinguish polymerase ε from other polymerases (*Table 1*).

Protocol 11. Assay conditions for polymerase ε

Equipment and reagents

- Apparatus for carrying out TCA precipitation reactions (see *Protocol 3*)
- $(dA)_{4000}$ and oligo$(dT)_{12-18}$ annealed at a 10:1 ratio as described in *Protocol 2*
- 100 mM dTTP
- Radiolabelled dTTP (as in *Protocol 3*)
- 2 × reaction buffer: 100 mM Hepes–KOH pH 7.5, 30 mM $MgCl_2$, 200 mM K glutamate pH 7.8, 20 mM DTT, 0.06% (v/v) Triton X-100, 40% (v/v) glycerol, 0.4 mg/ml BSA
- Stop buffer (see *Protocol 3*)

Method

1. For each sample set-up a 30–50 μl reaction mixture containing 1 × reaction buffer, 50 μM radiolabelled dTTP, 40 μM $(dA)_{4000}$: 4 μM oligo$(dT)_{12-18}$,[a] and the sample to be assayed.

Protocol 11. *Continued*

2. Incubate at 37 °C for 10–20 min.
3. Terminate the reactions and process the insoluble DNA products as described in *Protocol 3* (step 3 onwards).

[a] These ratios yield a inter-primer gap of approx. 135 nucleotides.

9. DNA polymerase accessory proteins

Proliferating cell nuclear antigen (PCNA) and replication factor C (RF-C) are two accessory factors important for processive DNA synthesis by polymerase δ or ε (110, 143, 145, 147, 154–161). The biological functions of these two accessory factors are described in separate chapters of this book.

10. Two important considerations for the study of DNA polymerases: processivity and fidelity

10.1 Processivity

What is processivity? Processivity of a DNA polymerase is defined as the number of times that a polymerase repeatedly incorporates dNMP onto a primer:template before its dissociation from the primer terminus. A polymerase that dissociates from the primer:template after each dNMP incorporation is said to be distributive with processivity of one. Processivity measurements give a quantitative estimate of a polymerase's ability to replicate DNA. The quantitative measurement of the processivity of a polymerase requires that each primer:template interacts with a polymerase molecule only one time. Thus, the primer:template should always be in large excess over the polymerase in the reaction. In general, no more than 10% of the primer should be used by a polymerase and short incubation time is required to limit the possibility of re-synthesis on a previously used primer. Specific ways to measure processivity are described in detail in ref. 162. Readers who are interested in this subject are advised to read this reference.

Protocol 12. An example of how to measure calf thymus polymerase δ processivity with a DNA trap

Equipment and reagents

- 8% acrylamide gel in standard sequencing gel size—the gel should be pre-run for at least 30 min at 70 W with the upper and lower buffer containing 89 mM Tris borate pH 8.3, 2 mM EDTA
- Electrophoresis apparatus and power supply
- Gel dryer
- Apparatus for autoradiography or phosphorimaging

- Poly(dA)$_{300}$: oligo(dT)$_{16}$ with the oligo(dT)$_{16}$ end-labelled with phosphate at the 5′ end using [γ-^{32}P]ATP (at 10 mCi/ml and approx. 3000 Ci/nmol)[a]
- Stop buffer (see *Protocol 3*)
- 5 × reaction buffer: 200 mM bis–Tris pH 6.8, 10% glycerol, 200 µg/ml BSA, 5 mM DTT, 30 mM $MgCl_2$
- Calf thymus DNA with optimal gap size for polymerase δ as DNA trap

Method

1. Set-up a 10 µl reaction mixture containing: 100 mM primer:template, 0.5 U purified calf thymus DNA polymerase δ, and 0.2 µg PCNA.
2. Start the reaction by addition of dTTP at a final concentration of 40 µM.
3. Incubate for 1 min at room temperature.
4. Add the DNA trap and incubate for a further 5 min.[b]
5. Stop the reaction by addition of 10 µl stop buffer.
6. Load 5 µl of each sample onto the pre-run 8% acrylamide gel, and electrophorese until the fastest migrating dye is about two-thirds of the way down the gel.
7. Dry the gel and subject to autoradiography or phosphorimaging.
8. The processivity of the reaction can then be directly observed from the size range of the products observed.

[a] As controls set-up reactions with no dTTP substrate, and with dTTP substrate but no DNA trap.
[b] Methods for annealing are described in *Protocol 2*. Alternative methods for annealing and end-labelling methods are described in ref. 162, and in Chapter 11 of this book.

10.2 Fidelity

What is fidelity? Fidelity of a polymerase indicates the errors made during DNA synthesis. There are several types of errors that a polymerase can make during DNA polymerization. These include deletion or addition of a nucleotide, insertion of a nucleotide (dNMP) that is not complementary to the template nucleotide, the incorporation of a nucleotide on a non-complementary primer terminus, or skip and slip resulting in a frame shift. In the past decade, numerous methods have been developed for measuring the different kinds of polymerase errors. Measurements of misinsertion fidelity, mispaired primer extension fidelity, metal-induced infidelity, and frame shift fidelity, all have different methods. Furthermore, there are different approaches for measurement of polymerase fidelity during initial encountering a primer terminus for incorporation of a nucleotide and incorporation of a nucleotide during elongation synthesis. Description of these protocols is beyond the scope of this chapter, however the general principles for measuring polymerase fidelity by these two methods are briefly described below.

10.2.1 Forward mutation assay

The principle of this assay is to measure the polymerase errors that inactivate the non-essential α-complementation activity of the *LacZ* gene in bacteriophage M13mp2. The polymerase of choice is allowed to fill a gapped *M13mp2* DNA *in vitro* with the gap positioned within the *LacZ*α. A portion of the gap-filled DNA product is first analysed by agarose gel to ensure the completion of gap-filling. Another aliquot of the gap-filled DNA products is transfected into competent *E. coli* cells and plated onto agar containing 5-bromo-4-chloro-3-indolyl-β-D-galactoside (X-Gal) and a lawn of α-complementation host *E. coli* (CSH50). Error-free polymerase product will yield functional α-peptide which is able to hydrolyse the X-Gal, resulting in dark blue M13 plaques. In the event of polymerase errors, the gap-filling products will not yield a functional α-peptide to complement the activity of LacZ gene product to hydrolyse X-Gal, thus will yield light blue plaques (due to partially functional α-peptide) or colourless plaques (due to complete inactivation of α-peptide). This method measures for loss of function and is termed 'forward mutation assay'. A broad spectrum of DNA polymerase errors can be measured by this method. Sequence analysis of DNA recovered from mutants will define the precise mutation. Readers who are interested in understanding the details of this forward mutation assay protocol are advised to read ref. 163.

10.2.2 Gel kinetic method

The assay is carried out by incubating a polymerase of choice with a 5′ ^{32}P-end-labelled primer:template to measure nucleotide misinsertion, exonucleolytic proof-reading, and lesion bypass efficiencies (164–168). The single-stranded 5′ ^{32}P-end-labelled primer products is then analysed by polyacrylamide gel electrophoresis to measure the nucleotide insertion frequency. Detailed protocols in measuring insertion error during initial encountering of a polymerase with the primer end, named standing start fidelity analysis, errors during elongation of a primer named running start kinetics, and how to measure polymerase fidelity containing exonuclease by the gel kinetic method are described in detail in ref. 168.

Readers who are interested in understanding the principle and how to perform and mathematically analyse the polymerase fidelity by the gel kinetic method are advised to read ref. 168.

Acknowledgements

T. S.-F. Wang is supported by NIH grants CA14835 and CA54415 from the National Institutes of Health; K. L. Conger is supported by the Cancer Biology pre-doctoral training grant CA09302; M. P. Arroyo is supported by the Medical Scientist Training Program (MSTP, GM07365).

References

1. Wang, T. S.-F. (1991). *Annu. Rev. Biochem.*, **60**, 513.
2. Wang, T. S.-F. (1996). In *Cellular DNA polymerases* (ed. M. L. DePamphilis), p. 461. Cold Spring Harbor Laboratory Press, Cold Spring Harbor, New York.
3. Hurwitz, J., Dean, F. B., Kwong, A. D., and Lee, S.-H. (1990). *J. Biol. Chem.*, **265**, 18043.
4. Bambara, R. A. and Jessee, C. B. (1991). *Biochim. Biophys. Acta*, **1088**, 11.
5. Challberg, M. D. and Kelly, T. J. (1989). *Annu. Rev. Biochem.*, **58**, 671.
6. Stillman, B. (1989). *Annu. Rev. Cell Biol.*, **5**, 197.
7. Stillman, B. (1996). *Science*, **274**, 1659.
8. Stillman, B. (1994). *J. Biol. Chem.*, **269**, 7047.
9. So, A. G. and Downey, K. M. (1992). *Crit. Rev. Biochem. Mol. Biol.*, **27**, 129.
10. Fisher, P. A. and Korn, D. (1977). *J. Biol. Chem.*, **252**, 6528.
11. Wahl, A. F., Geis, A. M., Spain, B. H., Wong, S. W., Korn, D., and Wang, T. S.-F. (1988). *Mol. Cell. Biol.*, **8**, 5016.
12. Pearson, B. E., Nasheuer, H.-P., and Wang, T. S.-F. (1991). *Mol. Cell. Biol.*, **11**, 2081.
13. Nasheuer, H.-P., Moore, A., Wahl, A. F., and Wang, T. S.-F. (1991). *J. Biol. Chem.*, **266**, 7893.
14. Moore, A. L. and Wang, T. S.-F. (1994). *Cell Growth Differ.*, **5**, 485.
15. Collins, K. L., Russo, A. A., Tseng, B. Y., and Kelly, T. J. (1993). *EMBO J.*, **12**, 4555.
16. Foiani, M., Marini, F., Gamba, D., Lucchini, G., and Plevani, P. (1994). *Mol. Cell. Biol.*, **14**, 923.
17. Copeland, W. C. and Wang, T. S.-F. (1993). *J. Biol. Chem.*, **268**, 26179.
18. Johnson, L. M., Snyder, M., Chang, L. M.-S., Davis, R. W., and Campbell, J. L. (1985). *Cell*, **43**, 369.
19. Wong, S. W., Wahl, A. F., Yuan, P. M., Arai, N., Pearson, B. E., Arai, K.-I., *et al.* (1988). *EMBO J.*, **7**, 37.
20. Miyazawa, H., Izumi, M., Tada, S., Takada, R., Masutani, M., Ui, M., *et al.* (1993). *J. Biol. Chem.*, **268**, 8111.
21. Damagnez, V., Tillit, J., de, R. A., and Baldacci, G. (1991). *Mol. Gen. Genet.*, **226**, 182.
22. Hirose, F., Yamaguchi, M., Nishida, Y., Masutani, M., Miyazawa, H., Hanaoka, F., *et al.* (1991). *Nucleic Acids Res.*, **19**, 4991.
23. Park, H., Francesconi, S., and Wang, T. S.-F. (1993). *Mol. Biol. Cell*, **4**, 145.
24. Cotterill, S., Lehman, I. R., and McLachlan, P. (1992). *Nucleic Acids Res.*, **20**, 4325.
25. Stadlbauer, F., Brueckner, A., Rehfuess, C., Eckerskorn, C., Lottspeich, F., Forster, V., *et al.* (1994). *Eur. J. Biochem.*, **222**, 781.
26. Foiani, M., Santocanale, C., Plevani, P., and Lucchini, G. (1989). *Mol. Cell. Biol.*, **9**, 3081.
27. Prussak, C. E., Almazan, M. T., and Tseng, B. Y. (1989). *J. Biol. Chem.*, **264**, 4957.
28. Bakkenist, C. J. and Cotterill, S. (1994). *J. Biol. Chem.*, **269**, 26759.
29. Hsi, K. L., Copeland, W. C., and Wang, T. S. (1990). *Nucleic Acids Res.*, **18**, 6231.
30. Park, H., Davis, R. E., and Wang, T. S.-F. (1995). *Nucleic Acids Res.*, **23**, 4337.
31. Foiani, M., Liberi, G., Lucchini, G., and Plevani, P. (1995). *Mol. Cell. Biol.*, **15**, 883.
32. Santocanale, C., Foiani, M., Lucchini, G., and Plevani, P. (1993). *J. Biol. Chem.*, **268**, 1343.

33. Bruckner, A., Stadlbauer, F., Guarino, L. A., Brunahl, A., Schneider, C., Rehfuess, C., *et al.* (1995). *Mol. Cell. Biol.*, **15**, 1716.
34. Wang, T. S.-F., Wong, S. W., and Korn, D. (1989). *FASEB J.*, **3**, 14.
35. Ito, J. and Braithwaite, D. K. (1991). *Nucleic Acid Res.*, **19**, 4045.
36. Delarue, M., Poch, O., Tordo, N., Morase, D., and Argos, P. (1990). *Protein Eng.*, **3**, 461.
37. Copeland, W. C. and Wang, T. S.-F. (1993). *J. Biol. Chem.*, **268**, 11028.
38. Copeland, W. C., Lam, N. K., and Wang, T. S.-F. (1993). *J. Biol. Chem.*, **268**, 11041.
39. Dong, Q., Copeland, W. C., and Wang, T. S. (1993). *J. Biol. Chem.*, **268**, 24163.
40. Dong, Q., Copeland, W. C., and Wang, T. S.-F. (1993). *J. Biol. Chem.*, **268**, 24175.
41. Dong, Q. and Wang, T. S.-F. (1995). *J. Biol. Chem.*, **270**, 21563.
42. Copeland, W. C., Dong, Q., and Wang, T. S.-F. (1995). In *Methods in enzymology* (ed. J. Campbell), Vol. 262, p. 294. Academic Press.
43. Campbell, J. L. (1993). *J. Biol. Chem.*, **268**, 25261.
44. Francesconi, S., Park, H., and Wang, T. S. (1993). *Nucleic Acids Res.*, **21**, 3821.
45. Dornreiter, I., Hoss, A., Arthur, A. K., and Fanning, E. (1990). *EMBO J.*, **9**, 3329.
46. Dornreiter, I., Erdile, L. F., Gilbert, I. U., von, W. D., Kelly, T. J., and Fanning, E. (1992). *EMBO J.*, **11**, 769.
47. Dornreiter, I., Copeland, W. C., and Wang, T. S.-F. (1993). *Mol. Cell. Biol.*, **13**, 809.
48. Park, P., Copeland, W., Yang, L., Wang, T., Botchan, M. R., and Mohr, I. J. (1994). *Proc. Natl. Acad. Sci. USA*, **91**, 8700.
49. Lee, S. K., Dong, Q., Wang, T. S.-F., and Lehman, I. R. (1995). *Proc. Natl. Acad. Sci. USA*, **92**, 7882.
50. Wong, S. W., Paborsky, L. R., Fisher, P. A., Wang, T. S., and Korn, D. (1986). *J. Biol. Chem.*, **261**, 7958.
51. Wang, T. S.-F., Hu, S.-Z., and Korn, D. (1984). *J. Biol. Chem.*, **259**, 1854.
52. Copeland, W. C. and Wang, T. S.-F. (1991). *J. Biol. Chem.*, **266**, 22739.
53. Rogge, L. and Wang, T. S. (1992). *Chromosoma*, **102 (suppl.)**, S114.
54. Wang, T. S.-F., Copeland, W. C., Rogge, L., and Dong, Q. (1995). In *Purification of mammalian DNA polymerases: DNA polymerase* α (ed. J. L. Campbell), p. 77. Academic Press.
55. Wahl, A., Kowalski, S. P., Harwell, L. W., Lord, E. M., and Bambara, R. A. (1984). *Biochemistry*, **23**, 1895.
56. Nasheuer, H.-P. and Grosse, F. (1987). *Biochemistry*, **26**, 8458.
57. Goulian, M. and Heard, C. J. (1989). *J. Biol. Chem.*, **264**, 19407.
58. Takada-Takayama, R., Hanaoka, F., Yamada, M., and Ui, M. (1991). *J. Biol. Chem.*, **266**, 15716.
59. Cotterill, S., Chui, G., and Lehman, I. R. (1987). *J. Biol. Chem.*, **262**, 16100.
60. Mitsis, P. G., Chiang, C.-S., and Lehman, I. R. (1995). In *Purification of DNA polymerase-primase (DNA polymerase* α*) and DNA polymerase* δ *from embryos of Drosophila melanogaster* (ed. J. L. Campbell), p. 62. Academic Press, Inc.
61. Plevani, P., Lucchini, G., Foiani, M., Valsasnini, P., Brandazza, A., Bianchi, M., *et al.* (1987). *Mol. Genet.*, **6**, 53.
62. Burgers, P. M. J. (1995). In *DNA polymerases from Saccharomyces cerevisiae* (ed. J. L. Campbell), p. 49. Academic Press, Inc.
63. Plevani, P., Francesconi, S., and Lucchini, G. (1987). *Nucleic Acids Res.*, **15**, 7975.
64. Lucchini, G., Francesconi, S., Foiani, M., Badaracco, G., and Plevani, P. (1987). *EMBO J.*, **6**, 737.

65. Gronostajski, R. M., Field, J., and Hurwitz, J. (1984). *J. Biol. Chem.*, **259**, 9479.
66. Hu, S. Z., Wang, T. S., and Korn, D. (1984). *J. Biol. Chem.*, **259**, 2602.
67. Kuchta, R. D., Reid, B., and Chang, L. M. (1990). *J. Biol. Chem.*, **265**, 16158.
68. Longhese, M. P., Jovine, L., Plevani, P., and Lucchini, G. (1993). *Genetics*, **133**, 183.
69. Francesconi, S., Longhese, M. P., Piseri, A., Santocanale, C., Lucchini, G., and Plevani, P. (1991). *Proc. Natl. Acad. Sci. USA*, **88**, 3877.
70. Marini, F., Pellecioli, A., Paciotti, V., Lucchini, G., Plevani, P., Stern, D. F., *et al.* (1997). *EMBO J.*, **16**, 639.
71. Copeland, W. C. (1996). *Protein Expression Purification*, **9**, 1.
72. Beard, W. A. and Wilson, S. H. (1995). In *Purification and domain mapping of mammalian DNA polymerase* β (ed. J. L. Campbell), p. 98. Academic Press, Inc.
73. Prasad, R., Widen, S. G., Singhal, R. K., Watkins, J., Prakash, L., and Wilson, S. H. (1993). *Nucleic Acids Res.*, **21**, 5301.
74. Leem, S. H., Ropp, P. A., and Sugino, A. (1994). *Nucleic Acids Res.*, **22**, 3011.
75. Budd, M. E. and Campbell, J. L. (1995). In *Purification and enzymatic and functional characterization of DNA polymerase* β*-like enzyme, POL4, expressed during yeast meiosis* (ed. J. L. Campbell), p. 108. Academic Press, Inc.
76. Abbotts, J., Sen, G. D., Zmudzka, B., Widen, S. G., Notario, V., and Wilson, S. H. (1988). *Biochemistry*, **27**, 901.
77. Wilson, S. H., Singhal, R. K., and Kumar, A. (1991). In *Structure and functional studies of mammalian DNA polymerase* β (ed. V. A. Bohr, K. Wassermann, and K. H. Kraemer), p. 343. Munksgaard, Copenhagen.
78. Casas, F. J., Kumar, A., Morris, G., Wilson, S. H., and Karpel, R. L. (1991). *J. Biol. Chem.*, **266**, 19618.
79. Davies, J. Z., Almassy, R. J., Hostomska, Z., Ferre, R. A., and Hostomsky, Z. (1994). *Cell*, **76**, 1123.
80. Sawaya, M. R., Pelletier, H., Kumar, A., Wilson, S. H., and Kraut, J. (1994). *Science*, **264**, 1930.
81. Pelletier, H., Sawaya, M. R., Kumar, A., Wilson, S. H., and Kraut, J. (1994). *Science*, **264**, 1891.
82. Pelletier, H. (1994). *Science*, **266**, 2025.
83. Tanabe, K., Bohn, E. W., and Wilson, S. H. (1979). *Biochemistry*, **18**, 3401.
84. Wang, T. S.-F. and Korn, D. (1980). *Biochemistry*, **19**, 1782.
85. Mosbaugh, D. W. and Linn, S. (1983). *J. Biol. Chem.*, **258**, 108.
86. Hirose, F., Hotta, Y., Yamaguchi, M., and Matsukage, A. (1989). *Exp. Cell Res.*, **181**, 169.
87. Singhal, R. K., Prasad, R., and Wilson, S. H. (1995). *J. Biol. Chem.*, **270**, 949.
88. Fornace, A. J., Zmudzka, B., Hollander, M. C., and Wilson, S. H. (1989). *Mol. Cell. Biol.*, **9**, 851.
89. Wiebauer, K. and Jiricny, J. (1990). *Proc. Natl. Acad. Sci. USA*, **87**, 5842.
90. Jenkins, T. M., Saxena, J. K., Kumar, A., Wilson, S. H., and Ackerman, E. J. (1992). *Science*, **258**, 475.
91. Shimizu, K., Santocanale, C., Ropp, P. A., Longhese, M. P., Plevani, P., Lucchini, G., *et al.* (1993). *J. Biol. Chem.*, **268**, 27148.
92. Wang, T. S.-F. and Korn, D. (1977). *Biochemistry*, **16**, 4927.
93. Wang, T. S.-F., Fisher, P. A., Sedwick, W. D., and Korn, D. (1975). *J. Biol. Chem.*, **250**, 5270.

94. Wang, T. S.-F. and Korn, D. (1982). *Biochemistry*, **21**, 1597.
95. Foury, F. (1989). *J. Biol. Chem.*, **264**, 20552.
96. Ropp, P. A. and Copeland, W. C. (1996). *Genomics*, **36**, 449.
97. Lewis, D. L., Farr, C. L., Wang, Y., Lagina, A., and Kaguni, L. S. (1996). *J. Biol. Chem.*, **271**, 23389.
98. Ye, F., Carrodeguas, J. A., and Bogenhagen, D. F. (1996). *Nucleic Acids Res.*, **24**, 1481.
99. Chiang, C. S. and Lehman, I. R. (1995). *Gene*, **166**, 237.
100. Heringa, J. and Argos, P. (1992). In *Evolution of viruses as recorded by their polymerase sequences* (ed. S. S. Morse), p. 87. Raven Press Ltd., New York.
101. Clayton, D. A. (1991). *Annu. Rev. Cell Biol.*, **7**, 453.
102. Clayton, D. A. (1992). *J. Inherit. Metab. Dis.*, **15**, 439.
103. Wallace, D. C. (1992). *Annu. Rev. Biochem.*, **61**, 1175.
104. Lewis, W. and Dalakas, M. C. (1995). *Nature Med.*, **1**, 417.
105. Wernette, C. M. and Kaguni, L. S. (1986). *J. Biol. Chem.*, **261**, 14764.
106. Mosbaugh, D. W. (1988). *Nucleic Acids Res.*, **16**, 5645.
107. Insdorf, N. F. and Bogenhagen, D. F. (1989). *J. Biol. Chem.*, **264**, 21491.
108. Gray, H. and Wong, T. W. (1992). *J. Biol. Chem.*, **267**, 5835.
109. Waga, S. and Stillman, B. (1994). *Nature*, **369**, 207.
110. Stillman, B. (1994). *Cell*, **78**, 725.
111. Boulet, A., Simon, M., Faye, G., Bauer, G. A., and Burgers, P. M. (1989). *EMBO J.*, **8**, 1849.
112. Pignede, G., Bouvier, D., Recondo, A.-M., and Baldacci, G. (1991). *J. Mol. Biol.*, **222**, 209.
113. Zhang, J., Chung, D. W., Tan, C.-K., Downey, K. M., Davie, E. W., and So, A. G. (1991). *Biochemistry*, **30**, 11742.
114. Zhang, J., Tan, C.-K., McMullen, B., Downey, K. M., and So, A. G. (1995). *Genomics*, **29**, 179.
115. Cullmann, G., Hindges, R., Berchtold, M. W., and Hubscher, U. (1993). *Gene*, **134**, 191.
116. Zhou, J.-Q., He, H., Tan, C.-K., Downey, K. M., and So, A. G. (1997). *Nucleic Acids Res.*,
117. Tan, C. K., Castillo, C., So, A. G., and Downey, K. M. (1986). *J. Biol. Chem.*, **261**, 12310.
118. Zhou, J.-Q., Tan, C.-K., So, A. G., and Downey, K. M. (1996). *J. Biol. Chem.*, **271**, 29740.
119. Arroyo, M. P., Downey, K. M., So, A. G., and Wang, T. S.-F. (1996). *J. Biol. Chem.*, **271**, 15971.
120. MacNeill, S. A., Moreno, S., Reynolds, N., Nurse, P., and Fantes, P. A. (1996). *EMBO J.*, **15**, 4613.
121. Zuo, S., Gibbs, E., Kelman, Z., Wang, T. S.-F., O'Donnell, M., Macneill, S. A., *et al.* (1997). *Proc. Natl. Acad. Sci. USA*, **94**, 11244.
122. Tsurimoto, T. and Stillman, B. (1990). *Proc. Natl. Acad. Sci. USA*, **87**, 1023.
123. Tsurimoto, T., Melendy, T., and Stillman, B. (1990). *Nature*, **346**, 534.
124. Waga, S., Bauer, G., and Stillman, B. (1994). *J. Biol. Chem.*, **269**, 10923.
125. Wang, Z., Wu, X., and Friedberg, E. C. (1993). *Mol. Cell. Biol.*, **13**, 1051.
126. Budd, M. E. and Campbell, J. L. (1995). *Mol. Cell. Biol.*, **15**, 2173.
127. Longley, M. J., Pierce, A. J., and Modrich, P. (1997). *J. Biol. Chem.*, **272**, 10917.

128. Downey, K. M. and So, A. G. (1995). In *Purification of mammalian polymerases: DNA polymerase* δ (ed. J. L. Campbell), p. 84. Academic Press, Inc.
129. Bauer, G. A., Heller, H. M., and Burgers, P. M. (1988). *J. Biol. Chem.*, **263**, 917.
130. Wintersberger, U. and Wintersberger, E. (1970). *Eur. J. Biochem.*, **13**, 11.
131. Chang, L. M. S. (1977). *J. Biol. Chem.*, **252**, 1873.
132. Crute, J. J., Wahl, A. F., and Bambara, R. A. (1986). *Biochemistry*, **25**, 26.
133. Nishida, C., Reinhard, P., and Linn, S. (1988). *J. Biol. Chem.*, **263**, 501.
134. Wahl, A. F., Crute, J. J., Sabatino, R. D., Bodner, J. B., Marraccino, R. L., Harwell, L. W., *et al.* (1986). *Biochemistry*, **25**, 7821.
135. Syvaoja, J. and Linn, S. (1989). *J. Biol. Chem.*, **264**, 2489.
136. Syvaoja, J., Suomensaari, S., Nishida, C., Goldsmith, J. S., Chui, G. S., Jain, S., *et al.* (1990). *Proc. Natl. Acad. Sci. USA*, **87**, 6664.
137. Syvaoja, J. E. (1990). *Bioessays*, **12**, 533.
138. Morrison, A., Araki, H., Clark, A. B., Hamatake, R. K., and Sugino, A. (1990). *Cell*, **62**, 1143.
139. Kesti, T., Frantti, H., and Syvaoja, J. E. (1993). *J. Biol. Chem.*, **268**, 10238.
140. Weiser, T., Gassmann, M., Thommes, P., Ferrari, E., Hafkemeyer, P., and Hubscher, U. (1991). *J. Biol. Chem.*, **266**, 10420.
141. Chui, G. S.-J. and Linn, S. (1995). In *Purification of mammalian DNA polymerases: DNA polymerase* ε (ed. J. L. Campbell), p. 93. Academic Press, Inc.
142. Chui, G. and Linn, S. (1995). *J. Biol. Chem.*, **270**, 7799.
143. Burgers, P. M. (1991). *J. Biol. Chem.*, **266**, 22698.
144. Lee, S. H., Pan, Z. Q., Kwong, A. D., Burgers, P. M., and Hurwitz, J. (1991). *J. Biol. Chem.*, **266**, 22707.
145. Lee, S. H., Kwong, A. D., Pan, Z. Q., and Hurwitz, J. (1991). *J. Biol. Chem.*, **266**, 594.
146. Podust, V. N. and Hubscher, U. (1993). *Nucleic Acids Res.*, **21**, 841.
147. Podust, V. N., Georgaki, A., Strack, B., and Hubscher, U. (1992). *Nucleic Acids Res.*, **20**, 4159.
148. Maga, G. and Hubscher, U. (1995). *Biochemistry*, **34**, 891.
149. Nasmyth, K. and Nurse, P. (1981). *Mol. Gen. Genet.*, **182**, 119.
150. Nurse, P., Thuriaux, P., and Nasmyth, K. (1976). *Mol. Gen. Genet.*, **146**, 167.
151. Araki, H., Ropp, P. A., Johnson, A. L., Johnston, L. H., Morrison, A., and Sugino, A. (1992). *EMBO J.*, **11**, 733.
152. D'Urso, G. and Nurse, P. (1997). *Proc. Natl. Acad. Sci. USA*, **94**, 12491.
153. Aboussekhra, A., Biggerstaff, M., Shivji, M. K. K., Vilpo, J. A., Monocollin, V., Podust, V., *et al.* (1995). *Cell*, **80**, 859.
154. Prelich, G., Kostura, M., Marshak, D. R., Mathews, M. B., and Stillman, B. (1987). *Nature*, **326**, 471.
155. Tan, C.-K., So, M. J., Downey, K. M., and So, A. G. (1987). *Nucleic Acids Res.*, **15**, 2269.
156. Tsurimoto, T. and Stillman, B. (1989). *EMBO J.*, **8**, 3883.
157. Lee, S. H. and Hurwitz, J. (1990). *Proc. Natl. Acad. Sci. USA*, **87**, 5672.
158. Tsurimoto, T. and Stillman, B. (1991). *J. Biol. Chem.*, **266**, 1950.
159. Pan, Z. Q., Chen, M., and Hurwitz, J. (1993). *Proc. Natl. Acad. Sci. USA*, **90**, 6.
160. Burgers, P. M. and Yoder, B. L. (1993). *J. Biol. Chem.*, **268**, 19923.
161. Hubsher, U., Maga, G., and Podust, V. N. (1996). In *DNA replication accessory*

proteins (ed. M. L. Depamphilis), p. 525. Cold Spring Harbor Laboratory Press, Cold Spring Harbor, NY.
162. Bambara, R. A., Fay, P. J., and Mallaber, L. M. (1995). In *Methods in enzymology* (ed. J. Campbell), Vol. 262, p. 270. Academic Press.
163. Bebenek, K. and Kunkel, T. A. (1995). In *Analyzing fidelity of DNA polymerases* (ed. J. L. Campbell), p. 217. Academic Press.
164. Boosalis, M. S., Petruska, J., and Goodman, M. F. (1987). *J. Biol. Chem.*, **262**, 14689.
165. Boosalis, M. S., Mosbaugh, D. W., Hamatake, R., Sugino, A., Kunkel, T. A., and Goodman, M. F. (1989). *J. Biol. Chem.*, **264**, 11360.
166. Mendelman, L. V., Petruska, J., and Goodman, M. F. (1990). *J. Biol. Chem.*, **265**, 2338.
167. Creighton, S., Huang, M.-M., Cai, H., Arnheim, N., and Goodman, M. F. (1992). *J. Biol. Chem.*, **267**, 2633.
168. Creighton, S., Bloom, L. B., and Goodman, M. F. (1995). In *Methods in enzymology* (ed. J. Campbell), Vol. 262, p. 232. Academic Press.

4

Identification and characterization of DNA helicases

DANIEL W. BEAN and STEVEN W. MATSON

1. Introduction

DNA helicases catalyse disruption of the hydrogen bonds between the two strands of duplex DNA to generate single-stranded DNA (ssDNA) products (1, 2). This reaction is generally referred to as an unwinding reaction. The ssDNA product(s) of the unwinding reaction is then available for use as a template for DNA replication or repair, or as a substrate in recombination. Enzymes that catalyse the unwinding of duplex RNA, RNA secondary structure, and DNA:RNA hybrids have also been described (3–7). These enzymes have important roles in mRNA biogenesis, transcription, and translation. Thus, helicase catalysed unwinding is not confined to the unwinding of duplex DNA. In fact, there are examples of enzymes that catalyse the unwinding of more than one of these substrates (8–10). The well characterized T-antigen, for example, unwinds both duplex DNA and RNA:DNA hybrids (9). Presumably these substrate preferences are related to the biological function of the protein, although these relationships are not always immediately obvious. None the less, it is clear that all transactions involving either DNA or RNA require the (transient) existence of a single-stranded intermediate, and thus helicases play key roles in all aspects of nucleic acid metabolism (11).

DNA helicases have been isolated from a variety of organisms including prokaryotes, eukaryotes, and viruses. A list of currently known DNA helicases from the prokaryote *Escherichia coli* and from the single celled eukaryote *Saccharomyces cerevisiae* is provided to illustrate the number and variety of these enzymes found in a single cell (*Table 1*). In each case the gene encoding the protein is known, and the enzyme has been purified and shown to have DNA unwinding activity *in vitro*. In addition, genetic studies have provided information regarding the cellular function of most of these proteins. It is clear that DNA helicases are involved in a wide range of DNA metabolic pathways both in prokaryotes and eukaryotes. In *E. coli* the precise role played by many of the helicases is well defined, while in yeast the roles

Table 1. DNA helicases

***E. coli* helicases**	**Gene**	**Polarity**	**Cellular role**
DnaB protein	*dnaB*	5′ to 3′	DNA replication
PriA protein	*priA*	3′ to 5′	DNA replication
Rep protein	*rep*	3′ to 5′	DNA replication
UvrAB complex	*uvrA* *uvrB*	5′ to 3′	Nucleotide excision repair
Helicase II	*uvrD*	3′ to 5′	Nucleotide excision repair, mismatch repair, recombination
Helicase IV	*helD*	3′ to 5′	Recombination
RecQ protein	*recQ*	3′ to 5′	Recombination
RecBCD complex	*recB* *recC* *recD*	*[a]	Recombination
RuvAB complex	*ruvA* *ruvB*	5′ to 3′	Recombination
Helicase I	*traI* (F plasmid)	5′ to 3′	Bacterial conjugation
***S. cerevisiae* helicases**			
Ssl2p	*SSL2*	3′ to 5′	Transcription (TFIIH subunit), nucleotide excision repair
Rad3p	*RAD3*	5′ to 3′	Transcription (TFIIH subunit), nucleotide excision repair
Pif1p	*PIF1*	5′ to 3′	Mitochondrial DNA repair/recombination, telomere maintenance
Srs2p	*SRS2*	3′ to 5′	DNA repair/recombination?
Dna2p	*DNA2*	3′ to 5′	DNA replication
Upf1p	*UPF1*	5′ to 3′	mRNA decay and translation
scHel1	*HEL1*	5′ to 3′	Unknown

[a] The RecBCD helicase has a strong preference for unwinding blunt-ended duplex DNA substrates and does not appear to exhibit a polarity as defined for the other helicases.

for most of the proteins have not been defined in detail. In addition to the helicases listed here, other genes encoding helicases will certainly be uncovered in both of these organisms. The reader is referred to a recent review for additional details on the cellular roles of the DNA helicases (11).

The diversity of DNA helicases found in *E. coli* and *S. cerevisiae* likely reflects the fact that each enzyme plays a specific role or roles in the cell, and in most cases one helicase cannot substitute for another. A comparison of the known enzymes from these model organisms reveals both similarities and differences. For instance, the proteins involved in the nucleotide excision repair pathway in *E. coli* include helicase II while the same pathway in yeast and other eukaryotes uses two DNA helicases, Ssl2p and Rad3p. Yeast also possess helicases required to meet the needs of eukaryotic DNA metabolism such as telomere maintenance and mitochondrial DNA recombination. Further comparison reveals an intriguing absence of yeast counterparts for several important *E. coli* DNA helicases. These include the helicase involved in DNA mismatch repair that corresponds to helicase II, a DNA helicase at the replication fork with a function similar to DnaB, and a DNA helicase or helicases with well defined roles in recombination such as RecBCD in *E. coli*. The discovery and characterization of yeast helicases that participate in these important biochemical pathways is an exciting avenue for further investigation that will certainly shed light on eukaryotic DNA metabolism.

The detailed mechanism utilized by a helicase to disrupt the hydrogen bonds between base pairs of a duplex substrate is not known for any enzyme. It is clear that helicase catalysed unwinding requires energy, and all helicases isolated to date catalyse the hydrolysis of nucleoside 5′ triphosphates (NTPs), usually ATP, to produce nucleoside 5′ diphosphate and Pi. This activity of the enzyme is often used to monitor purification. The energy provided by NTP hydrolysis is used by the enzyme to fuel the unwinding reaction. In general, the NTPase reaction is either dependent on or greatly stimulated by the presence of a polynucleotide, usually ssDNA or RNA.

The mechanism by which NTP hydrolysis is coupled to unwinding is also not known. This has been a subject of much interest and intense research effort in the last few years and readers are referred to an excellent review for details (12). We briefly summarize a current view of the reaction here. Most helicases appear to act catalytically as either dimers or hexamers, although many of these enzymes are isolated as monomers in solution. While this statement appears to be true for enzymes that have been well studied, it is not possible to generalize this view to all helicases. It is possible that helicases that function as monomers will be found. There are several clear examples of enzymes that form dimers or hexamers either in solution or when bound on DNA (reviewed in ref. 12). The oligomeric form of the enzyme provides for multiple DNA and NTP binding sites since each protomer contains a binding site for each ligand. Thus, a multimeric helicase need not dissociate from the DNA substrate during the course of an unwinding reaction. One model for DNA unwinding envisions a dimer 'rolling' along the DNA substrate with a specific directionality to achieve hydrogen bond disruption (13). It is easy to imagine one protomer binding to ssDNA that has just been unwound while the second protomer binds duplex DNA just ahead of the ssDNA:duplex DNA junction. As duplex DNA is rendered single-stranded the protomer bound to ssDNA is released and binds to the duplex to catalyse another 'step' in the unwinding reaction. Many details consistent with this mechanism have been confirmed for the *E. coli* Rep helicase (12). ATP binding and hydrolysis serve to modulate the affinity of each protomer for a specific form of nucleic acid, either ssDNA or duplex DNA. Thus, the ATPase reaction provides the energy required to move through the duplex DNA substrate. Whether other helicases will conform to this model remains to be determined and it should be noted that alternative models for unwinding DNA have been proposed (14, 15).

This chapter presents an overview of the biochemical reactions catalysed by DNA helicases and the assays used to measure them, including DNA-dependent ATPase assays and DNA helicase assays. The second portion of the chapter details some of the methods used in the identification and purification of DNA helicases with emphasis on the model eukaryote, *S. cerevisiae*.

2. Measuring the unwinding of duplex nucleic acids *in vitro*

2.1 Assays used to measure DNA helicase activity

The biochemical assays used to measure helicase activity *in vitro* are based on the ability of these enzymes to disrupt the hydrogen bonds between two complementary DNA strands. Typically, a relatively short [^{32}P]DNA strand is annealed to a much longer DNA strand, often a circular ssDNA isolated from phage M13 or some other suitable phage, to produce a partial duplex helicase substrate (*Figure 1*). After incubation with a DNA helicase, in the presence of appropriate cofactors, the products of the reaction are resolved on a native polyacrylamide gel to distinguish ssDNA that has been unwound from the partial duplex DNA substrate. The unwound or displaced [^{32}P]DNA fragment migrates faster on the polyacrylamide gel than the [^{32}P]DNA substrate allowing a product of the reaction to be directly observed.

Use of a circular partial duplex helicase substrate molecule provides several advantages for the assay of a DNA helicase. First, the substrate molecule is larger than the unwound DNA fragment making it easy to resolve product from substrate on a polyacrylamide gel. Secondly, a variety of different substrates can be constructed differing only in the position on M13 to which the oligonucleotide is annealed and the length of the duplex region. Thirdly, since most helicases prefer to bind ssDNA, there is a large target for protein binding and

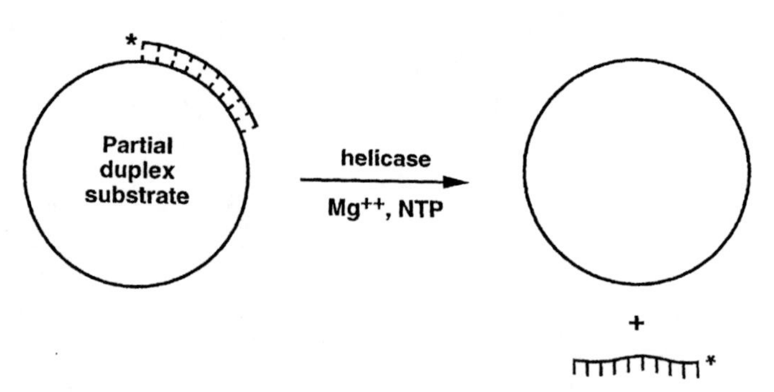

Figure 1. *In vitro* DNA helicase assay using partial duplex DNA substrates.

subsequent initiation of the unwinding reaction. Finally, most helicases unwind DNA with a specific polarity, either 5′-3′ or 3′-5′, defined with respect to the strand on which the protein is bound. A helicase of either polarity can unwind this substrate. This is particularly important when the polarity of a helicase is not known. A disadvantage of this assay is the fact that the assay only measures complete unwinding events, partially unwound substrate is not detected.

The source and length of the 'oligonucleotide' to be annealed to M13 ssDNA must be carefully considered. It is preferable to use an oligonucleotide of sufficient length to insure that a bona fide unwinding reaction is measured. If the oligonucleotide is too short, displacement is possible in the absence of helicase catalysed unwinding. In fact, substrates constructed using short oligonucleotides tend to be unstable under reaction conditions and when stored in low ionic strength buffers. Substrates with a duplex region of 50 base pairs (bp) or more are preferred. In specialized instances substrates with duplex regions of 20 bp have been used. In this case, care must be taken to insure the stability of the substrate during storage. Buffers containing 100 mM NaCl are recommended to prevent unwanted denaturation of the DNA. In addition, control reactions must be performed to demonstrate the dependence of the unwinding event on concomitant NTP hydrolysis.

In some cases it is necessary to modify the partial duplex substrate such that the fragment to be displaced contains a 'tail' (on either the 5′ end or the 3′ end) that is not annealed to the M13 ssDNA (*Figure 2*). A number of enzymes have been shown to require a 'tailed substrate' for *in vitro* measurement of helicase activity. For example, this type of substrate was used to detect the unwinding reaction catalysed by the phage T7 gene 4 protein (22). A partial duplex substrate with this property is prepared by insuring that the appropriate end of the oligonucleotide to be annealed to M13 ssDNA contains a region of DNA (usually 10–20 nucleotides in length) that is not complementary to M13 ssDNA.

Two different sources for the oligonucleotide to be annealed to M13 DNA have been used with equivalent results. An oligonucleotide can be

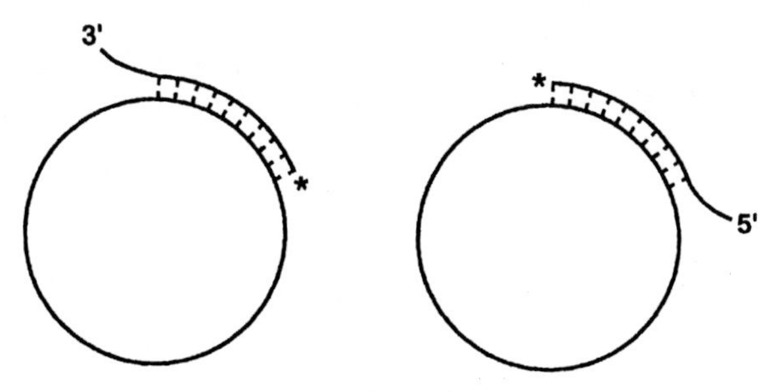

Figure 2. Partial duplex DNA substrates with ssDNA 'tails'.

synthesized that is complementary to a specific region on M13 DNA. Alternatively, a specific M13 replicative form I (RFI) DNA restriction fragment can be isolated and the strand complementary to the M13 ssDNA can be annealed to form the duplex region. *Protocol 1* is a method for the separation and isolation of M13 DNA restriction fragments suitable for substrate construction. A third source is the product of an asymmetric PCR reaction performed using linear M13 RFI DNA (23). The latter two approaches have the advantage of allowing the construction of much longer duplex regions than are possible using a synthetic oligonucleotide.

Protocol 1. Isolation of M13 RFI DNA restriction fragments

Equipment and reagents

- Horizontal gel electrophoresis apparatus
- Vertical gel electrophoresis apparatus
- Power supply
- Dialysis tubing (3500 M_r cut-off)
- Microcentrifuge (Eppendorf)
- UV transilluminator or hand-held UV light
- 1 × TBE buffer: 89 mM Tris, 89 mM borate, 1 mM EDTA pH 8.3
- *Hae*III restriction enzyme (New England Biolabs)
- TE buffer: 10 mM Tris–HCl pH 8, 1 mM EDTA
- STE buffer: 100 mM NaCl, 10 mM Tris–HCl pH 8, 1 mM EDTA
- Ethidium bromide (EtBr)
- Phenol (equilibrated with STE)
- Chloroform

Method

1. Isolate M13 RFI DNA using a standard protocol (24).
2. Digest 50 μg RFI DNA to completion with *Hae*III (or some other appropriate restriction enzyme).
3. Resolve the products of the restriction digest on a preparative 6% or 8% native polyacrylamide gel.[a]
4. Stain the gel with EtBr to visualize the DNA fragments.
5. Cut out the desired DNA fragment(s) under UV illumination.
6. Electroelute the DNA from the polyacrylamide gel slice by placing the gel slice in a dialysis bag with 0.4 ml 0.5 × TBE. Place the dialysis bag in a horizontal slab gel apparatus containing 0.5 × TBE with the gel slice perpendicular to the current flow. Electrophorese at 100 V until the DNA leaves the gel slice as observed under UV illumination. Reverse the current for 2 min to free DNA from the sides of the dialysis bag.
7. Recover the DNA-containing solution from the dialysis bag and pellet debris by centrifugation at 12 000 *g* for 5 min. Discard the pellet.
8. Extract with an equal volume of a 1:1 phenol:$CHCl_3$ mixture. Centrifuge to separate the phases and remove the upper (aqueous) phase.
9. Add 0.1 vol. 5 M NaCl and 2 vol. ethanol. Chill on ice for 30 min or at –20 °C until required.

10. Centrifuge at 4°C in a microcentrifuge for 30 min to pellet the DNA.
11. Wash the pellet with 70% (v/v) ethanol chilled to –20°C, vortex lightly, and centrifuge for 10 min at 4°C.
12. Dry the pellet and resuspend the precipitated DNA in an appropriate volume of TE buffer.
13. Quantify the isolated DNA by EtBr staining (25), or by running a known volume of the isolated DNA fragment on a native polyacrylamide gel with a set of DNA standards of known concentration.

[a]Use an 8% gel to isolate DNA fragments of 500 bp or less. Use a 6% gel to isolate DNA fragments of more than 500 bp.

Radioactive labelling of the oligonucleotide can be accomplished by extending the 3′ OH using the Klenow fragment of DNA polymerase I and an appropriate [α-^{32}P]dNTP (27) or by 5′ end-labelling using [γ-^{32}P]ATP and polynucleotide kinase following standard protocols (26). When restriction fragments are used as the source of the oligonucleotide it is often more convenient to label the 3′ OH by extension with a DNA polymerase in the presence of a [α-^{32}P]dNTP. This, in fact, is one reason that *Hae*III was chosen as the enzyme for preparation of restriction fragments (27). The first two residues incorporated at the 3′ OH of a *Hae*III restriction fragment are both dCMP. Thus, extension of the 3′ end using [α-^{32}]dCTP in the absence of the other three dNTPs insures relatively high specific activity labelling of the DNA substrate. Since *Hae*III produces a blunt-ended DNA fragment, the labelling reaction is performed after the DNA has been annealed on M13 ssDNA. The 5′ end-labelling method is often easier if a synthetic ssDNA oligonucleotide is being used to construct the helicase substrate. Oligonucleotides are supplied with a 5′ OH and therefore the 5′ PO_4 does not have to be removed prior to 5′ end-labelling. However, this substrate is subject to the action of 5′ phosphatases which remove the 5′ end-label.

Protocol 2a describes substrate construction using a *Hae*III cut restriction fragment annealed to M13 ssDNA. The fragment is labelled by extension with DNA polymerase I (Klenow fragment) after the annealing reaction. *Protocol 2b* describes the production of a helicase substrate using a synthetic oligonucleotide complementary to a region of M13 ssDNA. In this case the oligonucleotide is labelled, using polynucleotide kinase, prior to the annealing reaction.

Protocol 3 outlines a DNA helicase assay, using substrates constructed in either *Protocols 2a* or *2b*, and methods used for quantification of the reaction products. Controls are essential in the interpretation of the assay, including a control for non-enzymatic DNA unwinding as well as NTP-independent DNA unwinding. If substrates are used that have short duplex regions these controls are particularly important for proper interpretation of the results.

Another consideration is the selection of the NTP used in the reaction. Some helicases use NTPs other than ATP, for instance phage T7 gene 4 protein uses dTTP in the DNA unwinding reaction (22).

Protocol 2a. Construction of a partial duplex helicase substrate using a DNA restriction fragment

Equipment and reagents

- Chromatography column
- Water-bath at 65°C
- M13 ssDNA (New England Biolabs)
- Purified restriction fragment (*Protocol 1*)
- Annealing buffer: 7 mM Tris–HCl pH 8, 50 mM NaCl, 7 mM $MgCl_2$
- STE buffer
- DNA polymerase I (Klenow fragment) (New England Biolabs)
- Bio-Gel A5M or Sephacryl S-200 or Sephadex G100 (Bio-Rad)
- dCTP (US Biochemicals Inc.)
- [α-^{32}]dCTP (Amersham Corp.)

Method

1. Mix M13 ssDNA with an appropriate amount of the purified restriction fragment in annealing buffer and denature the DNA at 95°C for 3–5 min.
2. Incubate at 65°C for 15–30 min to allow the DNA to anneal.
3. Slowly cool to room temperature.
4. Add 20–40 μCi [α-^{32}P]dCTP and DNA polymerase I (Klenow fragment). Incubate at room temperature for 15–30 min.[a]
5. Add 50 μM unlabelled dCTP and incubate for 10 min at room temperature to complete extension on all 3′ ends.
6. Increase the volume of the reaction mixture to 100 μl by adding STE buffer and extract with a 1:1 mixture of STE saturated phenol and $CHCl_3$ (*Protocol 1*).[b]
7. Load the aqueous phase on an appropriate 1–2 ml size exclusion column (e.g. Bio-Gel A5M or Sephacryl S-200 or Sephadex G100) and collect the void volume which contains the helicase substrate.[c]
8. Store the helicase substrate at −20°C.
9. Estimate the DNA concentration based on the known concentration of DNA in the original annealing reaction and assuming about 80% recovery of DNA.

[a] The buffer used in the annealing step is suitable for the DNA polymerase reaction.
[b] This step is important for inactivation and removal of the DNA polymerase.
[c] The column is used to remove both the unincorporated label and the complementary strand of the restriction fragment that does not anneal to M13 ssDNA. The column should be equilibrated and developed with STE. A column at least 15 × 0.5 cm is necessary if a 5′ end-labelled oligonucleotide is used to construct the substrate since it is critical that all [^{32}P]oligonucleotide be removed from the substrate.

Protocol 2b. Construction of a helicase substrate using a synthetic oligonucleotide

Reagents

- Synthetic oligonucleotide
- Polynucleotide kinase (New England Biolabs)
- [γ-^{32}P]ATP (Amersham)

Method

1. Label the 5′ OH on an oligonucleotide using [γ-^{32}P]ATP and polynucleotide kinase following an established protocol (26).
2. Anneal the [^{32}P]DNA to M13 ssDNA as described in *Protocol 2a*, steps 1–3.
3. Purify the helicase substrate as described in *Protocol 2a*, steps 6–9.

Protocol 3. Protocol for a helicase reaction

Equipment and reagents

- Vertical slab gel apparatus
- Power supply
- X-ray film, scintillation counter, or PhosphorImager
- EDTA pH 8
- Native polyacrylamide gel, 0.5 × TBE (see *Protocol 1*), 25% glycerol, 6% or 8% acrylamide

Method

1. Incubate the helicase substrate with the helicase under appropriate reaction conditions, i.e. in the presence of a suitable buffer, Mg^{2+}, a reducing agent such as 2-mercaptoethanol, salts if necessary, and ATP. Several controls are important. First, a no enzyme control is essential to evaluate the extent of unwinding. Secondly, a heat denatured control provides a background value for radioactivity that cannot be displaced from the substrate. This also serves to evaluate the integrity of the [^{32}P]DNA strand. Finally, reactions lacking a NTP cofactor, or reactions containing a non-hydrolysable ATP analogue, are essential to insure the reaction depends on NTP hydrolysis.
2. Stop the reaction with sufficient EDTA to chelate Mg^{2+}. Include EDTA in a stop mixture that also contains 10–15% glycerol for gel loading purposes, 0.5 × gel running buffer, and 0.1% gel tracking dyes (bromphenol blue and xylene cyanol). It is also useful to add SDS to 0.1% to help remove bound proteins from the unwound DNA fragment. While not essential for many helicases, this can dramatically improve results with enzymes like DNA helicase II from *E. coli* which tend to bind to ssDNA.

Protocol 3. *Continued*

3. Load the samples on a *native* polyacrylamide gel (5–8% polyacrylamide, 0.5 × TBE, 25% glycerol). Use a low percentage gel (5–6%) for DNA fragments > 150 nucleotides (nt) in length and high percentage gels (8%) for DNA fragments < 100 nt in length. These gel percentages can be modified to suit particular conditions. However, a native gel must be used to accurately measure unwinding due to helicase activity. If protein binding to the displaced DNA fragment is a problem include 0.1% SDS in the gel and running buffer.
4. Electrophorese to resolve product from substrate. The gel running time and running conditions vary with the length of the oligonucleotide. Short products can be resolved from the substrate very quickly and run times of 1–2 h at 100 V are sufficient. Long products require correspondingly longer run times and higher voltage. For example, an 851 nt fragment requires an overnight run at 200 V for good separation of the product and the substrate.
5. Visualize the gel using autoradiography or by PhosphorImage analysis.
6. Quantify the unwinding reaction. There are several suitable methods available. We will mention three we have used successfully.
   (a) After autoradiography the X-ray film can be scanned using a densitometer. For accurate analysis it is important that the signal on the film be within the linear range of the film.
   (b) After autoradiography cut the displaced and substrate bands from the gel and count in a scintillation counter. This is highly reproducible and is not subject to problems with linearity of film response.
   (c) PhosphorImage and quantify the image using software provided by the manufacturer. We use this method almost exclusively now due to the ease of obtaining the image and the quantified data, and the large linear range available.
7. Calculate the fraction of the substrate unwound. This is accomplished using a simple equation that takes into account the background in both the no enzyme control and the heat denatured control. The fraction of the DNA fragment unwound is calculated as follows:

$$\%\ \text{unwound} = \frac{P - B_p}{(S - B_s) + (P - B_p)} \times 100$$

where P = radioactivity in the product, B_p = background radioactivity at the position of the product determined from the no enzyme control reaction, S = radioactivity remaining in the substrate, and B_s = background radioactivity at the position of the substrate determined from the heat denatured control reaction.

2.2 Fluorescence-based unwinding assays

There are two fluorescence-based assays available for detection of helicase activity (28, 29). The first of these assays (28) utilizes two complementary oligonucleotides, but of differing length. A fluorescent probe is placed on the 5′ end of one oligonucleotide and on the 3′ end of the other oligonucleotide. Thus, the two fluorescent probes are in close proximity and there is energy transfer from one to the other when excited at the appropriate wavelength. When the duplex region is unwound by a helicase there is a large change in the fluorescence signal due to the loss of energy transfer between the two fluorescent probes. This assay has the advantage of measuring unwinding in real time but requires that the polarity of the helicase be known so that the appropriate substrate (i.e. with a 5′ ssDNA tail or a 3′ ssDNA tail) can be constructed. This assay is particularly useful for advanced kinetic measurements of properties of the helicase. This assay has not been applied to purification of a helicase.

The second assay utilizes fluorescent molecules that bind DNA (e.g. EtBr, acridine orange) to monitor the unwinding event (29). In this instance, the fluorescent probe is bound to the duplex DNA substrate and unwinding is measured as a loss in the fluorescence signal due to unwinding of the substrate taking advantage of the low affinity of these molecules for ssDNA. This assay can also be used to measure unwinding in real time and has the advantage of being able to measure partial unwinding events. The utility of this assay for purification has not been evaluated. The assay has been used to characterize the unwinding reaction catalysed by the RecBCD enzyme from *E. coli* (29).

2.3 Assays used to measure DNA:RNA unwinding *in vitro*

The assay described in Section 2.1 can also be used, with modification, to measure the unwinding of DNA:RNA hybrid substrates. For this application a specific RNA is hybridized to the DNA and unwinding is measured as described (*Protocol 3*). This substrate will, necessarily, have a 5′ end on the RNA strand this is not annealed to the M13 ssDNA unless:

(a) Efforts are made to remove this RNA specifically.

(b) The promoter sequence is included in the M13 ssDNA.

T7 RNA polymerase has been used to produce specific transcripts with good success. These transcripts are purified and annealed to M13 ssDNA. In principal, DNA:RNA hybrids of any length or sequence can be made provided the appropriate M13 ssDNA, containing a region of DNA to which the RNA will anneal, is available. We typically denature the substrate to be used as a control with 85% formamide and heating to remove secondary structure. If necessary, the gel loading buffer can be modified to contain formamide (40%) to reduce the formation of RNA secondary structure. In the absence of heating, this formamide concentration is not sufficient to denature the substrate.

Protocol 4 describes the construction of a DNA:RNA partial duplex substrate using a radiolabelled RNA transcript annealed to M13 ssDNA. The length of the transcript can be varied and, if desired, the transcript can be designed with non-complementary tails on the 5′, 3′, or both ends.

Protocol 4. Preparation of a DNA:RNA partial duplex substrate

Equipment and reagents

- Power supply
- Vertical slab gel apparatus
- Denaturing urea/polyacrylamide gel: 1 × TBE, 8 M urea, 6–8% polyacrylamide
- Native polyacrylamide gel (see *Protocol 3*)
- T7 RNA polymerase (US Biochemicals)
- M13 ssDNA
- [α-^{32}P]UTP (Amersham)
- RNase-free DNase (New England Biolabs)
- TBE buffer (see *Protocol 1*)

Method

1. Prepare an appropriate RNA transcript using T7 RNA polymerase, *E. coli* RNA polymerase, or some other appropriate RNA polymerase in a run-off transcription reaction in the presence of [α-^{32}P]UTP using reaction conditions recommended by the supplier.[a]
2. Remove the template DNA by digestion with a RNase-free DNase.
3. Extract the [^{32}P]RNA with phenol:$CHCl_3$, precipitate with ethanol, and suspend in an appropriate volume of TE (see *Protocol 1*). Store at −80 °C to preserve the integrity of the RNA.[b]
4. Evaluate the integrity and length of the RNA on a denaturing urea gel.
5. Anneal the [^{32}P]RNA to ssDNA by following *Protocol 2a*, steps 1–3.[c]
6. Prepare the DNA:RNA hybrid on a preparative scale using the information gained in step 5 and as directed in *Protocol 2a*, steps 1–3.
7. Purify the substrate on a size exclusion column as described in *Protocol 2a* to remove any [^{32}P]RNA that failed to anneal to the DNA. Store the prepared substrate at −80 °C. Use directly in helicase reactions.

[a] It is essential that the template used for this reaction direct the synthesis of a RNA transcript that will anneal to an available ssDNA molecule. For example, one of the *Hae*III restriction fragments from M13 can be cloned in the multiple cloning site in a vector such as pBluescript which contains an RNA polymerase promoter. The plasmid DNA is then restricted with an appropriate restriction enzyme to provide the linear DNA template used in the run-off transcription reaction.

[b] The [^{32}P]RNA can be gel purified using a urea denaturing gel at this step if desired.

[c] To empirically determine the amount of [^{32}P]RNA to use, titrate known concentrations of M13 ssDNA with a fixed volume of [^{32}P]RNA in small volume (10 μl) annealing reactions. The products of the annealing reaction are then resolved on a native polyacrylamide gel and a concentration of [^{32}P]RNA is chosen that results in complete annealing of the [^{32}P]RNA to the ssDNA as evidenced by migration on the gel as a DNA:RNA hybrid.

2.4 Inhibiting nuclease activity in partially purified helicase preparations

2.4.1 The use of poly dI:dC

Nuclease activity in partially purified protein preparations can mask a helicase catalysed unwinding reaction by degrading the unwound [^{32}P]DNA product. The addition of poly dI:dC to helicase reactions quenches the activity of some nucleases that degrade the unwound substrate by providing an alternative substrate for the contaminating nuclease. For example, standard helicase reactions (*Protocol 3*) using partially purified helicases from yeast yield only degraded partial duplex DNA substrate. The addition of poly dI:dC at a concentration of 50 ng/reaction decreased the amount of nucleolytic degradation of the unwound product and revealed previously undetected helicase activity. The addition of 500 ng of poly dI:dC per reaction further quenched nuclease activity without significant inhibition of the helicase activity.

2.4.2 Use of sulfur-substituted oligonucleotides

A second alternative for dealing with contaminating nuclease activity is the use of sulfur (S)-substituted oligonucleotides in the preparation of partial duplex helicase substrates. S-substituted oligonucleotides are commercially available and the S substitution in the phosphodiester backbone of the oligonucleotide inhibits the nuclease activity of several common nucleases (30). We have used oligonucleotides that contain S substitutions throughout the oligonucleotide. The preparation of a partial duplex substrate containing a S-substituted oligonucleotide is as described in *Protocol 2a* and this substrate has worked well in partially purified preparations of helicases from both yeast and *E. coli*. It is important to note that use of this substrate does not insure inhibition of all nucleases.

2.5 Determining the polarity of an unwinding reaction

Most helicases unwind duplex DNA with a specific polarity that is defined with respect to the strand of DNA on which the enzyme is presumed to be bound. Linear partial duplex substrates with a central region of ssDNA, flanked by duplex regions on both ends of the substrate, are used to determine the polarity of the unwinding reaction (*Figure 3*). The product of the unwinding reaction is one of the two annealed fragments plus the remaining partial duplex molecule. These products can be resolved, based upon size, with the result indicating direction of unwinding. The critical features of a direction-determining substrate are the central loading region consisting of a long interior region of ssDNA (> 100 bases), double-stranded regions on the termini of the substrate, and a significant size difference between the two end fragments such that they can be easily resolved on a polyacrylamide gel. Numerous substrates have been used to detect the direction of unwinding

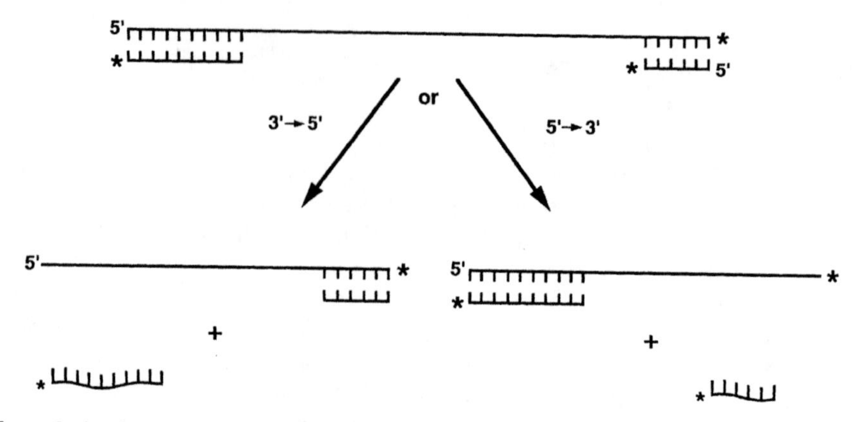

Figure 3. *In vitro* assay used to determine the polarity of an unwinding reaction.

with the major variable being the lengths of the duplex regions. We recommend that the duplex regions be as long as possible. Increased length increases the stability of the substrate and provides convincing evidence of DNA unwinding. Although short oligonucleotides (< 20 bp) have been used to construct direction-determining substrates, *Protocol 5* describes construction of a direction substrate that yields a 143 nt fragment if the helicase unwinds in a 3′-5′ direction, and a 202 nt fragment if the enzyme unwinds 5′-3′. This substrate is very stable and yields excellent results that are easy to interpret. However, some helicases are incapable of unwinding duplexes of these lengths and in those cases satisfactory direction substrates can be designed with shorter duplex termini.

Protocol 5. Preparation of a substrate to determine unwinding polarity

Equipment and reagents

- Chromatography column
- Chromatography resin (Bio-Gel A5M)
- *Cla*I restriction endonuclease
- Purified 341 bp *Hae*III restriction fragment from M13mp7 (see *Protocol 1*)

Method

1. Prepare the 341 bp *Hae*III RFI DNA restriction fragment from M13mp7 (*Protocol 1*).
2. Anneal the 341 bp *Hae*III restriction fragment from M13mp7 RFI DNA to M13mp7 ssDNA as described in *Protocol 2a*, steps 1–3.
3. Digest the partial duplex substrate with *Cla*I restriction endonuclease.[a]

4. Label all available 3′ OH termini using [α-^{32}P]dCTP and unlabelled dGTP as described (*Protocol 2a*).[b]
5. Phenol:chloroform extract and remove unincorporated nucleotide from the reaction as described in *Protocol 2a*, steps 6 and 7.

[a] This enzyme cuts once within the duplex region, producing a linear partial duplex molecule with 5′ GC overhangs at each end.
[b] Since the 5′ overhang is GC, dGTP must be included in the labelling reaction. This will result in 3′ end-labelling of all available 3′ OH ends in the partial duplex molecule. The substrate can then be used as a standard partial duplex substrate.

2.6 Stimulation of DNA unwinding by single-stranded DNA binding proteins (SSBs)

DNA helicases may be stimulated by the presence of SSBs in the unwinding reaction. It is presumed that SSBs bind to the single-stranded regions of the partial duplex substrate in addition to the newly formed single-stranded regions as the substrate is unwound. Some DNA helicases have been shown to be stimulated by the addition of SSBs while others appear to be inhibited. Striking results have been obtained with scHelI which requires SSBs in order to unwind long partial duplex substrates (31). Thus, one characteristic of DNA helicases is their catalytic capacity in the presence and absence of SSBs. An additional reason for the inclusion of SSBs, especially in partially purified preparations, is that SSB bound ssDNA is resistant to some forms of nuclease degradation, allowing the detection of unwinding where the substrate may otherwise be degraded.

Protocol 6. Unwinding reactions using SSBs

Reagents

- *E. coli* SSB or yeast RP-A protein
- Proteinase K (Boehringer Mannheim)

Method

1. Prepare an appropriate helicase substrate as described in *Protocol 2a*.
2. Assemble the components for a helicase reaction (*Protocol 3*) without the helicase and add SSB.[a]
3. Incubate on ice for 10 min prior to the addition of the helicase.
4. Add helicase and incubate at 30 °C for 10–60 min.
5. Stop the reaction as described in *Protocol 3* using a solution that contains proteinase K (100 μg/ml).[b]
6. Incubate at 37 °C for 20 min.

Protocol 6. *Continued*

7. Electrophorese and analyse as described in *Protocol 3*.[c]

[a] Helicase reactions are usually very sensitive to the concentration of SSB so it is important to calculate the amount of SSB required to bind all available ssDNA in the reaction and titrate SSB into the reaction at higher and lower concentrations. The binding site size of the *E. coli* SSB tetramer varies with reaction conditions (32) and must be taken into consideration when calculating the amount of SSB required to coat ssDNA. The eukaryotic SSB, RP-A, has a site size of 95 nucleotides (33).
[b] The SSB must be digested prior to resolution of the displaced strand on a polyacrylamide gel. Proteinase K has no effect on DNA unwinding but removes bound protein from the ssDNA, allowing resolution of unwound product on non-denaturing polyacrylamide gels.
[c] It is often useful to include 0.1% SDS in the native polyacrylamide gel and in the gel running buffer to insure that no protein is bound to the unwound DNA.

3. Measuring DNA-dependent ATPase activity *in vitro*

As noted above, all helicases isolated to date also catalyse a DNA stimulated NTP hydrolysis reaction. NTPase (ATPase) activity has been widely used to measure kinetic parameters of the reaction catalysed by a helicase and in purification of these proteins. The utility of the assay in the latter application is due to the ease with which it can be applied to column fractions, the quantitative nature of the assay, and the fact that contaminating nucleases do not destroy the substrate. This is especially important during the early stages of purification when nuclease contamination is a serious problem. It should be noted that contaminating nucleases can destroy the DNA cofactor added to DNA stimulated ATPase reactions. To circumvent this problem, very high concentrations of DNA can be used when column fractions are evaluated. A disadvantage of the ATPase assay is its lack of specificity. All helicases, and many other proteins in the cell, hydrolyse ATP. Thus, the presence of an active ATPase does not insure that a helicase, or the desired helicase, is being monitored. This issue is less problematic if the chromatographic properties of the protein are known. Alternatively, the use of Western blots with antibodies directed against the protein in combination with ATPase activity assays can be helpful. Under circumstances where little or nothing is known about the protein, we have successfully purified helicases by using helicase activity assays as soon as possible in the purification procedure to insure that the ATPase being monitored is, in fact, a helicase.

There are several dependable ATP hydrolysis assays available (34–37). Many of these can also be applied to other NTPs. We routinely use the assay described in *Protocol 7* which measures the conversion of [^{3}H]ATP to [^{3}H]ADP and Pi. The reactants and products are separated by ascending thin-layer chromatography. There are several advantages to this assay including its ease, the small amount of protein required, its quantitative nature, and the long half-life of [^{3}H]. However, this assay is not superior to the many other assays available. This assay will be described since it is the assay with which we have had most experience.

Protocol 7. DNA stimulated ATPase assay

Equipment and reagents

- Polyethyleneimine thin-layer chromatography plates (Brinkman)
- Thin-layer chromatography tank
- Hand-held UV (254 nm) light
- Scintillation counter
- [^{3}H]ATP or [^{32}P]ATP
- 20 mM ATP
- 20 mM ADP
- Chromatography buffer: 1 M formic acid, 0.8 M LiCl
- Scintillation fluid

Method

1. Assemble the reaction mixture containing appropriate salts, buffers, Mg^{2+}, DNA, and [^{3}H]ATP at an appropriate concentration.[a]
2. Add DNA to the reaction mixture to a final concentration of 30 μM (DNA phosphate).[b]
3. Add enzyme and incubate at 30°C for yeast enzymes or 37°C for *E. coli* enzymes for 10–60 min.[c]
4. Withdraw a sample (5 μl) of the reaction mixture and quench the reaction with an equal volume of 33 mM EDTA, 6 mM ADP, 6 mM ATP.[d]
5. Apply the entire 10 μl mixture (or some portion) to a polyethyleneimine plate at a pre-defined origin (usually 0.5 cm from the bottom of the plate) using a microcapillary or Hamilton syringe.[e]
6. Place the bottom edge of the thin-layer plate in water and allow the water to ascend on the thin-layer plate to a point just past the origin to remove dust and other contaminants.
7. Transfer the plate to chromatography buffer and develop (less than 1 h at 25°C).
8. Dry the plate on the bench or under a heat lamp.[f]
9. Identify the ADP and ATP markers in each lane by indirect UV illumination.[g]
10. Circle the ADP and ATP spots on the plate using a dull pencil.
11. Scrape the polyethyleneimine spots containing ATP and ADP off the plastic backing using a spatula.[h]

Protocol 7. *Continued*

12. Transfer the scrapings to a scintillation vial, add scintillation fluid, and determine the amount of radioactivity in each sample.[i]

[a] It is important that the ATP concentration be sufficiently high to support good ATPase activity for the enzyme being tested. This depends on the K_m for ATP which is not known initially. Use 0.5–1 mM [^{3}H]ATP when assaying column fractions since this is greater than the K_m for ATP of most helicases. Lower concentrations can be used, where appropriate, once a K_m for ATP has been determined.
[b] An appropriate DNA cofactor must be used in this reaction since most helicases are strongly stimulated by the presence of DNA. M13 ssDNA or poly(dT) are good sources of ssDNA. DNA is included at a concentration of 30 μM (nucleotide phosphate) to minimize the effects of contaminating nucleases. A lower concentration of DNA can be used if nucleases are not a problem.
[c] The appropriate time of incubation must be determined empirically.
[d] The EDTA serves to chelate Mg^{2+} ions and stop the reaction. The ATP and ADP serve as carriers of the reactants and products during chromatography and as convenient markers on the developed chromatogram.
[e] If the entire mixture is applied, two to three applications are necessary to spot the entire volume on the plate. Dry the spot after each application of quenched reaction mixture. We routinely use plates that are 10 cm in height and of variable width (depending on the number of samples to be analysed).
[f] It is important that the plate not be too dry. The desired time for drying must be determined empirically and will depend on the method of drying.
[g] The ADP migrates faster on the plate in this solvent system than does the ATP. We have used [^{32}P]ATP in this assay which enables the use of a PhosphorImager to quantify the reaction products, circumventing the need to scrape and count the spots in a scintillation counter.
[h] Do not contaminate ATP with ADP and vice versa.
[i] A no enzyme control is used to determine how much [^{3}H]ADP exists in the [^{3}H]ATP stock and to determine the specific activity of the [^{3}H]ATP used in the experiment. From this information it is possible to determine the specific activity (pmols of ATP hydrolysed per mg of protein) for the sample(s) assayed.

4. Purification of yeast helicases

Experimental determination of the biochemical properties of DNA helicases requires purified, enzymatically active preparations of these proteins. Purification is a difficult first step in the *in vitro* characterization of a helicase, but a number of strategies have proven successful, offering an array of options for purification of novel proteins. Most helicases are now purified from cell extracts that have been enriched for the target protein by the use of expression plasmids carrying the gene of interest. Expression plasmids usually provide a high concentration of the helicase in the initial extract and the use of affinity tags offers a convenient and specific purification step. Purified helicases have also been obtained from cell and tissue extracts without the benefit of expression plasmids. While this strategy has proven successful, a large amount of starting extract is required and yields of enzyme are low. Both of these approaches will be described below.

Often the sequence of a gene is known before biochemical function is determined, especially with the generation of vast amounts of sequence data by the genome sequencing projects. Because of this it is useful to consider what is known concerning sequence conservation among helicases prior to initiating the characterization of a putative helicase. The amino acid sequences of a large number of biochemically defined and putative helicases have been compared in computer-assisted homology searches (16–18). This analysis indicates the existence of several helicase superfamilies. Superfamilies I and II both demonstrate the presence of seven conserved amino acid regions (referred to as motifs I, Ia, II–VI), although not all seven motifs are identical between the two superfamilies. Two additional superfamilies, III and IV, have been identified (18). These proteins exhibit conserved amino acid sequences as well, but the regions of conservation are quite different than those found in superfamilies I and II, and their organization within the protein is also significantly different. The existence of these motifs in a protein sequence is not sufficient to allow one to conclude that the encoded protein is a helicase. There are several examples of proteins that share the seven motifs of superfamilies I and II and yet have no demonstrable helicase activity (19–21). Thus, caution is required in arriving at the conclusion that a particular protein catalyses the unwinding of duplex nucleic acids. Expression, purification, and biochemical characterization are required to definitively state that a protein has helicase activity.

The purification protocols discussed below are for proteins found in the budding yeast *S. cerevisiae* because we have had experience isolating enzymes from this organism and because the facile genetics and completed genome sequence make yeast a paradigm for the investigation of nucleic acid metabolism. Similar strategies may also be used for the isolation of enzymes from other organisms.

Using a combination of genetics and biochemistry seven DNA helicases have been characterized from yeast (see *Table 1*). The most complete characterizations have been of Rad3p and Ssl2p where DNA helicase activity has been demonstrated *in vitro* using purified proteins and both proteins have been shown to be necessary for reconstitution of the biochemical pathways of nucleotide excision repair and transcriptional activation. Genetic characterization preceded biochemical characterization and has been consistent with the *in vitro* biochemistry of these proteins. The role of DNA helicases in other metabolic pathways is much less well understood. For instance, as yet no helicase has been found to participate in DNA mismatch repair in a manner analogous to helicase II in *E. coli*. Perhaps more surprising, the helicase(s) responsible for unwinding DNA at the replication fork has not been identified. It is clear that many helicases are yet to be discovered in eukaryotes. The purification and biochemical characterization of helicases, as well as reconstitution of complex biochemical reactions *in vitro*, remains an active area of investigation. This section describes several strategies for the isolation of yeast DNA helicases, including isolation of DNA helicases fused with additional amino acids that facilitate affinity purification, isolation of native

DNA helicases expressed from plasmid-based systems, and isolation of DNA helicases from whole cell extracts.

4.1 Purification using affinity tags

Two approaches have been used to express and purify proteins from genes cloned in expression plasmids. One approach is to express the protein in its native state and develop a purification protocol using standard chromatographic steps. The second approach is to engineer the gene of interest such that the expressed protein contains an added 'affinity tag' that enables simplification of the purification protocol. We will discuss tagged helicases first since these are easily purified by standard protocols.

Affinity tags include epitopes for antibody affinity purification, histidine tags that greatly increase affinity for nickel-containing resins, fusions with maltose binding protein that allow purification on amylose resins, and other fusions, such as glutathione-*S*-transferase. Several tagged yeast DNA helicases have been purified. Two critical observations concerning these proteins are the relative ease of purification compared with native proteins and the fact that DNA helicase activity as well as ATPase activity was retained in spite of the added amino acids.

The *DNA2* gene is required for cell viability and the gene product appears to be involved in DNA replication although probably not as the replicative helicase (40). The purification and biochemical characterization of Dna2p was achieved by fusion of the *DNA2* coding sequence with a haemagglutin (HA) epitope tag consisting of 15 amino acids fused at the amino terminus of the protein (41). The HA tagged gene was expressed in yeast under control of the *GAL1* promoter. A reasonably pure, enzymatically active protein was obtained with only three chromatographic steps, including a final step on a HA monoclonal antibody affinity column.

Sufficient protein was obtained to perform an initial biochemical characterization including measurement of ATPase activity, the determination of a 3′-5′ polarity of unwinding, as well as the demonstration of a requirement for a 5′ tail on the DNA strand that is displaced. These results are encouraging considering the large size of Dna2p (172 kDa) and the modest amount of cell paste required for the purification (3–4 g).

The *UPF1* gene has been shown to be essential for nonsense-mediated decay of mRNA (42). Purification and characterization of the encoded polypeptide from yeast cell extracts was facilitated by fusing the gene with coding sequence for an epitope tag. *UPF1* was fused to a FLAG epitope, placed behind a strong constitutive promoter, and transformed into yeast. The gene product was partially purified from a cell lysate on a DEAE Sephacel column, followed by purification to near homogeneity using an immunoaffinity column with a monoclonal antibody against the FLAG epitope (10). The purification protocol yielded sufficient protein for biochemical characterization from two litres of cultured yeast cells. The resulting fusion polypeptide retained high DNA unwinding, RNA unwinding, and ATPase specific activity.

The Srs2 protein was expressed in *E. coli* using the pET-3c expression plasmid (43) engineered to add a short peptide, containing six histidine residues, to the carboxyl terminus of the protein. The overexpressed protein was purified by lysing the *E. coli* cells harbouring the overexpression plasmid in 6 M guanidine. The lysate was passed through a nickel column which bound the His tagged Srs2 protein. This purification step was followed by passage through a Sephacryl S-200 size exclusion column, and the final step was renaturation of the polypeptide by slowly lowering the guanidine concentration. The renatured protein regained DNA helicase and ATPase activity and possessed sufficient catalytic activity to allow an initial biochemical characterization of the gene product.

An affinity tag strategy that utilizes a maltose binding protein (MBP) fusion was employed recently to isolate a novel DNA helicase (A. Leitzel and S. W. Matson, unpublished results). The gene encoding this DNA helicase was previously identified as an open reading frame and cloned by PCR into a commercially available expression plasmid (New England Biolabs). The expressed protein is a N terminal fusion of the protein of interest with maltose binding protein. The fusion polypeptide was expressed in *E. coli* and isolation was accomplished by following *Protocol 8*. The DNA helicase activity of the eluate was directly measured using partial duplex substrates. Low levels of nuclease contamination necessitated the use of poly dI:dC and S-substituted unwinding substrates (see Sections 2.4.1 and 2.4.2).

Protocol 8. Purification of MBP–helicase fusion protein

Equipment and reagents

- Amylose resin (New England Biolabs)
- *E. coli* cells expressing a MBP–helicase fusion protein
- 10 mg/ml lysozyme (Sigma)
- Buffer A: 20 mM Tris–HCl pH 8.3, 20% glycerol, 1 mM EDTA, 0.5 mM EGTA, 15 mM 2-mercaptoethanol
- 10 mM maltose (Sigma)

Method

1. Resuspend cell paste at 2 g cells/ml using buffer A plus 0.2 M NaCl.
2. Freeze cell suspension at –80 °C.
3. Thaw the cells on ice and add lysozyme to 200 μg/ml and PMSF to 0.1 mM.
4. Heat to 20 °C in a water-bath and sonicate briefly to shear DNA.
5. Clear the lysate by centrifugation at 30 000 *g* for 30 min.
6. Equilibrate amylose resin in buffer A plus 0.2 M NaCl.
7. Incubate resin with cleared lysate (0.5 ml resin/3 g cell paste equivalent) at 4 °C for 2 h with gentle shaking.

Protocol 8. *Continued*

8. Pour a column with the resin plus bound proteins.
9. Wash the column with three column volumes of buffer A plus 0.2 M NaCl, followed by an extensive wash with buffer A plus 1 M NaCl until the protein concentration of the eluate is below detectable levels.
10. Elute the protein with buffer A plus 0.2 M NaCl plus 10 mM maltose.

4.2 Biochemical purification

The use of fusion proteins offers the advantage of ease of purification but the disadvantage of altering the primary amino acid sequence of the polypeptide by adding the tag. It is encouraging that enzymatic activity has been present in a number of these fusion proteins. However, the possibility of subtle biochemical effects rendered by the addition of the tag makes it necessary to isolate the native protein for rigorous biochemical characterization. Three DNA helicases; Rad3p, Sls2p, and Pif1p, have been isolated from yeast following expression of the native gene cloned in plasmids that contain yeast promoters. For detailed descriptions of the expression constructs and purification protocols see refs 44–46 for Rad3p, Sls2p, and Pif1p respectively. In all three cases sufficient pure enzyme was obtained to perform extensive biochemical characterization. Pif1p purification offers an interesting example of using subcellular localization to facilitate purification. Pif1p is targeted to the mitochondria and the majority of the polypeptide resides there, even in overexpressing strains. The protocol uses the mitochondrial fraction of yeast cell homogenates as a starting point in the chromatographic purification of the protein. Following preparation of the mitochondrial fraction, two chomatographic fractionation steps were required to obtain purified protein.

Purification of yeast DNA helicases, without the benefit of protein expression plasmids, has been accomplished in several laboratories (31, 38, 39). The yields are low and the amount of cell paste required as starting material is high. In spite of these drawbacks, it is possible to use purified protein to identify the gene encoding the polypeptide either by screening protein expression libraries with antibodies raised to the polypeptide or through chemical sequencing of the purified polypeptide.

Protocol 9 describes a large scale lysis procedure for obtaining a cleared lysate from yeast cells. This procedure has been used as a first step in isolation of DNA helicases without the benefit of expression plasmids. The protocol is also useful, in a scaled down version, for the isolation of helicases expressed from plasmids.

Whole cell protein extracts contain an array of ssDNA-dependent ATPases as well as nucleases capable of degrading the substrates used to measure DNA helicase activity. These are fractionated on a series of chromatographic columns with the goal of isolating individual helicases as well as the elimination of nuclease activities that interfere with the assay of DNA helicase activity. It is important to minimize the number of columns and to carefully monitor specific activity, usually expressed as NTPase activity per milligram of protein, to evaluate the efficacy of each purification step. It is also beneficial to monitor helicase activity using partial duplex substrates (*Protocols 2a, 2b*, and *3*) to detect not only helicase activity but also to reveal nucleases that will interfere with subsequent helicase characterization.

Protocol 9. Preparation of whole cell yeast lysates

Equipment and reagents

- Bead beater fitted with a 350 ml mixing chamber (Biospec)
- Yeast cells
- Lysis buffer: 0.2 M Tris–HCl pH 8, 8 mM EDTA, 6 mM 2-mercaptoethanol, 10% glycerol, 0.1% Brij-35
- 0.5 mm glass beads (Sigma)
- Protease inhibitors cocktail (1 ×): 10 mM benzamidine, 1 mM PMSF, 1 μg/ml pepstatin A, 1 μg/ml leupeptin, 10 μg/ml soybean trypsin inhibitor, 10 μg/ml TAME, 5 μg/ml aprotinin
- STE buffer: 10 mM Tris–HCl pH 8, 1 mM EDTA, 100 mM NaCl

Method

1. Rinse 1–2 kg (wet weight) yeast cells in H_2O and recover by centrifugation at 2500 *g*.[a]
2. Suspend the cell paste in an equal volume of lysis buffer that includes a cocktail of proteinase inhibitors.
3. Disrupt the cells using a bead beater. Fill the chamber to two-thirds capacity with 0.5 mm glass beads and then to full capacity with cell suspension to avoid blending air into the mixture. Disruption is accomplished with 6 × 30 sec bursts separated by 2 min cooling periods.[b]
4. Assess cellular disruption by viewing broken cells under a microscope or quantitate breakage by suspending an aliquot of the lysate in STE buffer, sedimenting cells and cell debris with 5 min spin at 12 000 *g* in a microcentrifuge, and measuring soluble protein in the supernatant.[c]
5. Clear the crude lysate by centrifugation for 90 min at 30 000 *g*.
6. Measure the total protein in the cleared supernatant.[d]

[a] Select cell strains that carry mutations in two genes encoding cellular proteinases, *PRB1* and *PEP4*.
[b] The bead beater mixing heats the lysate so care must be taken to keep the mixture cool.
[c] An efficient disruption gives approx. 3–5% soluble protein per wet wt of the cell paste although up to 10% soluble protein per wet wt cell paste has been reported (39).
[d] An efficient breakage should yield at least 20 mg protein/ml and a typical yield is 35–40 g soluble protein per kg yeast.

The isolation of scHelI required six chromatographic columns and a fractionation step using preparative ultracentrifugation (31). All buffers contained 0.1 mM PMSF as a protease inhibitor and 0.01% Nonidet P-40 which was essential for stability of scHelI. Two other DNA helicases have been isolated from yeast cell extracts; replication factor C associated helicase (38) and ATPase III (39). RF-C associated helicase was purified from 2.2 kg of yeast by fractionation on seven different chromatographic columns followed by glycerol gradient ultracentrifugation (38). The final yield was apparently not measured, but prior to the final purification step only 30 μg of protein remained. The first DNA helicase to be isolated from yeast cell extracts was ATPase III (39). The yield of the 60 kDa polypeptide was 150 μg per 500 g wet weight of cells and the purification required six chromatographic fractionation steps. Neither ATPase III nor RF-C associated helicase have been genetically characterized, perhaps owing to the low amounts of available polypeptide.

References

1. Geider, K. and Hoffmann-Berling, H. (1981). *Annu. Rev. Biochem.*, **50**, 233.
2. Matson, S.W. and Kaiser-Rogers, K.A. (1990). *Annu. Rev. Biochem.*, **59**, 289.
3. Dalbadie-McFarland, G. and Abelson, J. (1990). *Proc. Natl. Acad. Sci. USA*, **87**, 4236.
4. Lee, C.-G. and Hurwitz, J. (1992). *J. Biol. Chem.*, **267**, 4398.
5. Hirling, H., Scheffner, M., Restle, T., and Stahl, H. (1989). *Nature*, **339**, 562.
6. Pause, A. and Sonenberg, N. (1992). *EMBO J.*, **11**, 2643.
7. Brennan, C.A., Dombroski, A.J., and Platt, T. (1987). *Cell*, **48**, 945.
8. Matson, S.W. (1989). *Proc. Natl. Acad. Sci. USA*, **86**, 4430.
9. Scheffner, M., Knippers, R., and Stahl, H. (1989). *Cell*, **57**, 955.
10. Czaplinski, K., Weng, Y., Hagan, K.W., and Peltz, S.W. (1995). *RNA*, **1**, 610.
11. Matson, S.W., Bean, D.W., and George, J.W. (1994). *BioEssays*, **16**, 13.
12. Lohman, T.M. and Bjornson, K.P. (1996). *Annu. Rev. Biochem.*, **65**, 169.
13. Lohman, T.M. (1993). *J. Biol. Chem.*, **268**, 2269.
14. Geiselmann, J., Wang, Y., Seifried, S.E., and von Hippel, P.H. (1993). *Proc. Natl. Acad. Sci. USA*, **90**, 7754.
15. Lohman, T.M. (1992). *Mol. Microbiol.*, **6**, 5.
16. Gorbelenya, A.E., Koonin, E.V., Donchenko, A.P., and Blinov, V.M. (1989). *Nucleic Acids Res.*, **17**, 4713.
17. Hodgeman, T.C. (1988). *Nature*, **335**, 22.
18. Bork, P. and Koonin, E.V. (1993). *Nucleic Acids Res.*, **21**, 751.
19. Selby, C.P. and Sancar, A. (1995). *J. Biol. Chem.*, **270**, 4890.
20. Selby, C.P. and Sancar, A. (1997). *J. Biol. Chem.*, **272**, 1885.
21. Johnson, R.E., Prakash, S., and Prakash, L. (1994). *J. Biol. Chem.*, **269**, 28259.
22. Matson, S.W., Tabor, S., and Richardson, C.C. (1983). *J. Biol. Chem.*, **258**, 14017.
23. Naegeli, H., Bardwell, L., and Friedberg, E.C. (1992). *J. Biol. Chem.*, **267**, 392.
24. Greenstein, D. and Besmond, C. (1998). In *Current protocols in molecular biology* (ed. F.M. Ausubel, R. Brent, R.E. Kingston, D.D. Moore, J.G. Seidman, J.A. Smith, and K. Struhl), pp. 1.15.3. John Wiley and Sons, Inc.
25. Moore, D. and Chory, J. (1998). In *Current protocols in molecular biology* (ed.

F.M. Ausubel, R. Brent, R.E. Kingston, D.D. Moore, J.G. Seidman, J.A. Smith, and K. Struhl), pp. 2.6.9. John Wiley and Sons, Inc.
26. Tabor, S. (1998). In *Current protocols in molecular biology* (ed. F.M. Ausubel, R. Brent, R.E. Kingston, D.D. Moore, J.G. Seidman, J.A. Smith, and K. Struhl), pp. 3.10.3. John Wiley and Sons, Inc.
27. Matson, S.W. and George, J.W. (1987). *J. Biol. Chem.*, **262**, 2066.
28. Bjornson, K.P., Amaratunga, M., Moore, K.J.M., and Lohman, T.M. (1994). *Biochemistry*, **33**, 14306.
29. Eggleston, A.K., Rahim, N.A., and Kowalczykowski, S.C. (1996). *Nucleic Acids Res.*, **24**, 1179.
30. Eckstein, F. and Gish, G. (1989). *Trends Biochem. Sci.*, **14**, 97.
31. Bean, D.W., Kallam, W.E., and Matson, S.W. (1993). *J. Biol. Chem.*, **268**, 21783.
32. Bujalowski, W., Overman, L.B., and Lohman, T.M. (1988). *J. Biol. Chem.*, **263**, 4629.
33. Alani, E., Thresher, R., Griffith, J.D., and Kolodner, R.D. (1992). *J. Mol. Biol.*, **227**, 54.
34. Siegal, G., Turchi, J.J., Jessee, C.B., Myers, T.W., and Bambara, R.A. (1992). *J. Biol. Chem.*, **267**, 13629.
35. Lohman, T.M., Chao, K., Green, M.J., Sage, S., and Runyon, G.T. (1989). *J. Biol. Chem.*, **264**, 10139.
36. Matson, S.W. and Richardson, C.C. (1983). *J. Biol. Chem.*, **258**, 14009.
37. Lanzetta, P.A., Alvarez, L.J., Reinach, P.S., and Candia, O.A. (1979). *Anal. Biochem.*, **100**, 95.
38. Li, X., Yoder, B.L., and Burgers, P.M.J. (1992). *J. Biol. Chem.*, **267**, 25321.
39. Sugino, A., Ryu, B.H., Sugino, T., Naumovski, L., and Friedberg, E.C. (1986). *J. Biol. Chem.*, **261**, 11744.
40. Budd, M.E. and Campbell, J.L. (1995). *Proc. Natl. Acad. Sci. USA*, **92**, 7642.
41. Budd, M.E., Choe, W.-C., and Campbell, J.L. (1995). *J. Biol. Chem.*, **270**, 26766.
42. Leeds, P., Peltz, S.W., Jacobson, A., and Culbertson, M.R. (1991). *Genes Dev.*, **5**, 2303.
43. Rong, L. and Klein, H. (1993). *J. Biol. Chem.*, **268**, 1252.
44. Sung, P., Prakash, L., Matson, S.W., and Prakash, S. (1987). *Proc. Natl. Acad. Sci. USA*, **84**, 8951.
45. Guzder, S.W., Sung, P., Bailly, V., Prakash, L., and Prakash, S. (1994). *Nature*, **369**, 578.
46. Lahaye, A., Leterme, S., and Foury, F. (1993). *J. Biol. Chem.*, **268**, 26155.

5

Functional analysis of DNA replication accessory proteins

ULRICH HÜBSCHER, ROMINA MOSSI, ELENA FERRARI, MANUEL STUCKI, and ZOPHONÍAS O. JÓNSSON

1. Introduction: physiological importance of DNA replication accessory proteins

This chapter summarizes practical experiments that can be performed with DNA replication accessory proteins. The DNA, in preparation for DNA synthesis, has to become single-stranded to serve as a template for the replicative DNA polymerases (pols). For this a set of proteins have been identified that support the replicative and repair pols in performing processive, accurate, and rapid DNA synthesis. Furthermore these proteins prevent damage to the transient single-stranded DNA. The three best known are the proliferating cell nuclear antigen (PCNA), replication factor C (RF-C), and replication protein A (RP-A). DNA replication accessory proteins provide particular functions that are mandatory for pols. Such functions include the recruitment of particular pols when needed, the facilitation of pol binding to the primer terminus, the increase in pol processivity, the prevention of non-productive binding of the pol to single-stranded DNA, the release of the pol after DNA synthesis, and the bridging of pol interactions to other replication proteins. Thus it is not surprising that proteins carrying out such functions have been universally found in nature (reviewed in ref. 1). The eukaryotic auxiliary proteins RF-C and PCNA appear to be functional homologues of the bacteriophage T4 proteins gp44/62 and gp45, and subassemblies of the *E. coli* pol III holoenzyme γ-complex and β-subunit, respectively (reviewed in ref. 2).

Besides describing briefly the replication activities and the isolation protocols of PCNA, RF-C, and RP-A, we especially focus on some unique assays that were developed with these proteins. Such techniques include the *in vitro* ^{32}P-labelling of modified PCNA (*Protocol 4*), a subunit exchange assay to estimate the monomer exchange rate between PCNA trimers (*Protocol 5*), the kinase protection assay to measure interaction of PCNA with other proteins (*Protocol 9*), and the RF-C-dependent loading of PCNA onto DNA (*Protocol 10*).

2. Proliferating cell nuclear antigen (PCNA)

2.1 PCNA purification: general remarks

PCNA is an abundant protein in dividing eukaryotic cells. Its purification from various organisms has been described (3), e.g. human PCNA (4), yeast PCNA (5), and plant PCNA (6). The PCNA genes of various organisms have been cloned and used for expression in *E. coli* using several different expression systems (7–10). In general PCNA is expressed at a high level, as a soluble and biologically active protein from bacteria. In addition, heterologous expression allows an easy way of constructing various fusion derivatives of PCNA and amino acid-specific mutants. Fusing short sequences to either the N or C termini does not significantly alter the properties of PCNA as seen in *in vitro* DNA replication assays. This has for example allowed the construction of artificially phosphorylatable forms of PCNA and histidine tagged PCNA derivatives (11–13). Since PCNA is in most cases obtained from expression strains, we will describe in this section the behaviour of PCNA during isolation from overexpressing *E. coli* strains.

Human PCNA (hPCNA) is a highly acidic protein with a pI of 4.5 (3) and a calculated molecular weight of 28.8 kDa. However, due to its charge hPCNA will run as an approximately 36 kDa protein on SDS–PAGE. Yeast PCNA is somewhat less acidic with pI of 5.3 and runs closer to its expected position on SDS–PAGE (5). Nevertheless the chromatographic behaviour of PCNA from different organisms is very similar. In aqueous solution PCNA forms a stable trimer (5, 11, 14), although some dissociation and thus subunit exchange will take place. PCNA will therefore behave as a trimer in native PAGE, glycerol gradient centrifugation, and gel filtration.

2.2 Assay for PCNA

PCNA has been identified as a processivity factor for pol δ. The easiest way to quantify active PCNA is to determine the stimulation of pol δ in a poly(dA)/oligo(dT) assay (so called RF-C-independent assay, see also RF-C-dependent pol δ assay in Section 3.2). Since PCNA and pol δ are very conserved in structure in all eukaryotic cells they can be used interchangeably for *in vitro* assays from a variety of sources.

Protocol 1. Assay for PCNA

Reagents

- 5 × reaction buffer: 250 mM bis–Tris pH 6.5, 1.25 mg/ml BSA, 5 mM DTT, 30 mM $MgCl_2$, 50 mM KCl
- [^{3}H]dTTP (e.g. 1.1 Ci/mmol)
- Poly(dA)/oligo(dT) (base ratio 10:1)
- Pol δ

Method

1. Set-up a 25 μl reaction containing the following components: 1 × reaction buffer, 25 μM [^{3}H]dTTP, 0.5 μg poly(dA)/oligo(dT), 0.3 U pol δ, and PCNA to be assayed. Usually 50–100 ng homogeneous PCNA will saturate such an assay.
2. Incubate at 37 °C for 20 min.
3. Collect and measure the trichloroacetic acid insoluble material as described in Chapter 3, *Protocol 3*, or ref. 15.

2.3 Isolation of recombinant hPCNA

2.3.1 A short protocol for the expression and purification of hPCNA from *E. coli*

This protocol describes the expression of cloned hPCNA from a plasmid containing a cDNA encoding hPCNA, under the control of a T7 RNA polymerase promoter, in an *E. coli* host strain carrying the T7 RNA polymerase gene regulated by an IPTG-inducible promoter. This expression system was originally developed by Studier *et al.* (16, 17) and has been proven excellent for PCNA expression. The *E. coli* strain BL21(DE3)pLysS carrying the plasmid pT7/hPCNA (kindly provided by Bruce Stillman, Cold Spring Harbour Laboratory) is used for the PCNA expression. The purification scheme leads to highly concentrated and pure PCNA, free of contaminating nucleases and pol activity.

Protocol 2. Induction and purification of recombinant hPCNA

Equipment and reagents

- Sonicator
- 15 ml phosphocellulose column
- 10 ml Q-Sepharose (Pharmacia)
- 4 ml ceramic hydroxylapatite, type II (Bio-Rad) column
- BL21(DE3)pLysS:pT7/hPCNA
- IPTG
- Lysis buffer: 20 mM Tris–HCl pH 6.8, 1 mM EDTA, 10 mM $NaHSO_3$, 0.01% (v/v) NP-40, 1 mM DTT, 1 mM PMSF
- LB containing 50 μg/ml ampicillin and 34 μg/ml chloramphenicol (LAC)
- Gel loading buffer: 60 mM Tris–HCl pH 6.8, 2% (w/v) SDS, 2% (v/v) glycerol, 0.005% (w/v) bromphenol blue, 2% (v/v) 2-mercaptoethanol
- Buffer A: 20 mM Tris–HCl pH 6.8, 1 mM EDTA, 10 mM $NaHSO_3$, 0.01% (v/v) NP-40, 10% (v/v) glycerol
- Buffer B: 10 mM Tris–HCl pH 6.8, 5% (v/v) glycerol

Method

1. Start an overnight pre-culture of BL21(DE3)pLysS:pT7/hPCNA in 25 ml LAC (either from freshly grown colonies from a plate or a glycerol stock). Grow at 37 °C with shaking.
2. Next morning dilute the pre-culture into 1 litre of LAC and continue to shake at 37 °C. Increased yields are observed if the culture is well

Protocol 2. *Continued*

aereated and it is therefore recommended to split the culture into several culture flasks.

3. When the OD_{600} of the culture has reached 0.6, IPTG is added to 0.5 mM and shaking continued for 4–5 h. Take a 1 ml sample before induction and store at 4°C for SDS–PAGE analysis of the induction.
4. After induction remove a 1 ml sample for SDS–PAGE analysis and harvest the rest of the culture by spinning at 5000 *g*. The harvested cells can be stored at –70°C for several months without loss of PCNA activity.
5. To monitor the induction, spin down cells from 100 μl of the uninduced and induced cultures and resuspend in gel loading buffer. Boil for 2 min and load 40 μl of each onto a 12% SDS–PAGE gel. The overexpressed PCNA should be clearly visible at around 36 kDa.
6. Add 50 ml of lysis buffer and lyse the bacteria by freezing–thawing followed by brief sonication to reduce viscosity.[a]
7. Spin the lysate for 15 min at 25 000 *g* and discard the pellet.
8. Assemble a 15 ml phosphocellulose column and a 10 ml Q-Sepharose column so that the outlet of the phosphocellulose column is connected directly to the Q-Sepharose column. Equilibrate the columns in buffer A and load the supernatant from step 2 onto the columns at a flow rate of 0.5 ml/min. PCNA will pass through the phosphocellulose column and bind to the Q-Sepharose.
9. Finish loading by washing the columns with 20 ml buffer A with the columns still connected. Remove the phosphocellulose column and wash the Q-Sepharose column with 50 ml buffer A containing 0.1 M NaCl.
10. Elute PCNA with a 100 ml NaCl gradient from 0.1–0.7 M collecting 2 ml fractions. Identify PCNA-containing fractions by loading 10 μl of each (or every second) fraction onto SDS–PAGE and pool the peak fractions containing the 36 kDa band.
11. Equilibrate a 4 ml ceramic hydroxylapatite column in buffer B containing 300 mM NaCl.
12. Load directly the pool from the Q-Sepharose step at a flow rate of 0.5 ml/min.
13. Wash with 40 ml buffer B containing 1 M $MgCl_2$ followed by 20 ml buffer B with 50 mM NaCl.
14. Elute PCNA with a gradient from 0–250 mM $NaPO_4$ pH 6.8 in buffer B and monitor PCNA-containing fractions as before.

15. The purified PCNA can be stored at 4 °C for several weeks, or brought to 50% glycerol by dilution or dialysis and stored at −20 °C.
16. Finally, the activity in DNA synthesis can be determined as described in Section 2.2 (assay for PCNA). A yield of 50 mg of purified PCNA can be expected from 1 litre of culture.

[a] Alternatively, the cells can by lysed by passage through a French press.

2.3.2 Construction, expression, and purification of his–PCNA

Plasmids expressing histidine tagged PCNA derivatives are easily produced using commercially available expression vectors. A his–tag fusion allows rapid purification of recombinant PCNA and additionally provides a convenient tool for various biochemical assays such as pull-down assays. An example of the construction, purification, and use of his–PCNA is provided in ref. 11. In this case human PCNA is expressed with 23 additional amino acids (Met Gly $(His)_{10}$ Ser Ser Gly His Ile $(Asp)_4$ Lys His–) prior to the first Met of the authentic PCNA. Adding a $(His)_6$ tag to the C terminus of PCNA also results in active PCNA with similar properties (Z. O. Jónsson and U. Hübscher, unpublished results). The expression is performed as for untagged PCNA and his–PCNA is purified by a simple single column protocol.

Protocol 3. Induction and purification of histidine tagged hPCNA

Reagents

- Buffer A: 20 mM KPO_4 pH 7.8, 300 mM NaCl
- Ni-NTA resin (Qiagen)
- Imidazole–HCl pH 7.2
- Storage buffer of choice: e.g. 50 mM Tris–HCl pH 7.5, 50 mM NaCl, 1 mM EDTA, 1 mM DTT, 50% (v/v) glycerol

Method

1. Induce and harvest cells as outlined in *Protocol 1*, steps 1–5.
2. Add 50 ml buffer A and lyse the bacteria by freezing–thawing, followed by a brief sonication to reduce viscosity.
3. Spin the lysate for 15 min at 25 000 *g* and discard the pellet.
4. Mix the supernatant with 8 ml Ni-NTA resin and stir for 1 h at 4 °C.
5. Wash the resin once with 40 ml buffer A containing 10% (v/v) glycerol and pack it into a column.
6. Wash the column with 80 ml buffer A containing 10% (v/v) glycerol.
7. Elute his–PCNA with a 100 ml gradient from 0.01–0.5 M imidazole–HCl pH 7.2 in 10% (v/v) glycerol.
8. Monitor his–PCNA-containing fractions by SDS–PAGE. $(His)_{10}$–PCNA elutes at 0.25–0.3 M imidazole–HCl pH 7.2.
9. Remove imidazole by dialysis against storage buffer.

2.3.3 Artificially phosphorylated PCNA (ph–PCNA)

Addition of artificial phosphorylation sites to recombinant PCNA is a convenient way to obtain *in vitro* ^{32}P-labelled PCNA. Its use has already improved our understanding of the mechanism of assembly and action of eukaryotic replication auxiliary proteins. The recognition sequence of heart muscle kinase (Met Arg Arg Ala Ser Val) can be fused to either terminus of PCNA without affecting its activity in replication assays, even when phosphorylated (13, 18). We have constructed plasmids with phosphorylation sites on either the N or the C termini of PCNA (for details see ref. 12 and 13, respectively). The resulting fusion proteins can be expressed and purified according to *Protocol 1*. *Protocol 4* describes *in vitro* phosphorylation of modified ph–PCNA.

Protocol 4. *In vitro* phosphorylation of modified ph–PCNA

Reagents

- Protein kinase, catalytic subunit from bovine heart (HMK; Sigma, P-2645): 250 U of the kinase can be suspended in 50 μl 40 mM DTT (5 U/μl), and stored in 0.5 μl aliquots at –80 °C
- 10 × HMK buffer: 200 mM Tris–HCl pH 7.5, 10 mM DTT, 1 M NaCl, 120 mM $MgCl_2$
- [γ-^{32}P]ATP (10 μCi/ml)
- ph–PCNA (PCNA containing the phosphorylation site for HMK)

Method

1. Set-up a 10 μl labelling mixture containing 1 × HMK buffer, 1 μl [γ-^{32}P]ATP, 0.3 μl HMK (5 U/μl), and 250 ng ph–PCNA.
2. Incubate at 37 °C for 20 min.
3. Labelled PCNA can be kept on ice until use (not more than three days) and can be used in loading (see *Protocol 10* and *Figure 3*) or DNA replication (see Section 2.2) assays without additional purification.

2.4 A subunit exchange assay to estimate the exchange rate of PCNA monomers into trimers

This assay is based on the binding of his–PCNA subunits to Ni-NTA resin. When his–PCNA is incubated with wild-type PCNA at 37 °C a slow exchange of subunits takes place, leading to the formation of chimeric molecules. This assay has been used to measure the subunit exchange rate of PCNA and can be adapted to test the influence of PCNA mutations or other proteins and ingredients on the trimerization of PCNA. *Protocol 5* summarizes the measurement of subunit exchange kinetics of wild-type PCNA (see also *Figure 1*).

Protocol 5. Method to measure subunit exchange kinetics of wild-type PCNA

Reagents

- wt PCNA and his–PCNA
- 5 × buffer A: 2000 mM Tris–HCl pH 7.5, 1 mg/ml BSA, 50 mM $MgCl_2$
- Ni-NTA resin
- Imidazole–HCl

Method

1. Mix 30 μg wt PCNA and 30 μg his–PCNA in a final volume of 500 μl buffer A.
2. Incubate at 37 °C, remove aliquots of 50 μl after 0, 5, 10, 20, 30, 40, 50, and 60 min, and chill them on ice.
3. Adsorb the protein samples to 20 μl of a 50% slurry of Ni-NTA resin equilibrated in buffer A.
4. Incubate the samples at 4 °C for 15 min with gentle stirring.
5. Wash the resin three times in batch with 400 μl ice-cold 30 mM imidazole–HCl pH 7.2, containing 10% (v/v) glycerol.
6. Remove the supernatant carefully and resuspend the resin in 50 μl elution buffer containing 400 mM imidazole-HCl pH 7.2 and 10% (v/v) glycerol.
7. Centrifuge for 5 min in a Eppendorf centrifuge.
8. Mix the supernatant with 15 μl of 4 × SDS gel loading buffer (*Protocol 2*).
9. Boil the samples for 3 min, then run them on a 12.5% SDS–PAGE gel, and stain with Coomassie blue.
10. Quantify the two PCNA bands by densitometry.

The amounts of wt PCNA co-adsorbed with his–PCNA after varying times of incubation can be quantified by densitometry of SDS gels and normalized to the amount of his–PCNA in the same lane. In theory, the exchange mixture at equilibrium consists of wt PCNA homotrimer, 2 wt PCNA/1 his–PCNA, 1 wt PCNA/2 his–PCNA, and his–PCNA homotrimer in a quantitative ratio of 1:3:3:1, respectively. Only the forms containing at least one subunit of his–PCNA can adsorb to the Ni-NTA resin and therefore the maximal amount of wt subunits in the adsorbed pool is 75% of the input. Taking this into consideration, the normalized amount of wt PCNA must be amended with the coefficient 4/3 before plotting against time. For the calculations the minor portion of PCNA present as dimer or monomer is neglected. The exchange reaction fits to first order kinetics implying that a monomolecular reaction, presumably opening of the trimers, may be the rate limiting step of

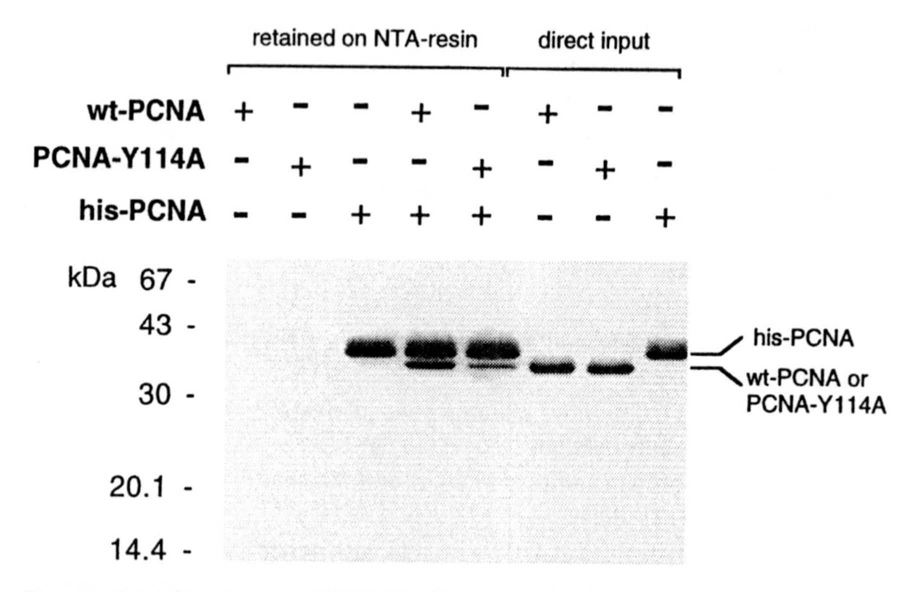

Figure 1. Subunit exchange of PCNA. The figure shows the subunit exchange of wt PCNA or mutant PCNA-Y114A with his–PCNA (11). 3 μg PCNA were mixed in 50 μl 40 mM Tris–HCl pH 7.5, 0.2 mg/ml BSA, and 10 mM $MgCl_2$ and incubated for 1.5 h at 37 °C. Then the proteins were adsorbed to Ni-NTA resin and processed as described in *Protocol 5*. The rightmost three lanes contained 0.6 μg wt PCNA, PCNA-Y114A, or his–PCNA loaded directly onto the polyacrylamide gel and served as a quantitative control (direct input). Reproduced with permission from ref. 11.

the exchange. By using this assay we obtained the value 80 min for the half-time of subunit exchange between wt PCNA and his–PCNA (11).

3. Replication factor C (RF-C)

3.1 RF-C purification: general remarks

RF-C is a heteropentameric protein essential for DNA replication and DNA repair (reviewed in ref. 1). It is a molecular matchmaker required for loading of PCNA onto double-stranded DNA and, thus, for PCNA-dependent DNA elongation by pols δ and ϵ. RF-C can be isolated by conventional techniques over several chromatographic steps from cells and tissues (19–21). Due to the duality of its matchmaker functions, namely to bind DNA and PCNA, it was found that the RF-C complex has peculiar physicochemical properties, thus leading to many problems during isolation such as stability, solubility in low salt, sensitivity to freezing–thawing, and heterogeneity in its behaviour on various chromatographic steps. Recent works by various groups indicated that RF-C can now be isolated from expression vectors, since all five subunits of yeast RF-C have been successfully cloned (22) and finally been expressed in *S. cerevisiae* in an operon containing all five subunits under the control of the galactose-inducible GAL-010 promotor (23). Furthermore, two reports describe the successful co-expression of the five subunits of human RF-C in the baculovirus expression system (24, 25).

It has been found that RF-C from *S. cerevisiae* can bind to PCNA–agarose beads, provided Mg-ATP was added to the binding buffer; this allowed binding of RF-C even at a concentration of 0.8 M NaCl. On the other hand binding did not occur at 0.3 M NaCl in the absence of Mg-ATP. These two binding modes were used to develop an improved purification method for RF-C (23). Yeast RF-C purified in this way was essentially homogeneous and had a several-fold higher specific activity than earlier preparations that were purified over several column procedures.

3.2 Assay for RF-C

RF-C is usually assayed by exploiting its stimulatory effects on pol δ under certain experimental conditions. This is done by carrying out an RF-C-dependent replication assay as described in *Protocol 6*.

Protocol 6. RF-C-dependent replication assay

Reagents

- 5 × reaction buffer: 200 mM Tris–HCl pH 7.5, 1 mg/ml bovine serum albumin, 5 mM DTT, 50 mM $MgCl_2$
- ATP
- All four dNTPs
- [α-^{3}H]dTTP (50 Ci/μmol)
- Primed M13 single-stranded DNA (*must* be circular DNA)
- *E. coli* SSB (or RP-A)
- Pol δ

Method

1. Set-up a reaction mixture of 25 μl containing 1 × reaction buffer, 1 mM ATP, 40 μM each of dATP, dGTP, dCTP, 15 μM [α-^{3}H]dTTP, 100 ng primed M13 single-stranded DNA, 100 ng PCNA, 350 ng *E. coli* SSB (or 500 ng RP-A), 0.2 U pol δ, and RF-C fractions to be tested. About 100 ng purified RF-C saturates the reaction under these conditions.
2. Incubate the reactions for 30 min at 37 °C.
3. Assay the trichloroacetic acid insoluble material as described in Chapter 3, *Protocol 3* or ref. 15.

3.3 Isolation of RF-C from HeLa cell nuclei

The first published protocol for RF-C purification (19) was modified to isolate active and stable RF-C from a 10 litre exponential HeLa cell culture. It has been found that nuclear extracts contain more RF-C than cytosolic extracts (19). However, even in nuclear extracts, RF-C activity is masked by other factors present, which makes it difficult to detect RF-C in the initial fractions by measuring activity with the previously described assay alone. Therefore, a protocol was developed, using an immunological method to trace RF-C containing fractions at the beginning of the purification procedure.

Protocol 7. Purification of RF-C from HeLa cell nuclei

Equipment and reagents

- 5 ml ceramic hydroxylapatite column (Bio-Rad)
- Pellet of 10 litre HeLa cell culture
- Lysis buffer: 20 mM Hepes pH 7.8, 1.5 mM $MgCl_2$, 5 mM KCl, 1 mM DTT, 0.1 mM PMSF
- Extraction buffer: 25 mM Tris–HCl pH 7.5, 200 mM NaCl, 1 mM EDTA, 1 mM DTT, 10 mM sodium bisulfite, 0.1 mM PMSF
- 2 ml ssDNA cellulose column (Pharmacia)
- Buffer A: 25 mM Tris–HCl pH 7.5, 1 mM EDTA, 10% (v/v) glycerol, 1 mM DTT, 10 mM sodium bisulfite, 0.01% (v/v) NP-40, 0.1 mM PMSF
- Buffer B: potassium phosphate pH 7.5 (variable concentrations as indicated below), 1 mM EDTA, 10% (v/v) glycerol, 1 mM DTT, 10 mM sodium bisulfite, 0.1 mM PMSF
- Antibody against RF-C that reacts in immunoblots
- Nitrocellulose membrane (Schleicher & Schuell)
- 1 ml Mono Q column (Pharmacia)
- Phosphocellulose (P11) column material
- RF-C storage buffer: 25 mM Tris–HCl pH 7.5, 150 mM NaCl 80% (v/v) glycerol, 0.75 mg/ml BSA, 1 mM DTT

Method

1. Resuspend the pellet of a 10 litre exponential HeLa cell culture in 100 ml of ice-cold lysis buffer and keep on ice for 15 min. Break the swollen cells with seven strokes of a tightly-fitting Dounce homogenizer and keep the lysate on ice for another 30–60 min.
2. Collect the nuclei by centrifugation at 500 *g* for 15 min and resuspend the crude nuclear pellet in 50 ml extraction buffer. Add pepstatin, aprotinin, chymostatin, and leupeptin to a final concentration of 1 μg/ml each. Stir slowly on ice for at least 1 h.
3. Centrifuge at 10 000 *g* for 30 min and load the supernatant slowly on a 30 ml phosphocellulose column, previously equilibrated with buffer A, containing 30 mM NaCl. Collect the flow-through and wash the column subsequently with 100 ml and 50 ml of buffer A, containing 330 mM NaCl and 660 mM NaCl, respectively.
4. The main portion of RF-C should be in the 660 mM NaCl wash, which can be confirmed by dotting 3 μl of each fraction on a piece of nitrocellulose membrane. Dry the membrane with a hair drier and block it with 5% milk in TBST.
5. Detect RF-C in the dots on the membrane with an antibody against RF-C, by using a standard Western blot procedure.
6. Dilute the RFC-containing fraction with buffer A, to a final NaCl concentration of 300 mM and load it on a 5 ml hydroxylapatite column, equilibrated with buffer A, containing 300 mM NaCl. Wash the column with 30 ml of buffer B, containing 100 mM potassium phosphate.
7. Elute RF-C with a 50 ml linear potassium phosphate gradient from 100–600 mM in buffer B. Collect 1 ml fractions and dot 3 μl of each fraction on a nitrocellulose membrane. Detect RF-C as described in step 5. Pool the RF-C containing fractions, eluting around 300 mM potassium phosphate.
8. Dialyse this pool against 1 litre of buffer A, containing 100 mM NaCl and load it very slowly on a 2 ml ssDNA cellulose colum (0.2 column volumes per hour), which has previously been equilibrated with buffer A containing 100 mM NaCl. Wash with 10 ml of equilibration buffer.
9. Elute RF-C with a 20 ml linear NaCl gradient from 100–600 mM in buffer A. Collect 0.25 ml fractions. RF-C in these fractions can now be detected using

the previously described complementation assay, because the inhibiting factors are purified away after this step. Pool the active fractions.

10. Dilute the ssDNA cellulose pool with buffer B to a final NaCl concentration of 100 mM and inject it onto a 1 ml Mono Q column, equilibrated with the same buffer. Elute RF-C with a 5 ml linear NaCl gradient from 100–400 mM. RF-C should elute at approx. 180 mM NaCl.
11. Pool the active fractions and freeze them by dripping them directly into liquid nitrogen. Store RF-C in liquid nitrogen. For short-term storage, thaw one frozen drop on ice and dilute it immediately 1:1 with storage buffer. In this buffer, RF-C is stable at −20 °C for at least one week.

3.4 Isolation of the PCNA binding region of the large RF-C subunit

Two functional regions of the large subunit of human RF-C have recently been identified (26), the DNA binding (fragment A) and the PCNA binding regions (fragment B). The DNA binding activity was localized between aa 366–477, where the conserved box I (restricted to the large subunit) is situated (22). This region shows striking similarity to the N terminal regions of all bacterial DNA ligases sequenced to date (see ref. 13 for references) and somewhat less, but still significant, similarity to the automodification domain of eukaryotic poly(ADP-ribose) pols. The PCNA binding region, which is adjacent to but does not overlap with the DNA binding region includes aa 478–712. This region of the protein has been shown to inhibit SV40 DNA replication *in vitro*, as well as RF-C-dependent loading of PCNA onto DNA and RF-C-dependent DNA synthesis. It also showed a dominant negative phenotype when expressed in mammalian cells (26).

The gene encoding the fragment B of the large RF-C subunit was cut out of the original vector (26) with *Eco*RI and cloned into the vector pET23a+ to gain a fragment with a his–tag at its C terminus. The resulting clone was sequenced to confirm the reading frame and the presence of the his–tag and finally expressed in *E. coli* BL21(DE3) (Z. O. Jónsson and U. Hübscher, unpublished results). Isolation of the PCNA binding fragment of the large RF-C subunit (fragment B) is described in *Protocol 8*.

Protocol 8. Isolation of the PCNA binding region of the large RF-C subunit (fragment B)

Reagents

- *E. coli* BL21(DE3)pLysS cells with the pET23/his-RFC-B plasmid containing the coding sequence for fragment B with a his–tag at its C terminus
- LAC (see *Protocol 2*)
- IPTG
- Buffer A: 20 mM Tris–HCl pH 7.5, 300 mM NaCl, 0.4 mM PMSF
- Buffer B: 25 mM Tris–HCl pH 7.5, 0.5 mM NaCl, 1 mM PMSF, 10 mM sodium bisulfite
- Buffer C: 10 ml 0.1 M $NaPO_4$ pH 8, 10 mM Tris–HCl, 8 M urea
- Buffer D: 20 mM Tris–HCl pH 7.5, 1 mM DTT, 1 mM EDTA, 100 mM NaCl, 0.4 mM PMSF, 0.5% (v/v) NP-40

Protocol 8. *Continued*

Method

1. Grow the cells in 1 litre (LAC) to OD_{600} of 0.6 and add IPTG to a final concentration of 1 mM.
2. Induce for 3.5 h at 37 °C and harvest the cells by centrifugation at 5000 *g* for 20 min.
3. Lyse the pellet in 10 ml buffer A by sonication for 2 min and incubation at 37 °C for 10 min. Centrifuge the lysate for 20 min at 18 000 *g*.
4. Wash the pellet with 2 × 10 ml each of buffer B plus 1 mM EDTA; buffer B plus 1 mM EDTA and 0.5% (v/v) Triton X-100; and buffer B plus 0.5 M urea. Resuspend the pellet each time by sonification and centrifuge 20 min at 13 000 *g*.
5. Solubilize the extensively washed pellet in buffer C. Centrifuge for 20 min at 18 000 *g*.
6. Analyse the fragment B of the large RF-C subunit in the supernatant by 12% SDS–PAGE.[a]
7. Pool the fragment B-containing fractions, dilute to a concentration of 0.2 mg/ml, and sequentially dialyse against buffer D containing 8 M, 5 M, 3 M, 2 M, 1.5 M, 1 M, 0.8 M, 0.6 M, 0.4 M, 0.2 M, and finally 0 M urea.[b] A yield of 25 mg purified fragment B can be expected from a 1 litre culture.

[a] The fragment B, appearing around 35 kDa, is generally more than 80% pure. If the protein has to be homogeneous further purification is performed over a Ni-NTA column (Qiagen) according to the manufacturer's standard procedure.

[b] If necessary, concentrate the purified protein over a Ni-NTA column. In this case leave EDTA and DTT out of the last dialysis buffer.

3.5 Kinase protection assay (KPA) to measure RF-C interaction with PCNA

The principle of the KPA is that an artificial phosphorylation site engineered on a protein is phosphorylated in the presence of another putative interacting protein thus reducing the accessibility of the phosphorylation site to the kinase (27). For example, ph–PCNA and a twofold molar excess of RF-C are pre-incubated with [γ-^{32}P]ATP for 2 min at 37 °C to allow interaction between the proteins. After addition of the heart muscle kinase (HMK) the kinetics of the phosphorylation reaction is assayed. If RF-C interacts with ph–PCNA, the accessibility of the artificial phosphorylation sites to the kinase is either reduced or the sites are not at all available. This results in a reduction of the amount of radioactivity incorporated into the ph–PCNA monomers. *Figure 2* shows an example of the KPA assay and demonstrates that the pentameric calf thymus RF-C can protect the C termini of PCNA from phosphorylation (for more details see ref. 13). *Protocol 9* details the kinase protection assay (KPA) to measure RF-C interaction with PCNA.

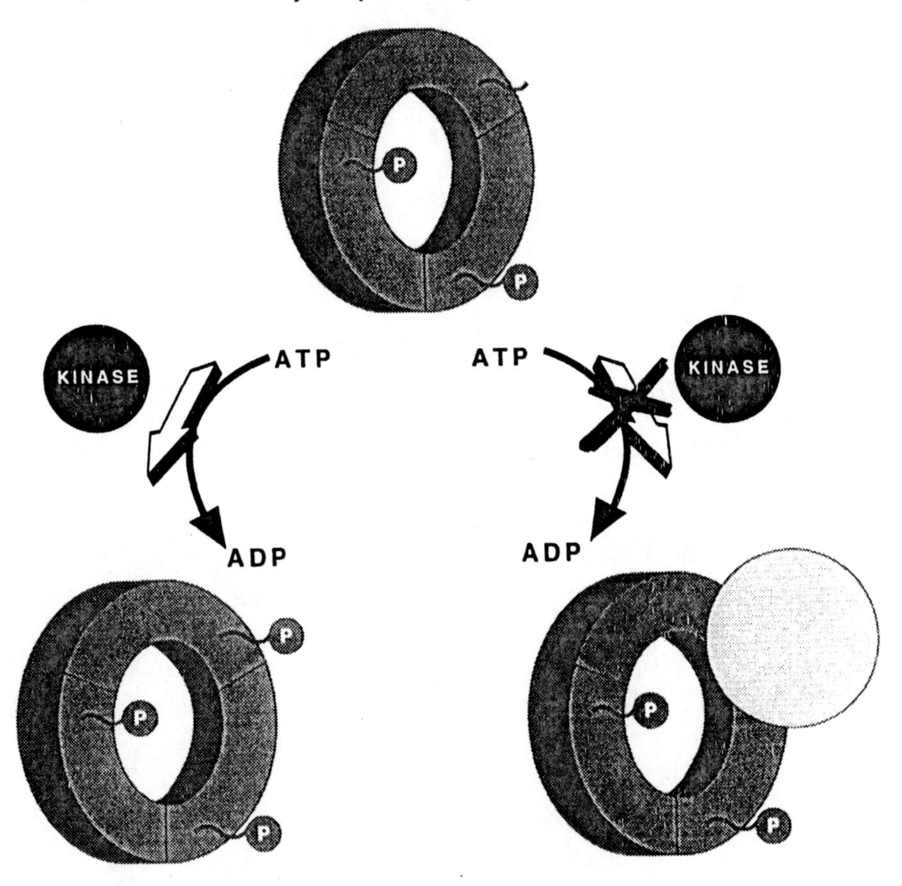

Figure 2. Kinase protection assay (KPA). For details see text and *Protocol 9*.

Protocol 9. Kinase protection assay (KPA) to measure interaction of PCNA with other proteins

Reagents

- N or C terminal phosphorylatable PCNA (nph– or cph–PCNA, respectively) as described in Section 2.3.3 (artificially phosphorylated PCNA)
- [γ-^{32}P]ATP
- 5 × KPA buffer: 100 mM Tris–HCl pH 7.5, 5 mM DTT, 60 mM $MgCl_2$, 1 mg/ml BSA
- HMK: heart muscle kinase (see *Protocol 4*)
- Gel fix: 10% trichloroacetic acid followed by 10% acetic acid, 12% methanol

Method

1. Mix 250 ng (3 pmol) of the phosphorylatable PCNA with an excess of the protein to be tested for interaction with PCNA (e.g. RF-C subunits, p21, pol δ, GADD 45) in a total volume of 10 μl KPA buffer.

Protocol 9. *Continued*

2. Add 3 μCi [γ-^{32}P]ATP and incubate the mix for 2 min at 37 °C.
3. After 2 min start the phosphorylation reaction by addition of 0.75 U HMK.
4. Incubate the reaction mix at 37 °C.
5. After 1, 2, 5, and 10 min take 2 μl aliquots and stop the reaction with 8 μl 2% SDS.
6. Analyse the samples by electrophoresis on a 12% SDS–PAGE gel.
7. Fix and dry the gel, then autoradiograph or quantify with a PhosphorImager.

3.6 RF-C-dependent loading of PCNA onto DNA

The radiolabelled PCNA can be used to quantify the loading onto DNA. PCNA once loaded onto the DNA can be cross-linked with glutaraldehyde and subsequently analysed on SDS–agarose gels (12). *Figure 3* shows an example of PCNA loading onto circular DNA.

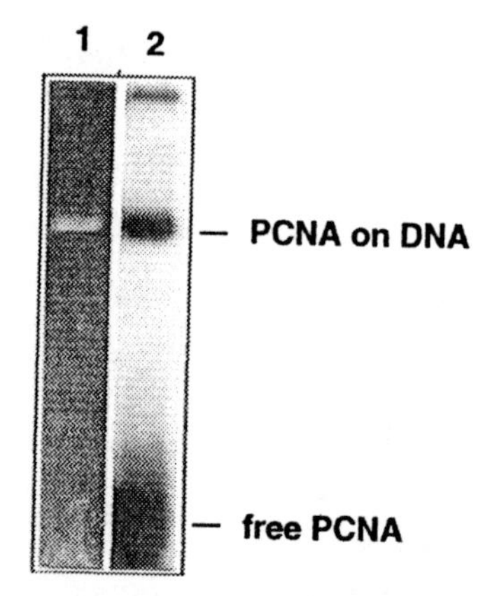

Figure 3. Loading of phosphorylated ph–PCNA onto DNA. Details of this experiments are described in *Protocol 10*. *Left*: ethidium bromide stain of the gel. *Right*: autoradiogram of the same gel indicating the position of PCNA on the DNA and free PCNA.

Protocol 10. RF-C-dependent loading of PCNA onto DNA

Reagents

- ^{32}P-phosphorylated N or C terminal ph–PCNA as described in *Protocol 4*
- 10 × reaction buffer: 400 mM triethanolamine–HCl pH 7.5, 2 mg/ml BSA, 10 mM DTT, 100 mM $MgCl_2$
- ATP
- Gapped circular DNA
- RF-C
- 1% (v/v) glutaraldehyde (freshly diluted from a 25% stock)

- Agarose gel running buffer: 45 mM Tris–borate pH 8.3, containing 1 mM EDTA and 0.1% SDS
- 6 × Ficoll 400 gel loading buffer (LB)
- Gel fix (see *Protocol 9*)

Method

1. Set-up 25 μl containing 1 × reaction buffer, 0.4 mM ATP, 4 μg/ml gapped circular DNA, 30 ng RF-C, and 50 ng ^{32}P-labelled ph–PCNA. (Also carry out a minus RF-C control.)
2. Incubate at 37 °C for 3 min.
3. Add 2.5 μl 1% (v/v) glutaraldehyde.
4. Incubate at 37 °C for 10 min.
5. Add 5.5 μl of a 6 × Ficoll 400 LB and a standard marker dye for an agarose gel, and separate the complexes on a 0.8% agarose gel.
6. Fix, dry, and quantitate the gel (as for *Protocol 9*).

4. Replication protein A (RP-A)

4.1 RP-A purification: general remarks

RP-A is an essential protein that participates in DNA replication, DNA repair, and homologous DNA recombination (reviewed in ref. 1). The protein was first discovered in human cells in the *in vitro* SV40 replication system (28) and has a heterotrimeric structure with polypeptides of molecular weights of 70 kDa (called RP-A1), 32–34 kDa (called RP-A2), and 11–14 kDa (called RP-A3). All three subunits of RP-A are essential for viability of the cell and have functions in DNA replication, DNA repair (nucleotide excision repair), and homologous DNA recombination. While several procedures are described in the literature to obtain an essential homogeneous three subunit RP-A preparation (28–30), the most commonly used method is to use the *E. coli* expression clone p11d-tRPA that co-expresses all three subunits together (31).

4.2 Assay for RP-A

The fate of RP-A during isolation and the determination of its biochemical activity can be followed in three different ways:

(a) By testing individual fractions with either monoclonal (32) or polyclonal (33) antibodies.
(b) By complementation of a RF-C-dependent DNA replication assay as outlined in Section 3.2 (assay for RF-C).
(c) With an unwinding assay (30, 33) as described in *Protocol 11*. This assay is very similar to the commonly used DNA helicase assays (see ref. 34 or Chapter 4 in this book).

Protocol 11. Unwinding assay for RP-A[a]

Reagents

- Single-stranded M13 circle onto which a short (e.g. 24mer) radioactively labelled oligonucleotide has been hybridized (3000 c.p.m./pmol)
- 5 × reaction buffer: 100 mM Tris–HCl pH 7.5, 20% (w/w) sucrose, 40 mM DTT, 400 mg/ml BSA

Method

1. Set-up a 25 µl reaction containing 1 × reaction buffer, 10 ng ^{32}P-labelled DNA substrate, and the protein fraction to be tested.
2. Incubate for 60 min at 37 °C.
3. Stop the reactions and analyse the products by PAGE (34).
4. Under these conditions the substrate stays at the top of the gel, while the unwound oligonucleotide migrates in. Fix and dry the gel and then quantitate the extent of unwinding.

[a] This assay is very sensitive to salt and $MgCl_2$ (30).

4.3 Isolation of recombinant human RP-A

As mentioned above the heterotrimeric RP-A can be isolated from the *E. coli* expression vector p11d-tRPA (31). The devised operon correctly induces all the three subunits. In order to confirm that active RP-A is expressed, activity should be followed at least in the last steps of the purification procedure. This includes three column steps: Affi-Gel Blue, hydroxylapatite, and FPLC Mono Q. Two modifications are critical to get good yield of active RP-A. First, the *E. coli* strain BL21(DE3) should be used as expression host for the p11dtRPA plasmid. We also tested strains carrying the pLysS and pLysE plasmids which expressed RP-A poorly. Secondly, the bacterial growth should be started by the inoculation of a 1 litre flasks of LB medium containing 100 µg/ml ampicillin with one bacterial colony, and growth of the cells should be allowed overnight at 30 °C without shaking.

Protocol 12. Isolation of recombinant human RP-A

Equipment and reagents

- Affi-Gel Blue column
- 1 ml Mono Q column
- French press
- HAP (hydroxylapatite) column
- LB medium containing 100 µg/ml amp
- IPTG
- HI buffer: 30 mM Hepes pH 7.8, 1 mM DTT, 0.25 mM EDTA, 0.25% (w/v) inositol, 0.01% (v/v) NP-40, and 1 mM PMSF
- NaSCN

Method

1. Inoculate six 1 litre flasks of medium with one colony each and let them grow overnight at 30 °C without shaking.
2. Next morning shake until the cells reach an OD_{600} of 0.6 and induce with 0.4 mM IPTG for 2 h.
3. Collect the cells by centrifugation at 1500 *g* for 20 min and resuspend the pellet in 30 ml HI buffer.
4. Freeze the resuspended cells at –80 °C, thaw, and lyse by three passages through a French press.
5. Spin at 18 000 *g* for 30 min at 4 °C.
6. Load the cell lysates immediately (without freezing) on a 60 ml Affi-Gel Blue column previously equilibrated with 5 vol. HI buffer containing 50 mM KCl.
7. Wash the column sequentially with 5 vol. HI buffer containing 50 mM KCl, 4 vol. HI buffer containing 0.8 M KCl, and 3 vol. HI buffer containing 0.5 M NaSCN.
8. Elute RP-A with 5 vol. HI buffer containing 1.5 M NaSCN.
9. Load the eluate directly on a 20 ml HAP column, previously equilibrated in HI buffer. It is better not to couple the two columns together because of their difference in size.
10. Wash the HAP column with 2 vol. HI buffer.
11. Elute RP-A with five column volumes HI buffer containing 80 mM potassium phosphate; collect 5 ml fractions.
12. Analyse the fractions by SDS–PAGE and pool the RP-A-containing fractions; avoid the *E. coli* SSB contaminating fractions (band at 18 kDa). Check the conductivity of the pool and adjust it to the same conductivity (by diluting with HI buffer) as the subsequent Mono Q equilibration buffer.
13. Load the HAP eluate onto a 1 ml Mono Q column equilibrated in HI buffer containing 100 mM KCl.
14. Wash the column with HI buffer containing 200 mM KCl and elute with a linear gradient of 200–1000 mM KCl (50 fractions of 200 μl). RP-A elutes at about 400 mM KCl.
15. Test the activity of the final RP-A preparation in a replication (see Section 3.2; assay for RF-C) or in an unwinding (see Section 4.2; assay for RP-A) assay. A yield of 0.3 mg homogeneous active RP-A can be expected from a 1 litre culture.

Acknowledgements

The work performed in the authors laboratory has been supported by the Swiss National Science Foundation (grant 31–43138.95) to R. M. and Z. O. J., by the Swiss Cancer League to M. S., and by the Kanton of Zürich to E. F. and U. H.

References

1. Hübscher, U., Maga, G., and Podust, V. N. (1996). In *DNA replication in eukaryotic cells* (ed. M. L. De Pamphilis), p. 525. Cold Spring Harbor Laboratory Press, Cold Spring Harbor, NY.
2. O'Donnell, M., Onrust, R., Dean, F. B., Chen, M., and Hurwitz, J. (1993). *Nucleic Acids Res.*, **21**, 1.
3. Tan, C.-K., Castillo, C., So, A. G., and Downey, K. M. (1986). *J. Biol. Chem.*, **261**, 12310.
4. Prelich, G., Kostura, M., Marshak, D. R., Mathews, M. B., and Stillman, B. (1987). *Nature*, **326**, 471.
5. Bauer, G. A. and Burgers, P. M. J. (1988). *Proc. Natl. Acad. Sci. USA*, **85**, 7506.
6. Ball, K. L. and Lane, D. P. (1996). *Eur. J. Biochem.*, **237**, 854.
7. Fien, K. and Stillman, B. (1992). *Mol. Cell. Biol.*, **12**, 155.
8. Matsumoto, T., Hata, S., Suzuka, I., and Hashimoto, J. (1994). *Eur. J. Biochem.*, **223**, 179.
9. Biswas, E. E., Chen, P.-h., and Biswas, S. B. (1995). *Protein Expression Purification*, **6**, 763.
10. Zhang, P., Zhang, S. J., Zhang, Z., Woessner, J. F. Jr., and Lee, M. Y. (1995). *Biochemistry*, **34**, 10703.
11. Jónsson, Z. O., Podust, V. N., Podust, L. M., and Hübscher, U. (1995). *EMBO J.*, **14**, 5745.
12. Podust, L. M., Podust, V. N., Sogo, J. M., and Hübscher, U. (1995). *Mol. Cell. Biol.*, **15**, 3072.
13. Mossi, R., Jónsson, Z. O., Allen, B. L., Hardin, S. H., and Hübscher, U. (1997). *J. Biol. Chem.*, **272**, 1769.
14. Brand, S. R., Bernstein, R. M., and Mathews, M. B. (1994). *J. Immunol.*, **153**, 3070.
15. Hübscher, U. and Kornberg, A. (1979). *Proc. Natl. Acad. Sci. USA*, **76**, 6284.
16. Studier, F. W. and Moffatt, B. A. (1985). *J. Mol. Biol.*, **189**, 113.
17. Studier, W. F., Rosenberg, A. H., Dunn, J. J., and Dubendorff, J. W. (1990). In *Methods in enzymology* (ed. D. V. Goeddel), Vol. 185, p. 60. Academic Press.
18. Yao, N., Turner, J., Kelman, Z., Stukenberg, P. T., Dean, F., Schechter, D., *et al.* (1996). *Genes Cells*, **1**, 101.
19. Tsurimoto, T. and Stillman, B. (1989). *Mol. Cell. Biol.*, **9**, 609.
20. Yoder, B. L. and Burgers, P. M. (1991). *J. Biol. Chem.*, **266**, 22689.
21. Podust, V. N., Georgaki, A., Strack, B., and Hübscher, U. (1992). *Nucleic Acids Res.*, **20**, 4159.
22. Cullmann, G., Fien, K., Kobayashi, R., and Stillman, B. (1995). *Mol. Cell. Biol.*, **15**, 4661.
23. Gerik, K. J., Gary, S. L., and Burgers, P. M. J. (1997). *J. Biol. Chem.*, **272**, 1256.

24. Cai, J. S., Uhlmann, F., Gibbs, E., Floresrozas, H., Lee, C. G., Phillips, B., *et al.* (1996). *Proc. Natl. Acad. Sci. USA*, **93**, 12896.
25. Podust, V. N. and Fanning, E. (1997). *J. Biol. Chem.*, **272**, 6303.
26. Fotedar, R., Mossi, R., Fitzgerald, P., Rousselle, T., Maga, G., Brickner, H., *et al.* (1996). *EMBO J.*, **15**, 4423.
27. Kelman, Z., Naktinis, V., and O'Donnell, M. (1995). In *Methods in enzymology* (ed. J. Campbell), Vol. 262, p. 430. Academic Press, NY.
28. Wobbe, C. R., Weissbach, L., Borowiec, J. A., Dean, F. B., Murakami, Y., Bullock, P., *et al.* (1987). *Proc. Natl. Acad. Sci. USA*, **84**, 1834.
29. Wold, M. S., Weinberg, D. H., Virshup, D. M., Li, J. J., and Kelly, T. J. (1989). *J. Biol. Chem.*, **264**, 2801.
30. Georgaki, A., Strack, B., Podust, V., and Hübscher, U. (1992). *FEBS Lett.*, **308**, 240.
31. Henricksen, L. A., Umbricht, C. B., and Wold, M. S. (1994). *J. Biol. Chem.*, **269**, 11121.
32. Kenny, M. K., Schlegel, U., Furneaux, H., and Hurwitz, J. (1990). *J. Biol. Chem.*, **265**, 7693.
33. Georgaki, A. and Hübscher, U. (1993). *Nucleic Acids Res.*, **21**, 3659.
34. Thömmes, P., Ferrari, E., Jessberger, R., and Hübscher, U. (1992). *J. Biol. Chem.*, **267**, 6063.

6

Isolation and characterization of lagging strand processing activities

GLENN A. BAUER and THOMAS MELENDY

1. Introduction

As a result of the bidirectional DNA replication process, lagging strand synthesis results in the formation of Okazaki fragments, hybrid polynucleotide molecules consisting of a short 5′ RNA primer and a long DNA extension. Lagging strand processing involves the removal of the RNA moiety from these fragments and the completion of DNA replication to synthesize a continuous DNA daughter molecule on the lagging strand at the replication fork.

1.1 Lagging strand processing in *E. coli*

Two enzyme activities are involved in the removal of the RNA primer from prokaryotic Okazaki fragments: RNase H and a 5′-3′ exonuclease. RNase H has the biochemical property of being an RNA endonuclease which digests an RNA strand that is hybridized to DNA. The 5′-3′ exonuclease activity, while not highly processive, has the ability to remove a ribonucleotide that is esterified to a deoxyribonucleotide (1). Single and double mutants in DNA polymerase I and RNase H show that the 5′-3′ exonuclease activity of DNA polymerase I plays the most essential role in lagging strand processing, whereas RNase H functions in an auxiliary role in removing the RNA moiety from Okazaki fragments (2). With the removal of the RNA primers from the Okazaki fragments, the resultant gaps are resynthesized by DNA polymerase I. DNA ligase connects each DNA fragment to the neighbouring fragment (3).

1.2 Lagging strand processing in eukaryotes

As has been discussed in other chapters in this book, identification of DNA replication proteins in higher eukaryotes has been more difficult due to the lack of an easily manipulable genetic system. Fortunately, the use of an *in vitro* SV40 DNA replication system as a complementation assay has provided the means for identifying the basal cellular factors required for the major synthetic stages of DNA replication (for reviews see refs 4–7). However, since little polynucleotide synthesis is involved, the identification of those factors required to remove the RNA primers and produce contiguous replicated daughter molecules has lagged. Therefore, biochemical studies, designed to identify activities characteristic of known *E. coli* lagging strand processing enzymes, have been used to identify eukaryotic enzymes predicted to be involved in this process during eukaryotic DNA replication.

Nucleotide synthesis on the lagging strand of the simian virus 40 DNA replication forks is initiated by DNA polymerase α/primase, which is responsible for laying down the RNA primer and a short length of initiator DNA (iDNA) (8–12). DNA polymerase α is non-processive and likely dissociates from the iDNA after synthesizing a short 30–40 nucleotide RNA:DNA primer (13). Replication factor C (RF-C) binds to the 3′ end of the iDNA and loads first proliferating cell nuclear antigen (PCNA), and then DNA polymerase δ, onto the DNA, thereby preventing reassociation of DNA polymerase α (8, 10, 14). The synthesis of the Okazaki fragment is completed by the RF-C–PCNA–polymerase δ holoenzyme. The RNA moiety of the Okazaki fragment is removed by the joint action of RNase H1 and the 5′-3′ nuclease, FEN1/RTH1 (15). The RFC–PCNA–polymerase δ holoenzyme proceeds to fill in the resulting gap which is ultimately sealed by DNA ligase I. Although studies using the SV40 *in vitro* DNA replication system appear to indicate that it is DNA polymerase δ that carries out the majority of the DNA synthesis during eukaryotic DNA replication (14), as is discussed elsewhere in this book, DNA polymerase ε may carry out some of these functions during cellular DNA replication (16, 17).

Much of the above model is based upon results from the *in vivo* SV40 DNA replication system. There is still uncertainty in the exact *in vivo* roles that RNase H1 and FEN1/RTH1 play in the removal of the RNA moiety of the Okazaki fragment. This chapter explores the biochemical analysis and purification of the enzymatic activities involved in eukaryotic lagging strand processing and discusses their possible *in vivo* roles.

2. Enzymes involved in eukaryotic lagging strand processing

This section will explore in more detail the three molecular players that carry out the maturation of the lagging strand Okazaki fragments into a single strand of DNA: RNase H1, the 5′-3′ nuclease (known variously as FEN1, MF1, and RTH1), and DNA ligase I.

2.1 RNase H1

RNase H enzymes have been defined as enzymes capable of degrading RNA chains hybridized to DNA (for review see refs 18 and 19). This has been shown to take place through an endonucleolytic mechanism. That is, RNase H enzymes randomly degrade RNA molecules annealed to DNA. The resulting small oligoribonucleotides (5–20 nucleotides long) can then dissociate from the DNA template. The primary function of RNase H enzymes is believed to be the removal of RNA primers from newly synthesized Okazaki fragments. This would appear to be an essential process, since these RNA primed fragments must have their RNA primers removed before the subsequent Okazaki fragment can be ligated to the DNA. Indeed, RNase H enzymes have been shown to be an integral part of the DNA replication machinery in a wide variety of biological systems from bacteriophage, to *E. coli*, to human cells. Interestingly, mutations in yeast that destroy RNase H1 function are not lethal (R. Crouch as reported in ref. 20), suggesting either that other enzymes exist that can carry out this function or that RNase H activity is not absolutely essential for Okazaki fragment processing.

Eukaryotic RNase H enzymes have been divided into two classes: the type 1 enzymes, which exhibit relatively large native molecular weights (68–90 kDa) and can utilize either Mg^{2+} or Mn^{2+} as a divalent cation cofactor; and type II enzymes, which are smaller (~ 30 kDa) and can only use Mg^{2+} as their divalent cation cofactor. Both types of RNase H break down RNA strands that are hybridized to DNA (see *Protocol 1*). It is the first class of enzymes, RNase H1 enzymes, that have been shown to be the RNase enzymes involved in the removal of RNA primers from Okazaki fragments. The purification of RNase H1 from calf thymus is given in *Protocol 2*. It should be noted that although this purification results in a preparation of the bovine type 1 RNase H, the extreme lability of this protein usually results in the purification of a 32 kDa proteolysed form of RNase H1. Experience with purification of other DNA replication enzymes from sources varying from yeast to human suggests that this same protocol should be useful for the purification of RNase H1 from other eukaryotes. However, RNase H1 is noteworthy for its lack of consistency during purification. Hence, care should be taken to evaluate each purification step carefully for the presence of RNase H1 activity.

Protocol 1. RNase H assay[a]

The template consists of short segments of tritiated RNA (poly A) randomly annealed to longer stretches of DNA (poly dT). RNase H activity will digest the annealed poly(A) resulting in the release of tritiated, acid soluble monophosphate nucleotide residues.

Equipment and reagents

- Scintillation counter
- [^{3}H]poly(A): 100 μCi/ml, 500 μCi/μmole (Amersham)
- Poly(dT) (Pharmacia)
- Aquasol scintillation fluid (New England Nuclear)
- Calf thymus DNA (Sigma)
- Trichloroacetic acid (TCA; Sigma)

Protocol 1. *Continued*

A. *Preparation of template*

1. Combine 3500 pmol (nucleoside residue) of [^{3}H]poly(A) (1.75 μCi) and 7000 pmol (nucleoside residue) of poly(dT) in a total volume of 40 μl (use dH_2O for remaining volume).
2. Boil the solution for 1 min and allow to cool for 15 min at 37 °C.

B. *RNase H assay*

1. Combine 2 μmole (of ribonucleotide, ~ 0.05 μCi) of the heteroduplex template (part A) with 50 mM Tris–HCl pH 7.5, 5 mM 2-mercapto-ethanol, 10 mM $(NH_4)_2SO_4$, 10 mM $MgCl_2$, 200 μg BSA, and the enzyme fraction in a 50 μl reaction.
2. Incubate the reaction mixture for 15 min at 37 °C.
3. Add 50 μl calf thymus DNA (1 mg/ml) and 0.2 ml 10% TCA to the reaction mixture.
4. Incubate the resulting solution for 10 min at 4 °C, and centrifuge at 7000 r.p.m. for 10 min in a microcentrifuge.
5. Add 0.2 ml of the supernatant to 4 ml scintillation fluid and measure the soluble counts by scintillation counting.

[a] Adapted from ref. 21.

Protocol 2. Purification of RNase H1[a]

Equipment and reagents

- FPLC Fast Protein Liquid Chromatography system (Pharmacia) or compatible high pressure chromatography system
- Mono Q HR5/5 column, Mono S HR5/5 column (Pharmacia)
- Buffer A: 10 mM Tris pH 8.4, 0.5 mM EDTA, 0.5 mM EGTA, 1 mM DTT
- Buffer B: buffer A plus 15% (v/v) glycerol
- Fetal calf thymus (Pell Freez Biologicals)
- DEAE–Sepharose, CM–Sepharose, phenyl–Sepharose, heparin–Sepharose, Affi-Gel Blue–Sepharose (Pharmacia)
- Cheesecloth (Fisher Scientific)

Method

All procedures should be carried out at 4 °C unless otherwise indicated.

1. Homogenize 400 g fetal calf thymus in 1.2 litres buffer A containing 50 mM KCl.
2. Centrifuge the homogenate at 9000 *g* for 30 min at 4 °C.
3. Recover the supernatant and centrifuge at 100 000 *g* for 60 min at 4 °C.
4. Filter the supernatant (from step 3) through four layers of cheesecloth and apply onto a 500 ml DEAE–Sepharose column equilibrated with buffer A containing 50 mM KCl.

5. Directly apply the fraction that flows through the DEAE–Sepharose column to a 500 ml CM–Sepharose column equilibrated in buffer A containing 50 mM KCl.
6. Elute the protein with a 2.5 litre linear gradient from 50–400 mM KCl.
7. Assay the fractions for RNase H activity (*Protocol 1*), pool the active fractions, and precipitate the protein with $(NH_4)_2SO_4$ at 60% saturation.
8. Collect the protein pellet by centrifugation at 10 000 *g* for 30 min at 4°C and dissolve the pellet in buffer A containing 1 M $(NH_4)_2SO_4$.
9. Clarify the solution by centrifugation (as in step 8), apply the supernatant to a 40 ml phenyl–Sepharose column, and develop with a 200 ml decreasing salt gradient from 1–0.4 M $(NH_4)_2SO_4$ in buffer A.
10. Assay the fractions for RNase H activity (*Protocol 1*), pool the active fractions, and dialyse against ten volumes of buffer B containing 100 mM KCl.
11. Apply the dialysate to a Mono S HR5/5 FPLC column.
12. Apply the Mono S flow-through directly to a Mono Q HR5/5 FPLC column equilibrated in buffer B containing 100 mM KCl.
13. Apply the Mono Q flow-through to a 1 ml heparin–Sepharose column equilibrated in buffer A containing 10% (v/v) glycerol and 100 mM KCl.
14. Develop the column with a 15 ml gradient from 100–600 mM KCl in buffer A containing 10% (v/v) glycerol.
15. Assay the fractions for RNase H activity (*Protocol 1*), pool the active fractions, and dilute the sample with buffer A to a conductivity equal to buffer A containing 100 mM KCl.
16. Apply the diluted sample to a 1 ml blue–Sepharose column equilibrated in buffer A containing 100 mM KCl.
17. Elute the RNase H activity from the column with buffer A containing 500 mM KCl and 2 M $MgCl_2$.
18. Assay the fractions for RNase H activity (*Protocol 1*), pool, and dialyse the active fractions against buffer A containing 50% (v/v) glycerol. Aliquot the dialysate, freeze in liquid nitrogen, and store at –70°C.

[a] Adapted from refs 15 and 21.

It has been shown more recently that RNase H1 has an additional endonucleolytic activity. Using hybrid RNA:DNA chains hybridized to DNA, it was shown that RNase H1 cleaves the template just 5′ to the RNA:DNA junction. Cleavage takes place at the junction regardless of the length of the

RNA chain (15, 22). This suggests strongly that the cleavage is endonucleolytic. More recent studies have shown that the RNA that is removed from the RNA:DNA hybrid remains intact. This study also showed that the same type of cleavage site, just 5′ to the RNA:DNA junction, is selected whether or not the nucleotides around the junction, or anywhere on the entire RNA:DNA hybrid molecule, are base paired (23). This was a surprising finding since the digestion of RNA chains by RNase H had previously been shown to only occur when the RNA substrate was base paired to DNA. These studies indicate that RNase H1 has two distinct activities: a random endonucleolytic activity that removes RNA that is base paired to DNA, and a structure-specific endonucleolytic activity that cleaves just 5′ to the RNA:DNA junction of hybrid molecules and is independent of whether the substrate is base paired. The possible relevance of this activity to Okazaki fragment processing *in vivo* will be discussed in Section 4.

2.2 5′-3′ nuclease (FEN1/RTH1)

The 5′-3′ nuclease that is critical for removing the residual RNA:DNA junction from the RNA primers used for lagging strand DNA replication has been identified several times by laboratories in a number of different ways. Exonuclease assays were used to identify this enzyme from both murine and bovine sources (24, 25). This general biochemical assay measures the digestion of one strand of a duplex DNA template (see *Protocol 4*).

Ishimi *et al.* (26) used a partial SV40 *in vitro* DNA replication assay that required this 5′-3′ exonuclease for the generation of covalently closed circular DNA products. More recently, an SV40 *in vitro* DNA replication assay containing both leading and lagging strand DNA synthesis systems was also used to identify this enzyme (called MF1 in this study) as required for synthesis of covalently closed replicated molecules (27). A purification for human FEN1/RTH1 is provided in *Protocol 5*. As noted above for RNase H1, experience with purification of other DNA replication enzymes from a variety of sources suggests that this same protocol should be useful for the purification of FEN1/RTH1 from other eukaryotic sources.

This exonuclease also has a structure-specific endonuclease activity that will remove unannealed 5′ tails from an otherwise base paired DNA duplex (28). Interestingly, a related 'flap endonuclease' assay designed to identify a nuclease involved in recombination also identified this same enzyme (called FEN1 in this study) (29). It has been noted that the 5′-3′ exonuclease activity may actually be a result of the enzyme's endonucleolytic activity acting upon transiently unpaired 5′ ends of otherwise paired DNA (30).

FEN1 appears to be the same enzyme as the yeast rad 2 homologue 1 (RTH1). It is interesting that a null mutation in RTH1 is temperature-sensitive and hyper-recombinative but not lethal (31), suggesting that alternate pathways may exist for Okazaki fragment processing.

Protocol 3. Preparation of [^{3}H]poly(dA-dT) template for FEN1/RTH1 exonuclease activity[a]

Reagents

- Reaction mixture: 50 mM Tris–HCl pH 7.6, 50 mM potassium acetate, 10 mM $MgCl_2$, 0.1 mM EDTA, 10% (v/v) glycerol, 0.01% NP-40, 0.1 mg/ml BSA, 5 mM dATP, 5 mM dTTP
- Poly(dA-dT) (Pharmacia)
- Sephacryl S-200 (Pharmacia)
- [^{3}H]dATP: 30 Ci/mmol, 1 μCi/μl (Amersham)
- DNA polymerase I (Boehringer Mannheim)
- Proteinase K (Gibco BRL)

Method

1. Concentrate 50 μl (50 μCi) of [^{3}H]dATP down to 20 μl with a SpeedVac.
2. Mix 30 μg poly(dA-dT) with the reaction buffer components containing Tris, potassium acetate, $MgCl_2$, EDTA, and glycerol, and incubate for 5 min at 70 °C. Let sit at RT for 20 min.
3. Add [^{3}H]dATP, dATP, dTTP, BSA, NP-40, and 15 U of *E. coli* DNA polymerase I (in a total reaction volume of 600 μl).
4. Incubate the mixture for 16 min at 37 °C.
5. Add SDS to 0.2%, EDTA to 20 mM, and proteinase K to 0.5 mg/ml. Incubate the mixture for 60 min at 37 °C.
6. Apply 100 μl of the reaction mixture to a 1 ml Sephacryl S-200 spin column equilibrated in TE.
7. Combine the void fractions and extract once with an equal volume of 1:1 phenol:chloroform.
8. Extract sample with an equal volume of chloroform.
9. Ascertain that there is at least 0.2 M NaCl present in the sample and add 2 vol. of ethanol to the sample to precipitate the nucleic acids, mix, and let sit at RT for 15 min.
10. Collect the DNA template by centrifugation at 12 000 r.p.m. for 15 min in a microcentrifuge at RT.
11. Resuspend the DNA template pellet in 100 μl TE.

[a] Adapted from ref. 24.

This template consists of random length base paired segments of alternating dA-dT co-polymer. Following treatment by DNA polymerase I, a small percentage of the dAMP residues become tritiated. FEN1/RTH1 activity digests the annealed polynucleotide (starting at the 5′ end) and releases acid soluble monophosphate nucleotide residues, including some tritiated dAMP residues.

Protocol 4. Exonuclease assay for measuring activity of FEN1/RTH1[a]

Equipment and reagents

- Scintillation counter
- Reaction mixture: 25 mM Tris–HCl pH 7.6, 5 mM $MgCl_2$, 1 mM DTT, 0.1 mM EDTA, 0.1 mg/ml BSA (fraction V), 0.01% (v/v) NP-40, 10% (v/v) glycerol
- FEN1/Rad27 fractions
- Template: 5 pmol/μl (in nucleotides) [^{3}H]poly(dA-dT) (see *Protocol 3*)
- Aquasol scintillation fluid (New England Nuclear)
- Salmon sperm DNA (Sigma)
- TCA (Sigma)

Method

1. Assemble the reaction mixture with the template and add the FEN1/Rad27 fraction (1 μl) in a total reaction volume of 20 μl.
2. Incubate the reaction for 15–30 min at 37 °C.
3. Add 20 μl 1 mg/ml sheared salmon sperm DNA (in TE) and 80 μl ice-cold 10% (w/v) TCA to the reaction. Mix and let the reaction sit on ice for 10 min.
4. Spin the reaction in a microcentrifuge at 12 000 r.p.m. for 10 min at 4 °C.
5. Add 80 μl of supernatant to 4 ml scintillation fluid and measure radioactivity in a scintillation counter.

[a] From S. Waga, adapted from ref. 24.

Protocol 5. Purification of FEN1/MF1[a]

Equipment and reagents

- Ultracentrifuge and a Beckman SW50.1 rotor
- FPLC Fast Protein Liquid Chromatography system (Pharmacia) or compatible high pressure chromatography system
- Hydroxylapatite resin HPHT (Bio-Rad)
- Human 293 cells (human embryo kidney cells transformed with fragments of adenovirus type 5 DNA) propagated in suspension cultures in F-13 medium (Life Technologies) containing 10% calf serum
- Extraction buffer: 15 mM Tris-HCl pH 7.5, 1 mM EDTA, 10% (w/v) sucrose, 400 mM NaCl, 0.5 mM benzamidine, 2 μM pepstatin A , 0.1 mM PMSF (Boehringer Mannheim)
- Buffer S: 25 mM KPO_4 pH 7.2, 1 mM EDTA, 10% glycerol, 0.01% NP-40, 1 mM DTT, 0.5 mM benzamidine, 2 μM pepstatin A, 0.1 mM PMSF
- Buffer H: buffer S without the EDTA
- Buffer Q: 20 mM triethanolamine pH 7.4, 1 mM EDTA, 0.01% NP-40, 10% (v/v) glycerol, 1 mM DTT, 0.5 mM benzamidine, 2 μM pepstatin A, 0.1 mM PMSF
- S-Sepharose resin, Mono Q, and Mono S HR5/5 FPLC columns (Pharmacia)
- Glycerol gradient: 25 mM KPO_4 pH 7.2, 1 mM EDTA, 150 mM NaCl, 1 mM DTT, 0.5 mM benzamidine, 2 μM pepstatin A, 0.1 mM PMSF

Method

All procedures should be carried out at 4 °C unless otherwise indicated.

1. Extract the nuclei from 40 litres of human 293 cells (cell density of 5×10^5 cells/ml) using hypotonic dounce homogenization (30 strokes) and incubate in 200 ml extraction buffer for 30 min at 4 °C.
2. Remove the cellular debris by centrifugation at 12 000 *g* for 20 min.
3. Collect the supernatant and prepare an $(NH_4)_2SO_4$ precipitation at 25% saturation. Centrifuge for 30 min at 13 000 *g* at 4 °C. Recover supernatant and add $(NH_4)_2SO_4$ to 60% saturation. Collect precipitate by centrifugation at 13 000 *g* for 30 min.
4. Resuspend the resulting pellet in 60 ml buffer S and dialyse overnight against buffer S with 25 mM NaCl (fraction II).
5. Apply fraction II to a 2 × 7 cm S-Sepharose column pre-equilibrated in buffer S with 25 mM NaCl.
6. Develop the column with a linear 220 ml NaCl gradient 25–600 mM in buffer S. Assay the fractions for FEN1 nuclease activity (*Protocol 4*). FEN1 elutes at 250 mM NaCl.
7. Pool the active fractions (fraction III) and apply them directly to a 1.4 × 3.8 cm hydroxylapatite column pre-equilibrated in buffer H.
8. Develop this column with a 60 ml linear 25–500 mM KPO_4 gradient in buffer H. Assay the fractions for FEN1 nuclease activity (*Protocol 4*). The activity elutes at 240 mM KPO_4.
9. Pool the active fractions and dialyse them overnight against buffer Q (fraction IV).
10. Apply fraction IV onto a Mono Q HR5/5 column pre-equilibrated in buffer Q.
11. Dialyse the flow-through from the Mono Q column for 6 h against buffer S with 25 mM NaCl (fraction V).
12. Apply fraction V to a Mono S HR5/5 column pre-equilibrated in buffer S with 25 mM NaCl.
13. Develop the column with a 10 ml linear 25–400 mM NaCl gradient in buffer S. Assay the fractions for FEN1 nuclease activity (*Protocol 4*). The activity elutes at 200 mM NaCl (fraction VI).
14. Subject portions of fraction VI to glycerol gradient sedimentation using a 5 ml 15–40% (v/v) glycerol gradient (in buffer S with 150 mM NaCl) in a Beckman SW50.1 rotor for 24 h at 45 000 r.p.m. at 4 °C.

[a] Adapted from ref. 27.

2.3 DNA ligase I

DNA ligases are defined as enzymes that can act to join the 5′ monophosphate of one deoxynucleotide chain to the adjacent 3′ hydroxyl of a deoxynucleotide chain (for review see refs 32 and 33). The most rapid and common assay for ligase activity is to treat hybridized radiolabelled homopolymer with the enzyme or fraction in question, followed by treatment with phosphatase (*Protocol 7*). Ligase activity will, by fusing the 5′ phosphate to the adjacent 3′ hydroxyl, result in a protection of the labelled 5′ phosphate from hydrolysis by phosphatase. The results of this reaction can also be visualized by examination of the nucleic acid products. Following protease digestion and denaturation of the nucleic acids, it is a simple matter to subject the nucleic acids to gel electrophoresis and evaluate whether the shorter DNA chains have been fused to form longer chains (*Protocol 7*, part B).

DNA ligases require a source of energy to carry out each ligation. The *E. coli* replicative ligase utilizes NADH as an energy source. However, most other ligases utilize ATP. Like many ATPase enzymes, DNA ligases can be trapped in the act of hydrolysing ATP. This property can also be used as an assay for DNA ligases. The primary use of this assay is to identify the protein species that is carrying out DNA ligation. In this assay radiolabelled ATP is included in the reaction and all the proteins in the reaction are then denatured and separated on a denaturing protein gel. The denatured ligase can then be identified by autoradiography (*Protocol 8*).

Although several DNA ligases have been identified in eukaryotic cells, biochemical studies using artificial templates (14, 34) or the reconstituted *in vitro* SV40 DNA replication system (27) have indicated that DNA ligase I is the enzyme involved in eukaryotic Okazaki fragment processing.

Protocol 6. Preparation of DNA ligase template for phosphatase-resistance assay[a]

Equipment and reagents

- Sep-Pak C18 cartridge
- Oligo(dT) (average length 16 nucleotides) (Pharmacia)
- Poly(dA) (average length 290 nucleotides) (Pharmacia)
- [γ-^{32}P]ATP: 10 mCi/ml, 5000 Ci/mmol (Amersham)
- T4 polynucleotide kinase (New England Biolabs)

Method

1. Label oligo(dT) (25 pmol nucleotide) with [γ-^{32}P]ATP and T4 polynucleotide kinase as described by the supplier.
2. Purify the labelled oligonucleotide using a Sep-Pak C18 cartridge as described (38).
3. Combine the labelled oligonucleotide and poly(dA) (25 pmol nucleotide) in TE with 0.1 M NaCl in a final volume of 500 μl.
4. Anneal by heating mixture to 95°C for 5 min and allowing to slowly cool to RT.

[a] From S. Waga, adapted from refs 35–37.

The template consists of 5′ phosphate radiolabelled, short (~ 16 nt) dTMP oligonucleotides that are randomly annealed to longer (~ 290 nt) stretches of DNA (poly dA). DNA ligase activity will ligate the 5′ phosphate residues to the 3′ hydroxyl ends of adjacent dTMP oligonucleotides. If the template alone is treated with phosphatase, nearly all the 5′ labelled phosphate will be released as acid soluble product. If the template is treated with DNA ligase activity, those labelled 5′ phosphate residues that are ligated to adjacent oligonucleotides will be protected from digestion by phosphatase. This will result in the presence of additional radioactive material in the acid insoluble nucleic acid fraction.

Protocol 7. Phosphatase-resistance DNA ligase assay[a]

Equipment and reagents

- Scintillation counter
- Ligase reaction buffer: 60 mM Tris–HCl pH 8, 10 mM $MgCl_2$, 5 mM DTT, 1 mM ATP, 50 μg/ml BSA
- TCA (Sigma)
- Calf intestinal phosphatase (Boehringer Mannheim)
- tRNA (Sigma)

A. *Evaluation of phosphatase resistance*

1. Combine ligase fraction and 0.1 pmol of labelled oligo(dT):poly(dA) (see *Protocol 6*) in ligase reaction buffer in 20 μl final volume.
2. Incubate reaction for 15 min at 16°C.
3. Stop reaction by the addition of 20 μl 0.5 M NaCl and 40 μl phenol: chloroform.
4. Extract reaction and recover the aqueous layer.
5. Add 5 μl 10 mg/ml BSA and 1 ml ice-cold 5% (w/v) TCA to the aqueous supernatant.
6. Mix the TCA precipitation mixture and incubate on ice for 5 min.
7. Centrifuge sample at 12 000 r.p.m. for 10 min at 4°C in a micro-centrifuge and discard the supernatant.
8. Rinse the pellet with 500 μl 5% (w/v) TCA.
9. Centrifuge the sample at 12 000 r.p.m. for 10 min at 4°C in a micro-centrifuge and discard the supernatant.
10. Dissolve the pellet in 100 μl 100 mM Tris–HCl pH 8 containing 0.2 U of calf intestinal phosphatase.
11. Incubate the reaction at 37 °C for 30 min.
12. Add 1 μl 10 mg/ml tRNA and 120 μl ice-cold 10% (w/v) TCA to the reaction.
13. Mix the TCA precipitation mixture and incubate on ice for 5 min.

Protocol 7. *Continued*

14. Centrifuge the sample at 12 000 r.p.m. for 10 min at 4°C in a microcentrifuge and discard the supernatant.
15. Place the microcentrifuge tube in a scintillation vial and evaluate the Cherenkov counts in the pellet using a scintillation counter.

B. *Evaluation of ligated products*

1. Carry out ligation reaction as described above in part A, steps 1 and 2.
2. Mix 5 μl of each reaction with 10 μl 95% (v/v) formamide loading dye (38).
3. Heat reactions to 100°C for 5 min.
4. Subject 5 μl of each reaction to separation on a 10% polyacrylamide sequencing gel (38).
5. After electrophoresis fix, dry, and subject gel to autoradiographic analysis.

[a] From S. Waga, adapted from refs 35–37.

Protocol 8. DNA ligase assay—formation of covalent intermediate[a]

Reagents

- Covalent intermediate ligase reaction buffer: 50 mM Tris–HCl pH 7.5, 10 mM $MgCl_2$, 5 mM DTT, 50 μg/ml BSA
- [α-^{32}P]ATP: 3000 Ci/mmol, 10 mCi/ml (Amersham)
- 5 × stop solution: 10% (w/v) SDS, 250 mM Tris–HCl pH 6.8, 25% (v/v) glycerol, 100 mM DTT, 0.05% (w/v) bromphenol blue

Method

1. Combine ligase fraction and 0.1 μl (1 μCi) [α-^{32}P]ATP in covalent intermediate ligase reaction buffer in 20 μl final volume.
2. Incubate the reaction for 5 min at 25°C.
3. Stop the reaction by mixing in 5 μl 5 × stop solution.
4. Heat reactions to 100°C for 3 min.
5. Apply 10 μl of each reaction to a SDS–polyacrylamide minigel with a 10% (w/v) polyacrylamide resolving region (39).
6. Subject the samples to electrophoresis until the dye front reaches the bottom of the gel.
7. Fix and stain the gel with Coomassie brilliant blue using standard staining methods (39).

8. Dry the gel onto Whatman 3MM filter paper using a standard gel drying apparatus.
9. Expose X-ray film to the dried gel for various times until the desired exposure is obtained.

[a] From S. Waga.

These biochemical assays can be used as criteria to identify and purify DNA ligases. A purification for bovine DNA ligase I is provided in *Protocol 9*. As noted above for the other Okazaki fragment processing enzymes, experience with purification of other DNA replication enzymes from a variety of sources suggests that this same protocol should be useful for the purification of DNA ligase I from other eukaryotic sources.

There are several criteria that can be used to differentiate the two most common eukaryotic DNA ligases. The two enzymes have been shown to be immunologically distinct (40). In addition, DNA ligase I and II elute uniquely, at 0.09 M and 0.28 M PO_4 respectively, from hydroxylapatite (41). These ligases can also be differentiated functionally, since only DNA ligase I can effectively ligate blunt-ends (42).

Protocol 9. Purification of DNA ligase I[a]

Equipment and reagents

- Waring blender
- Mono Q HR5/5 FPLC column (Pharmacia)
- FPLC Fast Protein Liquid Chromatography system (Pharmacia) or compatible high pressure chromatography system
- Calf thymus (Pell Freez Biologicals)
- Homogenization buffer: 0.1 M NaCl, 50 mM Tris–HCl pH 7.5, 1 mM EDTA, 0.5 M DTT, with the protease inhibitors (Boehringer Mannheim) 1 mM PMSF, 1.9 μg/ml aprotinin, and 0.5 μg/ml each of leupeptin, pepstatin, chymostatin, and TLCK
- Buffer A: 20 mM NaCl, 50 mM Tris–HCl pH 7.5, 1 mM EDTA, 0.5 mM DTT
- Buffer B: 1 M NaCl, 50 mM Tris–HCl pH 7.5, 1 mM EDTA, 0.5 mM DTT
- Buffer C: 30 mM NaCl, 50 mM Tris–HCl pH 7.5, 1 mM EDTA, 0.5 mM DTT, 10% (v/v) glycerol
- P11 phosphocellulose (Whatman)
- Ultrogel AcA 34 (Pharmacia)
- Hydroxylapatite (Bio-Rad)
- Double-stranded DNA cellulose (Sigma)

Method

All procedures should be carried out at 4°C unless otherwise indicated.

1. Disrupt 2 kg calf thymus tissue by homogenizing small fractions for 3 × 30 sec in a Waring blender with a proportional amount (total 4 litres for 2 kg thymus) of ice-cold homogenization buffer.
2. Stir the resulting slurry for 1 h at 4°C and remove the cellular debris by subjecting to centrifugation at 10 000 *g* for 30 min at 4°C.
3. Dilute the supernatant (fraction I) to a final NaCl concentration of 20 mM by adding four volumes of 50 mM Tris–HCl pH 7.5, 1 mM EDTA, and 0.5 mM DTT.

Protocol 9. *Continued*

4. Absorb the diluted fraction I to 3.5 litres P11 phosphocellulose pre-equilibrated in buffer A and stir gently for 60 min at 4 °C.
5. Wash the phosphocellulose with 10 litres buffer A.
6. Elute the adsorbed proteins with 3 litres buffer A containing 0.5 M NaCl.
7. Slowly dissolve 231 g $(NH_4)_2SO_4$ per each litre of eluate volume adding 1 M Tris base to maintain a pH range of 7.1–7.5.
8. Gently stir the slurry for 30 min and then centrifuge at 10 000 *g* for 30 min at 4 °C.
9. Add additional $(NH_4)_2SO_4$ (160 g/litre) and neutralize as necessary with 1 M Tris base.
10. Collect the precipitate by centrifugation at 10 000 *g* for 30 min at 4 °C. Divide the pellet into four equal portions and freeze rapidly in liquid nitrogen. Store the pellets at –70 °C.
11. Thaw one portion of the second $(NH_4)_2SO_4$ pellet (step 10) and suspend with buffer B to obtain a thick slurry. Dialyse against buffer B for 4 h at 4 °C.
12. Remove insoluble material by centrifugation at 10 000 *g* for 30 min at 4 °C. Recover the supernatant (fraction III).
13. Apply fraction III to a 2.5 × 100 cm Ultrogel AcA 34 column pre-equilibrated with buffer B.
14. Elute the proteins with buffer B. Assay fractions for DNA ligase activity (*Protocol 7*, part A). DNA ligase elutes just before the major protein peak.
15. Pool the active fractions and add K_2HPO_4 to 1 mM (fraction IV).
16. Apply fraction IV to a 1.6 × 29 cm hydroxylapatite column pre-equilibrated in buffer C.
17. Develop the column with 120 ml step elutions of 50, 150, and 400 mM K_2HPO_4 pH 7.5 with 0.5 mM DTT. Assay fractions for DNA ligase activity (*Protocol 7*, part A).
18. Dialyse the active fractions (150 mM eluate, fraction V) against buffer C.
19. Apply fraction V to a 1.6 × 5 cm double-stranded DNA cellulose column, pre-equilibrated with buffer C.
20. Elute the proteins with a 100 ml 0–1 M NaCl gradient in buffer C. Assay fractions for DNA ligase activity (*Protocol 7*, part A).
21. Pool the active fractions (fraction VI), dialyse against buffer C with 50% (v/v) glycerol, aliquot, and store at –70 °C.

[a] Adapted from ref. 36.

3. Reconstitution of lagging strand processing

The currently accepted model of lagging strand synthesis is that DNA polymerase α-primase, with assistance from SV40 Tag and RP-A, synthesizes the RNA primer and the iDNA to prime each Okazaki fragment. Recent studies have indicated that the relatively error-prone DNA polymerase α-primase is likely not involved in the synthesis of a significant portion of the lagging strand (9, 11, 13, 14, 17). The original leading strand DNA polymerase switching model (8) has been invoked as the likely mechanism by which another DNA polymerase (either DNA polymerase δ or ε) becomes the major DNA polymerase involved in Okazaki fragment synthesis (14). RNase H1, FEN1/RTH1, and DNA ligase I then co-operate to remove the RNA primer and ligate each Okazaki fragment to the preceding fragment.

The first reconstitution of lagging strand processing was performed as part of a 'complete' SV40 *in vitro* DNA replication assay that generated covalently closed, newly synthesized plasmid molecules (26). In retrospect, it is now known that this artificial system mimics only lagging strand synthesis and is often referred to as the lagging strand assay. The first reconstitution of both leading and lagging strand replication and processing was accomplished through the use of an SV40 *in vitro* DNA replication system that was wholly dependent upon purified proteins for the generation of covalently closed replicated daughter molecules (27). In this study it was shown that the addition of DNA ligase I and MF1 (FEN1/RTH1) were sufficient to complement the previously described reconstituted SV40 DNA replication initiation and elongation system (8) for the generation of fully replicated covalently closed daughter molecules. This result initially appeared to be in conflict with earlier results from Turchi and Bambara (34). In the Turchi and Bambara (34) study, which used an artificial template assay to mimic Okazaki fragment processing, the results clearly indicated that RNase H1 was also required for this process. Although this discrepancy was initially attributed to possible contaminating RNase H1 activity in other protein preparations, it is also possible that different reaction conditions may have favoured differing Okazaki fragment processing pathways, one of which may not require RNase H1 activity (see Section 4).

Another important observation in the Waga *et al.* (27) study was that levels of DNA ligase I and MF1 (FEN1/RTH1) sufficient for the complete SV40 DNA replication assay, were incapable of ligating Okazaki fragments generated by DNA polymerase α-primase in the artificial lagging strand assay (14). To address this an artificial substrate was designed that reproduces the structural characteristics of two adjacent Okazaki fragments on the lagging strand template. This artificial lagging strand template consists of two primers hybridized to one longer single-stranded DNA template. The first DNA primer is fully hybridized and displays a free 3′ hydroxyl available for the continuation of synthesis by a DNA polymerase. The second, downstream, primer is a fully hybridized RNA:DNA hybrid molecule that has a short

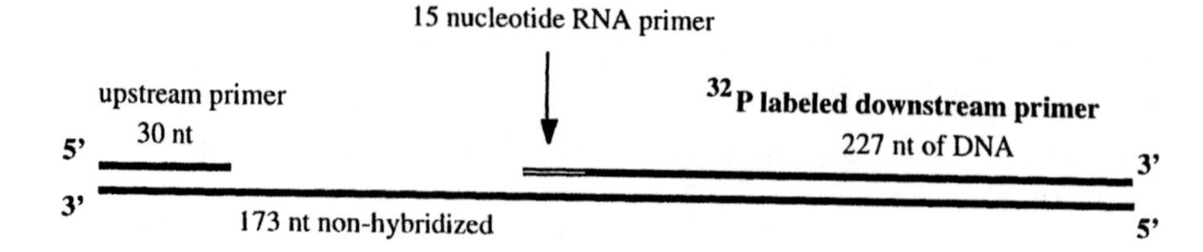

Figure 1. Artificial lagging strand template. The template whose production is described in *Protocol 10* is shown. DNA strands are depicted as solid black lines and RNA strands are depicted as a thin double line. Analysis upon denaturing PAGE will reveal a single labelled entity of 242 nt. If the RNA is specifically removed by chemical or nuclease treatment a 227 nt labelled product will be generated. Extension of the upstream primer followed by ligation will result in a full-length (445 nt) labelled product. If the RNA primer has been properly removed this product will be resistant to digestion by RNase and hydroxide. Extension of the upstream primer will result in an unlabelled 203 nt nascent strand. If the 15 nt RNA primer is removed by nuclease action the primer can be extended to 218 nt. Ligation of this 218 nt strand to the 227 nt labelled DNA strand generates the 445 nt labelled ligation product. Alternative strategies, such as labelling the upstream primer at the 5′ end and synthesizing an unlabelled downstream primer, could be used to evaluate other aspects of Okazaki fragment processing (14).

(15 nucleotide) 5′ RNA segment that is attached to a radiolabelled 3′ DNA segment (see *Figure 1*, construction of this template is described in *Protocol 10*). This template was used to demonstrate that Okazaki fragment processing is highly efficient when the RF-C–PCNA–polymerase δ holoenzyme extends Okazaki fragments but is inefficient when DNA polymerase α extends these fragments (14). Although inefficient, some ligation is seen when DNA polymerase α-primase is the only synthetic DNA polymerase and high levels of FEN1/RTH1 and DNA ligase I are present. Differing levels of these enzymes in the two studies could explain why Ishimi *et al.* (26) observed that DNA ligase I and FEN1/RTH1 were capable of generating covalently closed daughter molecules in the artificial lagging strand assay while Waga *et al.* (27) did not see this result.

Protocol 10. Preparation of artificial lagging strand template[a]

Reagents

- pBluescript SK(–) (Stratagene)
- T3 RNA polymerase (Stratagene)
- T4 DNA polymerase (New England Biolabs)
- Proteinase K (Gibco BRL)
- Glycogen (Gibco BRL)
- Helper phage M12 K07 (Stratagene)
- [α-^{32}P]ATP: 3000 Ci/mmol, 10 mCi/ml (Amersham)
- [α-^{32}P]CTP: 3000 Ci/mmol, 10 mCi/ml (Amersham)
- *Pvu*II (New England Biolabs)
- Superscript reverse transcriptase (Gibco BRL)
- A synthesized 40 nucleotide oligonucleotide designed to hybridize to pBSK(–) ssDNA from nucleotide 948–987

A. *Preparation of template for RNA synthesis*

1. Digest 20 μg pBluescript SK(–) with *Sac*I.
2. Add TE to bring the volume to 100 μl and extract with an equal volume of 1:1 phenol:chloroform.
3. Ascertain that there is at least 0.2 M NaCl present in the sample and add 2 vol. of ethanol to the sample to precipitate the nucleic acids, mix, and let sit at RT for 15 min.
4. Collect the DNA by centrifugation at 12000 r.p.m. for 15 min in a microcentrifuge at RT. Dry the pellet under vacuum for 5 min.
5. Resuspend the pellet and treat the DNA with T4 DNA polymerase (New England Biolabs) and dNTPs to create blunt-ends as per manufacturer's instructions.
6. Extract with phenol:chloroform and ethanol precipitate as described in steps 2–4.
7. Resuspend the DNA in TE and separate on a 0.8% (w/v) agarose gel.
8. Recover the pBSK(–)/*Sac*I blunt-end fragment from the gel.
9. Treat sample with proteinase K to remove RNases.
10. Add TE to bring the volume to 100 μl, extract the DNA first with an equal volume of phenol, then with an equal volume of chloroform.
11. Ethanol precipitate as described in steps 3 and 4 and resuspend the pellet in 20 μl TE.

B. *Synthesis and preparation of RNA primer*

1. Treat approx. 5 μg pBSK(–)/*Sac*I blunt-end fragment in 0.2 ml with 100 U T3 RNA polymerase (Stratagene) as per manufacturer's instructions for 1 h at 37 °C.
2. Extract with phenol:chloroform and ethanol precipitate as described in part A, steps 2–4.
3. Resuspend the pellet in 80% (v/v) formamide in 0.5 × TBE.
4. Heat sample to 100 °C for 3 min.
5. Apply sample to a 20% non-denaturing polyacrylamide gel adjacent to labelled markers and subject to electrophoresis.[b]
6. Expose wet gel to X-ray film for 1 min.
7. Cut out the sample lane of the gel containing the 15 nucleotide region and crush the gel into pieces.
8. Resuspend the gel pieces in several volumes of 1% (w/v) SDS, 2 M ammonium acetate.
9. Agitate the suspension at 37 °C for 4 h.
10. Collect the supernatant and add 1 μl 20 mg/ml glycogen.

Protocol 10. *Continued*

11. Ethanol precipitate as described in part A, steps 3 and 4.
12. Rinse the pellet with 100% ethanol and dry the pellet.
13. Dissolve the pellet in 100 μl H_2O.

C. *Annealing of RNA primer and extension with dNTPs*

1. Prepare single-stranded pBluescript SK(–) DNA, using helper phage M13 K07, as per manufacturer's (Stratagene) instructions.
2. Combine 4 μg single-stranded pBSK(–) DNA, 5 μl RNA primer (see part B), and water to 15 μl final volume.
3. Heat sample to 70°C for 2 min in a heating block and allow the sample and the block to cool to RT for 20 min.
4. Combine the sample with 10 μl (100 μCi) [α-^{32}P]dATP (Amersham), 5 μl 2.5 mM each dCTP, dGTP, and TTP, 500 U Superscript reverse transcriptase (Gibco BRL) in manufacturer's first strand buffer, in a final volume of 47.5 μl.
5. Incubate the reaction for 10 min at RT and 60 min at 37°C.
6. Add 5 μl 2.5 mM dATP to the reaction and incubate for 30 min at 37°C.
7. Extract with phenol:chloroform and ethanol precipitate as described in part A, steps 2–4.
8. Resuspend the pellet in 30 μl TE.

D. *Addition of upstream primer*

1. Combine the 30 μl of the dNMP extended RNA primer (product of part C) with 45 pmol of a 40 nucleotide ssDNA primer designed to hybridize to ssDNA pBSK(–) from nucleotide 948–987 in 40 μl final volume.
2. Heat this reaction to 70°C for 2 min in a heating block and allow block and sample to cool to RT for 20 min.
3. Digest the reaction products with the restriction enzyme *Pvu*II as per manufacturer's instructions.
4. Apply the sample to a 1 × TBE 3.5% polyacrylamide gel (29:1) in a loading buffer that contains xylene cyanol and separate the sample by electrophoresis at 4°C.
5. Expose the wet gel to X-ray film and develop the film.
6. Excise the radioactive band that migrates slightly slower than the xylene cyanol marker dye (remaining radioactive products should remain near the origin of the gel).
7. Crush the gel slice and extract with several volumes of 0.1% (w/v) SDS, 0.5 M ammonium acetate.

8. Agitate the suspension for 4 h at 37 °C.
9. Collect the supernatant, add 1 μl 20 mg/ml glycogen, and ethanol precipitate the nucleic acids as described in part A, steps 3 and 4.
10. Resuspend the pellet in 60 μl TE. Use 0.25 μl of this product for each 10 μl lagging strand DNA replication reaction.

[a] From S. Waga, as outlined in ref. 14.
[b] Radioactive markers can be synthesized using the same procedure as part B, steps 1–4 at 1/20 the volume, and including 2 μCi of [α-^{32}P]CTP in the reaction.

4. Alternative models for lagging strand processing

The apparent dispensability of RNase H1 for Okazaki fragment processing (27), and the reported lack of DNA replication defect in yeast cells without functional RNase H1 (R. Crouch as reported in ref. 20), both suggest that there may be more than one pathway for Okazaki fragment processing. Bambara *et al.* (20) have recently proposed two alternative models for Okazaki fragment processing.

The accepted mechanism for lagging strand processing is that RNase H1 first cleaves the RNA portion of the Okazaki fragment. Although this was previously believed to take place through a random endonucleolytic cleavage of RNA hybridized to DNA, recent work suggests that RNase H1 cleaves just 5′ to the RNA:DNA junction through a structure-specific endonucleolytic cleavage (15, 22). FEN1/RTH1 then would exonucleolytically remove the nucleotides around the RNA:DNA junction. The gap would then be filled in from the 3′ end of the preceding Okazaki fragment by the RFC–PCNA–polymerase δ holoenzyme (or possibly DNA polymerase ε) and the two DNA molecules ligated by DNA ligase I.

The alternative lagging strand processing pathway proposed by Bambara *et al.* (20) is that a DNA polymerase-associated DNA helicase may displace the 5′ end of the Okazaki fragment prior to nuclease action. The recently discovered novel endonuclease function of RNase H1 is highly active on non-base paired RNA:DNA hybrid molecules and could easily cleave the RNA region from the displaced 5′ end of the Okazaki fragment (23). Assuming the displaced region of the Okazaki fragment extended into the DNA region of the molecule, this would create a 5’ DNA flap structure. It was the cleavage of this type of 5′ flap structure that was initially used to identify FEN1/RTH1 as FEN1 (29). Therefore this structure could easily be cleaved by FEN1/RTH1, leaving the newly extended preceding Okazaki fragment (extended by the DNA polymerase associated with the displacing DNA helicase) in the proper position to be ligated to the newly cleaved fragment by the action of DNA ligase I. Since FEN1/RTH1 is capable of removing both the RNA and DNA regions from a helicase-displaced 5′ end (43), this latter Okazaki fragment processing model could explain the lack of a strict requirement for RNase H1, seen both genetically, and in previously described biochemical systems (14, 26, 27).

Although not essential in the alternative lagging strand processing pathway, RNase H1 activity is still likely to be important in this pathway. Rapid, endonucleolytic removal of the RNA region of these lagging strand Okazaki fragments would help prevent formation of occasional stable RNA:DNA hairpins in the displaced tail. This type of structure can effectively block FEN1/RTH1 from acting at the flap junction (44). Another interesting aspect of this alternative processing pathway is that it is more likely to result in the removal of the majority of the iDNA synthesized by the relatively error-prone DNA polymerase α (20).

Sequences in the yeast genome have been identified that encode proteins with some homology to these known processing enzymes, thereby hinting at potential alternative processing pathways. However, with our current understanding of this process, these alternative proteins, if functional in Okazaki fragment processing, would likely act through one of the two mechanisms described above. For a more in depth review of this subject matter, the reader is referred to the recent review by Bambara *et al.* (20).

5. Additional protein–protein interactions

As one of the final groups of cellular DNA replication proteins to be characterized, the Okazaki fragment processing activities are only now beginning to be evaluated with regards to their protein–protein interactions. Most protein interaction studies with the lagging strand processing enzymes relate to apparent specific effects of one protein on the biochemical activity of a second protein. Although more recently, studies in *S. cerevisiae* have begun demonstrating apparent genetic interactions with these factors.

RNase H enzymes have been reported to associate with and stimulate their respective DNA polymerases α from yeast and calf thymus (45, 46); however, the mechanistic reason for such an interaction is unclear. It has recently been shown that PCNA, part of the leading strand complex, stimulates the activity of FEN1/RTH1 tenfold (47). A temperature-sensitive mutant of the DNA2 helicase from *S. cerevisiae*, a protein which has been implicated in DNA replication, has been shown to be complemented by overexpression of FEN1/RTH1 (48). This result suggests that these two proteins may interact and that the DNA2 helicase may be involved in lagging strand DNA replication. Another DNA helicase, helicase E, has been shown to co-purify with both FEN1/RTH1 and DNA polymerase ε and therefore may also be involved in lagging strand DNA replication (49–51). As the Okazaki fragment processing proteins become increasingly available in the near future, it is likely that other important protein–protein interactions will be identified.

Acknowledgements

We would like to thank Drs S. Waga, R. Bambara, R. Murante, and L. Henricksen for scientific protocols, copies of submitted manuscripts, and scientific discussions. Work in the laboratory of T. Melendy is supported by National Institutes of Health Grant GM56406 and a research grant from the American Cancer Society (RPG-9807601-GMC).

References

1. Konrad, E. B. and Lehman, I. R. (1974). *Proc. Natl. Acad. Sci. USA*, **71**, 2048.
2. Ogawa, T. and Okazaki, T. (1984). *Mol. Gen. Genet.*, **193**, 231.
3. Konrad, E. B., Modrich, P., and Lehman, I. R. (1973). *J. Mol. Biol.*, **77**, 519.
4. Hurwitz, J., Dean, F. B., Kwong, A. D., and Lee, S. H. (1990). *J. Biol. Chem.*, **265**, 18043.
5. Melendy, T. and Stillman, B. (1992). In *Nucleic acids and molecular biology* (ed. F. Eckstein and D. M. J. Lilley), Vol. 6, p. 129. Springer–Verlag, Berlin.
6. Bambara, R. A. and Huang, L. (1995). *Prog. Nucleic Acid Res. Mol. Biol.*, **51**, 93.
7. Brush, G. S., Kelly, T. J., and Stillman, B. (1995). In *Methods in enzymology* (ed. J. Abelson and M. Simon), Vol. 262, p. 522. Academic Press, San Diego, CA.
8. Tsurimoto, T., Melendy, T., and Stillman, B. (1990). *Nature*, **346**, 534.
9. Bullock, P. A., Seo, Y. S., and Hurwitz, J. (1991). *Mol. Cell. Biol.*, **11**, 2350.
10. Murakami, Y., Eki, T., and Hurwitz, J. (1992). *Proc. Natl. Acad. Sci. USA*, **89**, 952.
11. Nethanel, T., Zlotkin, T., and Kaufmann, G. (1992). *J. Virol.*, **66**, 6634.
12. Denis, D. and Bullock, P. A. (1993). *Mol. Cell. Biol.*, **13**, 2882.
13. Wang, T. S. (1991). *Annu. Rev. Biochem.*, **60**, 513.
14. Waga, S. and Stillman, B. (1994). *Nature*, **369**, 207.
15. Turchi, J. J., Huang, L., Murante, R. S., Kim, Y., and Bambara, R. A. (1994). *Proc. Natl. Acad. Sci. USA*, **91**, 9803.
16. Linn, S. (1991). *Cell*, **66**, 185.
17. Zlotkin, T., Kaufmann, G., Jiang, Y., Lee, M. Y., Uitto, L., Syvaoja, J., *et al.* (1996). *EMBO J.*, **15**, 2298.
18. Crouch, R. J. and Dirksen, M. L. (1982). In *Nucleases* (ed. S. M. Linn and R. J. Roberts), p. 211. Cold Spring Harbor Press, Cold Spring Harbor, NY.
19. Crouch, R. J. (1990). *New Biol.*, **2**, 771.
20. Bambara, R. A., Murante, R. S., and Henricksen, L. A. (1997). *J. Biol. Chem.*, **272**, 4647.
21. Eder, P. S. and Walder, J. A. (1991). *J. Biol. Chem.*, **266**, 6472.
22. Eder, P. S., Walder, R. Y., and Walder, J. A. (1993). *Biochimie*, **75**, 123.
23. Murante, R. S., Henricksen, L. A., and Bambara, R. A. (1998). *Proc. Natl. Acad. Sci. USA*, **95**, 2244.
24. Goulian, M., Richards, S. H., Heard, C. J., and Bigsby, B. M. (1990). *J. Biol. Chem.*, **265**, 18461.
25. Siegal, G., Turchi, J. J., Myers, T. W., and Bambara, R. A. (1992). *Proc. Natl. Acad. Sci. USA*, **89**, 9377.
26. Ishimi, Y., Claude, A., Bullock, P., and Hurwitz, J. (1988). *J. Biol. Chem.*, **263**, 19723.

27. Waga, S., Bauer, G., and Stillman, B. (1994). *J. Biol. Chem.*, **269**, 10923.
28. Murante, R. S., Huang, L., Turchi, J. J., and Bambara, R. A. (1994). *J. Biol. Chem.*, **269**, 1191.
29. Robins, P., Pappin, D. J., Wood, R. D., and Lindahl, T. (1994). *J. Biol. Chem.*, **269**, 28535.
30. Wu, X. and Lieber, M. R. (1996). *Mol. Cell. Biol.*, **16**, 5186.
31. Sommers, C. H., Miller, E. J., Dujon, B., Prakash, S., and Prakash, L. (1995). *J. Biol. Chem.*, **270**, 4193.
32. Lindahl, T. and Barnes, D. E. (1992). *Annu. Rev. Biochem.*, **61**, 251.
33. Lindahl, T., Prigent, C., Barnes, D. E., Lehmann, A. R., Satoh, M. S., Roberts, E., *et al.* (1993). *Cold Spring Harbor Symp. Quant. Biol.*, **58**, 619.
34. Turchi, J. J. and Bambara, R. A. (1993). *J. Biol. Chem.*, **268**, 15136.
35. Olivera, B. M. and Lehman, I. R. (1968). *J. Mol. Biol.*, **36**, 261.
36. Tomkinson, A. E., Lasko, D. D., Daly, G., and Lindahl, T. (1990). *J. Biol. Chem.*, **265**, 12611.
37. Tomkinson, A. E., Roberts, E., Daly, G., Totty, N. F., and Lindahl, T. (1991). *J. Biol. Chem.*, **266**, 21728.
38. Sambrook, J., Fritsch, E. F., and Maniatis, T. (ed.) (1989). *Molecular cloning: a laboratory manual.* Cold Spring Harbor Press, Cold Spring Harbor, NY.
39. Harlow, E. and Lane, D. (1988). *Antibodies: a laboratory manual.* Cold Spring Harbor Press, Cold Spring Harbor, NY.
40. Soderhall, S. and Lindahl, T. (1975). *J. Biol. Chem.*, **250**, 8438.
41. Soderhall, S. and Lindahl, T. (1973). *Biochem. Biophys. Res. Commun.*, **53**, 910.
42. Elder, R. H., Montecucco, A., Ciarrocchi, G., and Rossignol, J. M. (1992). *Eur. J. Biochem.*, **203**, 53.
43. Murante, R. S., Rust, L., and Bambara, R. A. (1995). *J. Biol. Chem.*, **270**, 30377.
44. Murante, R. S., Rumbaugh, J. A., Barnes, C. J., Norton, J. R., and Bambara, R. A. (1996). *J. Biol. Chem.*, **271**, 25888.
45. Karwan, R., Blutsch, H., and Wintersberger, U. (1984). *Adv. Exp. Med. Biol.*, **179**, 513.
46. Hagemeier, A. and Grosse, F. (1989). *Eur. J. Biochem.*, **185**, 621.
47. Li, X., Li, J., Harrington, J., Lieber, M. R., and Burgers, P. M. (1995). *J. Biol. Chem.*, **270**, 22109.
48. Budd, M. E. and Campbell, J. L. (1997). *Mol. Cell. Biol.*, **17**, 2136.
49. Siegal, G., Turchi, J. J., Jessee, C. B., Myers, T. W., and Bambara, R. A. (1992). *J. Biol. Chem.*, **267**, 13629.
50. Turchi, J. J., Murante, R. S., and Bambara, R. A. (1992). *Nucleic Acids Res.*, **20**, 6075.
51. Turchi, J. J., Siegal, G., and Bambara, R. A. (1992). *Biochemistry*, **31**, 9008.

7

Analysis of telomere replication activities: telomerase and telomere binding proteins

LEA A. HARRINGTON

1. Introduction

This chapter outlines the types of assays currently available for the analysis of two major classes of proteins important for telomere replication: telomerase, and telomere binding proteins. Since many of the other activities important for telomere replication, including DNA polymerases and helicases, are not necessarily telomere-specific, they will not be discussed here.

There have been many recent advances in the detection of telomerase activities *in vitro*. The finding that telomerase activity is detected in a wide variety of human cancers, and is present at very low levels or undetectable in many normal, somatic tissues, has considerably raised interest in the study of telomerase activity regulation (reviewed in ref. 1). In addition, the recent identification and cloning of telomere binding proteins and telomerase components from several organisms has opened the way for a detailed biochemical examination of telomerase structure and function in many organisms (reviewed in refs 2–4).

The aim of this chapter is to direct someone interested in beginning an analysis of telomerase or telomere binding proteins to the techniques best suited to their needs. Most of these techniques are applicable to several organisms, however mammalian activities will be the focus, with reference to other organisms where possible.

2. Telomerase elongation activities

2.1 Considering which telomerase elongation assay to use

The choice of telomerase assay to use depends upon two key considerations:

(a) From which organism is the tissue or cell source you wish to use:
- is the activity abundant or rare?

- is the activity processive or non-processive *in vitro*?
- is the activity as yet uncharacterized?

(b) Is there a need to quantify the activity in your sample:

- do you need to compare total activity between different samples or cell sources?
- will the assay be used to biochemically purify the activity?
- will the assay be used on a crude cell lysate, or a partially purified extract?

The advantages and disadvantages of the 'standard elongation assay', the first *in vitro* assay developed for telomerase activity (5), will be compared to newer, more sensitive techniques such as the telomere repeat amplification protocol (TRAP) (6, 7) and modifications of this method (*Table 1*).

There is one last important consideration before you go ahead. Since other polymerase activities can give spurious results in the telomerase assay, it is important to determine that the reaction products observed are telomerase-specific. This can usually be confirmed if the following criteria apply to the products of the telomerase elongation assay:

(a) The elongation activity is sensitive to protease and ribonuclease digestion.
(b) The elongation products are dependent upon the input telomeric substrate.
(c) The elongation activity shows specific incorporation of telomeric nucleotides.

2.2 Analysis of telomerase activity using the standard elongation assay

In this assay, a single-stranded G-rich oligonucleotide serves as the *in vitro* substrate for telomerase. This assay was first used to identify telomerase in the ciliate *Tetrahymena thermophila* (5, 8, 9), but has subsequently been adapted to mammalian telomerase (10, 11). A typical standard elongation assay using partially purified *Tetrahymena* cell extract is shown in *Figure 1*. In *Tetrahymena*, for example, the telomeric repeat is TTGGGG (12). The enzyme is incubated with a telomeric primer such as d(TTGGGG)$_3$, in the presence of dTTP and [α-^{32}P]dGTP (5, 8). Using its telomerase RNA sequence 5′-CAACCCCAA-3′ as a template (9), the enzyme adds TTGGGG, one nucleotide at a time, to the 3′ end of the telomeric primer (13). Repeated realignment (termed 'translocation') between the primer and the telomerase RNA template results in the synthesis of hundreds of telomeric repeats in 1 h at 30 °C (13). After extraction of the telomerase products and resolution on a denaturing acrylamide gel, the ladder of telomerase extension products can be visualized by autoradiography (*Figure 1*). Since the assay is quantitative, and detects each nucleotide added by telomerase, this is the method of choice

Table 1. Comparison between assays used to detect telomerase activity *in vitro*

Method	Sensitivity	Quantitative	Processivity	Advantages	Applications (see text for refs)
SEA	~ 10^5 cells	Yes	Yes	• Detects quantitative activity differences • Detects differences in enzyme processivity • Detects each nucleotide added to primer	• Ciliates • Yeast • *Xenopus* • Mammals
TRAP	< 100 cells	No	No	• Detects low activity • Useful for non-processive enzymes	• Humans • Mouse • Yeast • Plants
TRAP–SPA	< 100 cells	No	No	• Handles large sample numbers (i.e. 96) • No gel required to detect activity	• Large scale screening of telomerase activity
TRAP–ITAS	< 100 cells	Semi	No	• More quantitative than TRAP • Can detect activity in embedded tissue sections	• Internal standard • Fluorescence-based • Embedded sections
Stretch–PCR	~ 1000 cells	Semi	Semi	• Semi-quantitative for activity levels and processivity	

Figure 1. Detection of *Tetrahymena* telomerase activity using the standard elongation assay. Lane 1: ^{32}P-5′-end-labelled input telomeric primer $(GGGGTT)_3$, indicated at left with an *arrow*. Lane 2: approx. 0.1 µg unlabelled telomeric primer was incubated with partially purified *Tetrahymena* telomerase (as in ref. 47) and assayed according to the standard elongation assay in *Protocol 3*.

for a detailed dissection of the biochemistry of the telomerase elongation mechanism (reviewed in ref. 4).

The standard elongation assay has been successfully used to characterize the telomerase elongation reaction and other associated telomerase activities, such as the ability of telomerase to recognize and elongate short telomeric sequences, and an endonucleolytic activity (reviewed in ref. 4). It has also been used in the purification and cloning of telomerase RNA and protein components (14–18). In addition, the assay has been employed by Autexier, Greider, and colleagues to develop an *in vitro* reconstitution assay for ciliate and human telomerases (19, 20).

The only major disadvantage of this technique is that it is not very sensitive, and therefore it can be difficult to detect telomerase activity in crude cell lysates where telomerase activity is rare, or is non-processive *in vitro*. Human telomerase activity, although less abundant than in ciliates, is processive *in vitro*, and can be detected in S100 cell extracts using the standard elongation assay (10, 21) (*Protocols 1* and *3*). In mouse, *Xenopus*, and yeast, however, the activity is usually non-processive, and only relatively

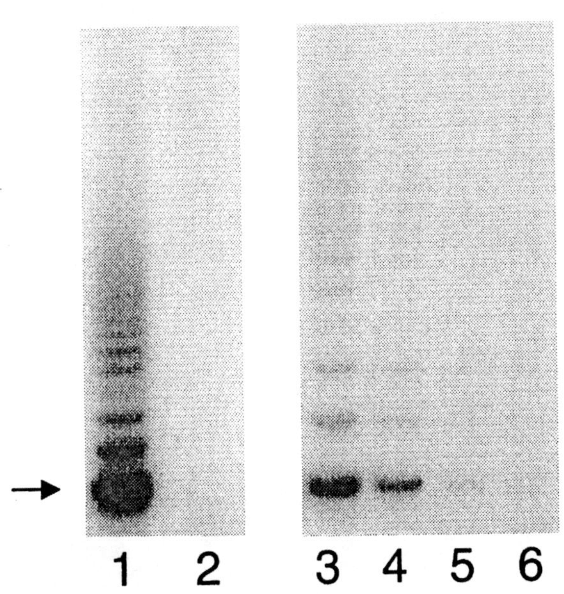

Figure 2. Detection of murine telomerase activity using the telomeric repeat amplification protocol (TRAP). S100 lysates from a mouse immortalized neuroblastoma cell line (N2A) were assayed according to the TRAP method in *Protocol 5*. Lanes 1 and 2: approx. 1 µg S100 cell lysate was incubated without (lane 1) and with (lane 2) 0.1 µg ribonuclease A prior to the TRAP reaction. Lanes 3–6 (a different experiment, therefore banding pattern varies slightly from that in lane 1): the TRAP protocol carried out on 1 µg, 0.5 µg, 0.1 µg, and 0.05 µg S100 cell lysate, respectively. The *arrow* at left indicates the position of the first telomeric repeat PCR amplification product.

short (~ 6–12 nt) extension products are detected (11, 22–26). In this case, partial purification of telomerase, for example by anion exchange chromatography (*Protocol 2*) or other purification steps has improved detection (11, 25, 26).

Protocol 1. Preparation of S100 cell extracts (5, 10, 21)

Equipment and reagents

- 7 ml Dounce homogenizer with B pestle (25–55 μm clearance)
- Rubber policeman (if cells to be harvested are adherent)
- Beckman Ti50 rotor and compatible tubes
- Beckman ultracentrifuge or comparable model
- Ice-cold PBS: 2.7 mM KCl, 1.5 mM KH_2PO_4, 137 mM NaCl, 8 mM Na_2HPO_4
- Ice-cold 2.3 × hypotonic buffer (HB): 1 × HB is 10 mM Hepes pH 8, 3 mM KCl, 1 mM $MgCl_2$, 1 mM DTT, 0.1 mM PMSF, 10 U/ml RNasin, 1 μM leupeptin, 10 μM pepstatin A

Method

1. Collect approx. 6×10^6 cells by centrifugation (if cells are adherent, scrape from plates using rubber policeman) for 10 min at 1800 r.p.m. (500 *g*) at 4 °C with no brake.
2. Rinse the cell pellet twice in ice-cold PBS by centrifugation for 3 min at 2000 r.p.m. at 4 °C.
3. Rinse the cells once in ice-cold 2.3 × HB, and centrifuge as in step 2.
4. Resuspend the pellet in 0.75 vol. (relative to volume of cell pellet) 2.3 × HB, incubate on ice for 10 min.
5. Transfer the cells to an ice-cold 7 ml Dounce homogenizer. Homogenize on ice using a B pestle.
6. Incubate the extract on ice for 30 min. Transfer the extract to a cold 1.5 ml Eppendorf tube, and centrifuge for 10 min at 10 000 r.p.m. at 4 °C.
7. Add 0.02 vol. 5 M NaCl, and transfer the extract to a 3.8 ml Beckman polycarbonate or polyallomer tube (or other tube compatible with a Beckman Ti50 rotor).[a]
8. Centrifuge for 1 h at 38 000 r.p.m. (100 000 *g*) at 4 °C in a Beckman Ti50 rotor. Add glycerol to a final concentration of 20% (v/v), aliquot to 1.5 ml screw-cap tubes (Starstadt), and store at –80 °C. A typical lysate concentration is approx. 4 mg/ml.

[a] If the extract volume is too small, 0.8 ml polyallomer tubes and plastic adaptors for the Ti50 rotor can be purchased from Beckman.

Protocol 2. Fractionation of telomerase activity by DEAE agarose chromatography[a] (11)

Reagents

- Diethylaminoethyl (DEAE) agarose (Bio-Rad)
- 2.3 × hypo buffer plus protease inhibitors and RNasin (2.3 × HB) (see *Protocol 1*)

Method

1. Pre-equilibrate a 1 ml DEAE agarose column in 0.1 M NaCl in 2.3 × HB.
2. Load 3–5 mg of an S100 extract onto the column, collect flow-through.
3. Wash the column with 3 ml 0.1 M NaCl in 2.3 × HB, collect wash fractions.
4. Elute the column with 3 ml each of 0.2 M NaCl in 2.3 × HB, 0.3 M NaCl in 2.3 × HB. Collect all fractions.
5. Save 2–20 µl for a telomerase elongation assay (*Protocol 3*), to determine which eluted fraction contains the majority of telomerase activity (for subsequent purification or study). Use the remainder immediately in subsequent purification steps, or quick-freeze in liquid nitrogen, and store at –80°C.

[a]Useful when enzyme activity is difficult to detect in crude S100 cell lysate. Other columns may be similarly used to partially purify telomerase.

Protocol 3. Analysis of telomerase activity using the standard elongation assay[a,b] (5, 10, 21)

Equipment and reagents

- 8% (w/v) denaturing acrylamide sequencing gel and sequencing gel apparatus
- 10 × standard elongation assay (SEA) buffer: 20 mM dATP, 20 mM dTTP, 10 mM $MgCl_2$, 10 mM spermidine, 50 mM 2-mercaptoethanol, 500 mM potassium acetate, 500 mM Tris–acetate pH 8.5
- d(TTAGGG)$_3$ DNA primer (available by order from any oligonucleotide synthesis service)
- [α-^{32}P]dGTP: 400 Ci/mmol (Amersham)
- Stop buffer: 20 mM EDTA, 10 mM Tris–HCl pH 7.5, 0.1 mg/ml ribonuclease A
- Proteinase K solution: 0.3 mg/ml proteinase K in 10 mM Tris–HCl pH 7.5, 0.5% (w/v) SDS (optional)
- Gel loading dye: 100% formamide, 0.25 mg/ml bromphenol blue and xylene cyanol
- 1 × TBE

Method

1. Mix 4 µl 10 × SEA buffer with 1 µM d(TTAGGG)$_3$ DNA primer, and 1–5 µl (10–50 µCi) [α-^{32}P]dGTP. Bring to 20 µl, and add 20 µl S100 cell extract (*Protocol 1*), or partially purified telomerase extract (*Protocol 2*).
2. Incubate the reaction at 30°C for 1 h. Terminate the reaction with 50 µl

Protocol 3. *Continued*

stop buffer, and incubate at 37 °C for 15 min. As an optional step, incubate with 50 μl proteinase K solution for 10 min at 37 °C.

3. Extract the DNA products with 1 vol. of 1:1 phenol:chloroform, and precipitate with 40 μl 2.5 M ammonium acetate, 4 μg tRNA, and 500 μl ethanol.
4. Precipitate the DNA by centrifugation at 10000 r.p.m. for 10 min. Remove the ethanol, dry the pellet, and resuspend in 3 μl gel loading dye.
5. Boil the samples for 1 min, chill on ice, and load onto a pre-run, 8% (w/v) denaturing polyacrylamide gel at 1500 V for approx. 2.5 h, or until bromphenol blue dye is approx. 1–2 inches from bottom edge of plate.
6. Dry the gel on a vacuum gel dryer, and expose to film (XAR-5, Kodak) for one to two days or a PhosphorImager screen overnight.

[a] To confirm the ribonuclease sensitivity of the elongation products, typically each test sample is assayed in the absence (as in step 2) and in the presence of DNase-free ribonuclease A (Boehringer Mannheim). Incubate 0.01–100 ng RNase A with 20 μl telomerase extract for 10 min at 30 °C, prior to the addition of SEA mixture. To confirm that the activity is also protease-sensitive, incubate 20 μl of extract with 0.3 mg/ml proteinase K in 0.5% SDS for 10 min at 37 °C prior to addition to the SEA mixture (5, 10, 21).

[b] The substrate specificity of the telomerase elongation activity can be tested using several primers of different sequence (reviewed in ref. 4). There should be little or no detectable telomerase activity upon omission of input primer, or incubation with C-rich oligonucleotide primers. Finally, the incorporation of nucleotides should be specific to the telomeric sequence of the organism used as the source of telomerase. For example, human telomerase activity should only give telomerase products when TTP, dATP, and dGTP is in the reaction mixture (one nucleotide, usually dGTP, is radiolabelled). For example, addition of radiolabelled dCTP, in the presence of other unlabelled nucleotides, should not yield radioactive extension products with wild-type telomerase. More detailed analyses using chain termination to determine the exact sequence of the nucleotides added in the standard elongation assay, are discussed elsewhere (4).

2.3 Analysis of telomerase activity using the polymerase chain reaction (PCR)

For most other applications, the telomeric repeat amplification protocol (TRAP) method is most commonly used (6, 7). Since telomerase can elongate oligonucleotides with non-telomeric 5′ ends (27, 28), the usual telomeric substrate (as in the standard elongation assay) can be replaced with a primer that contains unique 5′ sequence followed by a short telomeric sequence (6, 7). This modification allows the subsequent amplification of the telomerase elongation products using the unique 5′ sequence as an anchor primer (e.g. the TS primer in *Protocol 5*), and a complementary telomeric sequence as the other primer (e.g. the CX primer in *Protocol 5*). Because the initial telomerase extension products are subsequently amplified by the polymerase chain reaction (PCR), it is at least 1000 times more sensitive than the standard elongation assay. Mouse telomerase activity, for example, which is normally not processive and difficult to detect, can readily be detected in mouse S100 lysates by this technique (*Figure* 2). The TRAP method (*Protocol* 5), and modified versions that include an internal PCR amplification standard (*Protocol* 6), possess the following advantages:

(a) The method is very sensitive.
(b) It can detect activity in crude cell lysates or in purified fractions.
(c) The processivity of the enzyme is not a consideration.
(d) A semi-quantitative comparison can be made between samples or cell sources.

The disadvantages of this technique, and how they can be partly overcome, are outlined below:

(a) The spurious inhibition of PCR amplification in crude cell lysates can lead to a falsely negative result. This can be overcome by including an internal telomerase amplification standard (actually an internal standard for *Taq* polymerase), called ITAS (see *Protocol* 6). Amplification of the ITAS DNA may utilize one or more primers in common with the TRAP primers, or it may utilize novel primers that are included after the telomerase elongation reaction. Even in a sample without telomerase activity, the standard serves as a normalization control for the PCR amplification. Therefore, the amount of amplification of this product can be used to normalize the potential difference in *Taq* inhibition between samples. In addition, a titration of cell extract (for example, from 5–0.05 μg) can be compared for each sample: whereas the amount of ITAS amplification should remain the same, the amount of telomerase activity will decrease.
(b) The PCR amplification products do not accurately reflect the distribution of telomerase elongation products, or quantitatively represent the amount of telomerase activity in the lysate. The inclusion of an internal amplification standard, as discussed above, can be used to normalize the amplification of telomerase products between samples and different lysate concentrations. This has somewhat improved the linearity of the TRAP assay. Modifications of the TRAP assay (*Protocols 5–8*) have also incorporated mismatched sequence into the C-rich primer used in the subsequent PCR amplification; this modification has also reportedly increased both the linearity of the TRAP products, and better reflects the distribution of telomerase elongation products after the initial elongation step.
(c) The formation of dimers between the two oligos used in the PCR amplification (TS and CX) can lead to a PCR product that is not ribonuclease-sensitive, and therefore falsely positive. This can be overcome by either mismatches in the CX amplification primer (as in *Protocol* 5,

where the CX primer is intentionally mismatched at three residues), or by 'hot-start' PCR. The PCR amplification products should therefore be dependent on input lysate, and should also be sensitive to pre-incubation of the lysate with ribonuclease and proteinase K.

(d) The TRAP method does not directly assay the actual telomerase elongation products. If you are undertaking a detailed analysis of the biochemistry of the telomerase elongation reaction, the standard elongation assay is therefore the method of choice.

The preparation of crude telomerase lysates has also been modified in the TRAP assay, utilizing a detergent lysis to allow the preparation of lysates from relatively few cells (6, 7). *Protocol 4* outlines this modified S100 protocol; however, the telomerase lysates prepared in *Protocols 1* and *2* can also be directly assayed in the TRAP and TRAP–ITAS methods (*Protocols 5* and *6*).

Protocol 4. Preparation of a CHAPS cell lysate (6, 7)

Equipment and reagents

- Microultracentrifuge, or ultracentrifuge with 0.8 ml adaptors, as in *Protocol 1*
- PBS (see *Protocol 1*)
- Ice-cold wash buffer: 10 mM Hepes pH 7.5, 1.5 mM $MgCl_2$, 10 mM KCl, 1 mM DTT
- Ice-cold CHAPS lysis buffer: 10 mM Tris–HCl pH 7.5, 1 mM $MgCl_2$, 1 mM EGTA, 0.1 mM PMSF, 5 mM 2-mercaptoethanol, 0.5% (w/v) CHAPS (Pierce), 10% (v/v) glycerol

Method

1. Wash cells once with PBS by centrifugation at 1000 *g* for 1 min at 4 °C.
2. Wash cells once with ice-cold wash buffer, as in step 1.
3. Resuspend cells at 10^4–10^6 cells per 20 μl ice-cold CHAPS lysis buffer.
4. Incubate the suspension on ice for 30 min.
5. Transfer the suspension to a 1.5 ml microultracentrifuge tube or similar ultracentrifuge tube (Beckman 0.8 ml tube, if using Ti50 rotor as in *Protocol 1*), and centrifuge at 100 000 *g* for 30 min at 4 °C.
6. Aliquot supernatant, quick-freeze, and store at –80 °C.

Protocol 5. Analysis of telomerase activity using the telomere repeat amplification protocol (TRAP) (6, 7)[a]

Equipment and reagents

- Thermal cycler and 0.5 ml PCR amplification tubes
- 15% (w/v) non-denaturing acrylamide gel, in 0.5 × TBE, and electrophoresis apparatus
- Ampliwax (Perkin Elmer)[b]
- *Taq* DNA polymerase (Boehringer Mannheim)
- 0.5 × TBE

- TS oligonucleotide: 5′-AATCCGTCGAGCA-GAGTT-3′
- 10 × TRAP assay buffer: 200 mM Tris–HCl pH 8.3, 15 mM $MgCl_2$, 630 mM KCl, 0.05% (v/v) Tween 20, 10 mM EGTA, 500 μM each of dGTP, dATP, TTP, dCTP, 1 mg/ml BSA, 10 μg T4g32 protein (Boehringer Mannheim)[c]
- CX oligonucleotide: 5′-CCCTTACCCTTACC-CTTACCCTAA-3′
- Non-denaturing loading dye: 30% (v/v) glycerol, 0.25% (w/v) bromphenol blue, 0.25% (w/v) xylene cyanol
- [α-^{32}P]dCTP or [α-^{32}P]dGTP: 3000 Ci/mmol, 10 μCi/μl (Amersham)

Method

1. Lyophilize 0.1 μg CX primer onto the bottom of a 0.5 ml PCR amplification tube, and seal it with 7–10 μl molten Ampliwax. Allow the wax to solidify at room temperature.[b]
2. For each reaction, prepare a 50 μl mix containing 5 μl 10 × TRAP assay buffer, 0.1 μg TS oligo, 2 U *Taq* DNA polymerase in a 0.5 ml PCR tube.
3. Add 0.2–0.4 μl [α-^{32}P]dCTP or [α-^{32}P]dGTP.
4. Add 1–2 μl CHAPS cell lysate (0.1–5 μg). Incubate at room temperature for 10 min.
5. If not using an oil-free thermal cycler, layer mineral oil on top of sample. Carry out PCR amplification for 27 cycles of 94 °C for 30 sec, 50 °C for 30 sec, and 72 °C for 1.5 min.
6. Add non-denaturing load dye to 1 ×, load up to one-half the sample on a 15% (w/v) non-denaturing acrylamide gel until bromphenol blue dye has migrated 8–12 inches from the wells.

[a] For elongation by telomerases from both *Tetrahymena* and human, the substrate primer need only contain a few telomeric nucleotides at the primer 3′ end (27, 28). Thus, telomerases from other organisms may also utilize the TS primer (e.g. ref. 34).
[b] To avoid the requirement of an Ampliwax barrier to separate the CX and TX primers, omit step 1, perform steps 2–4, and add 0.1 μg CX primer just prior to step 5.
[c] T4g32 protein may be omitted.

Protocol 6. Analysis of telomerase activity using the TRAP assay and an internal telomerase amplification standard (TRAP–ITAS) (29)

Equipment and reagents

- See *Protocol 5*
- TS-overlap primer 5′-AATCCGTCGAGCAG-AGTTGTGAATGAGGCCTTC-3′
- CX-overlap primer 5′-CCCTTACCCTTACCC-TTACCCTAATAGGCGCTCAATGTA-3′

Method

1. Prepare a CHAPS cell lysate as described in *Protocol 4*, except increase the Tween 20 detergent to 0.5% (v/v).[a]

Protocol 6. *Continued*

2. Carry out the TRAP assay as described in *Protocol 5*, with the following modifications:[a]
   (a) Increase the incubation time of the extract with TS primer to between 30 min to 1 h.
   (b) Decrease the concentration of unlabelled nucleotides to 25 μM.
   (c) Increase the number of PCR cycles to 34 cycles.
   (d) Add 30 μM of TS-overlap and CX-overlap primers to the PCR mix above the wax barrier.[b,c]

[a] The modifications in steps 1 and 2 are designed to increase the sensitivity of the TRAP assay to as few as one to three cells (29).
[b] The TS-overlap and CX-overlap primers are added to the TRAP assay for the first use: the PCR product that is generated may be gel purified, and 5 attograms used in each subsequent TRAP assay.
[c] If not using a wax barrier, add the ITAS standard (or primers) after the telomerase elongation step, just prior to the PCR amplification.

At least two telomerase detection kits that employ a modification of the TRAP assay are now commercially available (Boehringer Mannheim, and Oncor Inc.).

A detailed protocol outlining the modification of TRAP plus ITAS using a fluorescently labelled TS primer has also been described (30). In this non-radioactive assay, the fluorescently tagged TRAP products are resolved on an ALF DNA sequencer II apparatus, and analysed using Fragment Manager software (Pharmacia). Finally, a modification of this assay has also be adapted for the analysis of telomerase activity in single 5 μm embedded cryostat sections (31). This method is particularly useful for tumours or other such samples where it is not possible to prepare a tissue homogenate in parallel with the embedded sample.

The scintillation proximity assay is a further improvement on the TRAP assay, taking advantage of the ability of low energy β-emitting isotopes, such as tritium, to cause emission of a nearby fluor (32). When adapted to the TRAP assay, as outlined in *Protocol 7*, this method allows the large scale analysis of telomerase activity in 96-well plates, without requiring resolution of the products on a gel, or radiographic imaging.

Similar to the controls for the specificity of the standard elongation assay, the TRAP assay should not yield telomerase products when the lysate has been pre-incubated with ribonuclease or proteinase K, or when either the TS or CX primer has been omitted. The amplification of telomerase reaction products should also be specific to the nucleotides present in the forward elongation reaction, as in the standard elongation assay. In this case, additional nucleotides are added after the initial telomerase elongation step, to allow for PCR amplification of the products. To confirm the primer specificity of the telomerase products, the TS oligo can be replaced with other G-rich primers. The relative banding pattern of the PCR products may vary with the length of the forward elongation primer and where the primer aligns with the RNA template. However, the spacing between the PCR products should NOT depend upon altering either the CX or TS primer. That is because the spacing between the products (i.e. 6 nt) is a function of the telomerase elongation products, and not the substrate or the amplification conditions. As an example, the TRAP protocol has recently been used to identify telomerase activity in plants (33–35). In the case of characterizing a completely new telomerase activity, these controls were helpful in determining the primer specificity of the plant telomerases from different species (34).

Protocol 7. Analysis of telomerase activity using the TRAP method and the scintillation proximity assay (TRAP–SPA) (32)

Equipment and reagents

- See *Protocol 5*
- MicroBeta scintillation counter (Wallac) or comparable scintillation counter
- 5′ biotinylated CX primer (Biot-CX)
- [Me-^{3}H]TTP: 114 Ci/mmol (Amersham)
- Streptavidin coated fluoromicrospheres, diluted 1:4 in 0.56 M EDTA

Method

1. Prepare the cell lysate as in *Protocol 4* (CHAPS), adjust the lysate to 5×10^3 cells/ml, and clarify by centrifugation for 30 min at 15 000 r.p.m.
2. Use approx. 1×10^4 cells (2 μl) per reaction.
3. Carry out the TRAP reaction in *Protocol 5* with the following modifications:[a,b]
   (a) Replace the CX primer with a 5′ biotinylated version (Biot-CX).
   (b) Reduce the concentration of TTP to 2 μM (dATP, dCTP, dGTP remain at 50 μM).
   (c) Replace [α-^{32}P]dGTP with 2 μCi [Me-^{3}H]TTP.
4. After the PCR amplification, mix the 40 μl reaction with 50 μl of streptavidin coated fluoromicrospheres and incubate for 10 min at 37 °C to allow binding of the biotinylated, ^{3}H-labelled reaction products to the streptavidin beads.
5. Count the microtitre plate (or 0.5 ml assay tube) using a MicroBeta scintillation counter (Wallac).

[a] The T4g32 protein may be omitted.
[b] To perform the PCR without a wax barrier (e.g. if TRAP–SPA assays are in a 96-well microtitre plate), add the Biot-CX primer just prior to the 31 cycle amplification, in an oil-free thermal cycler.

A recent adaptation of the PCR-based method for the amplification of telomerase extension products has been developed, called 'Stretch–PCR' (36). In principle, the assay is similar to TRAP, except that the 3′ amplification primer also contains unique sequences at its 5′ end. During the first PCR cycle, the 5′ unique overhang does not hybridize with the telomeric repeats, and the primer aligns randomly along the telomerase reaction products. However, because of the high annealing temperature (68 °C) during the PCR, in subsequent cycles the primer will selectively anneal at the ends of previously amplified products that already contain the unique 5′ sequence. This apparently results in a more accurate reflection of the distribution of telomerase products after the first elongation step. In addition, the linearity of the PCR products with varying lysate concentration is increased, with fairly linear results being obtained over a 10^5-fold cell dilution range (36).

Protocol 8. Analysis of telomerase activity using Stretch–PCR (36)

Equipment and reagents

- Thermal cycler
- 1 × hypo buffer (see *Protocol 1*)
- *n*-Nonanoyl-*N*-methyl-glucomide (MEGA-9)[a]
- 2 × elongation buffer: 100 mM Tris–potassium acetate pH 8.5, 1 mM dATP, 1 mM TTP, 1 mM dGTP, 10 mM 2-mercaptoethanol, 2 mM $MgCl_2$, 2 mM EGTA, 0.1 mg/ml BSA, 2 mM spermidine, 0.2 mM spermine
- TAG-U DNA primer, sequence 5′-GTAAAA-CGACGGCCAGTTTGGGTTGGGTTGGGTTG-3′[b]
- CTA-R DNA primer, sequence 5′-CAGGAA-ACAGCTATGACCCCTAACCCTAACCCTAACCCT-3′
- PCR buffer: 20 mM Tris–HCl pH 8.3, 75 mM KCl, 1.5 mM $MgCl_2$, 0.05% polyoxyethylene ether W-1
- dNTP mix: 10 mM each of dATP, dGTP, TTP, and 1 mM dCTP
- [α-^{32}P]dCTP: 3000 Ci/mmol
- *Taq* polymerase mix: 10 μl PCR buffer, 0.2 μl dNTP mix, 0.3 μl 5 U/μl *Taq* polymerase, 0.5 μl [α-^{32}P]dCTP

A. *Preparation of S100 cell extracts*[a]

1. Harvest 5 × 10^6 cells as in *Protocol 1*, steps 1–3.
2. Resuspend cells in 1 × HB containing 0.5% (w/v) MEGA-9 and incubate on ice for 20 min.
3. Clarify the lysate in a 1.5 ml Eppendorf tube at 10 000 r.p.m. for 10 min at 4 °C. Add 0.02 vol. 5 M NaCl, and incubate the extract at 4 °C for 20 min with mild shaking.
4. Prepare a 100 000 *g* supernatant (S100) as described in *Protocol 1*, step 8.

B. *Stretch–PCR*

1. Mix 20 μl of the 2 × elongation buffer with 20 pmol denatured TAG-U, and add 20 μl S100 extract as prepared in part A. Incubate at 30 °C for 60 min.

2. Terminate the reaction and isolate the DNA products as in *Protocol 3*, steps 2–4, except do not add tRNA carrier during the ethanol precipitation.
3. Redissolve the dried DNA pellet in 39 μl PCR buffer containing 5 pmol CTA-R primer. Overlay the sample with mineral oil, heat the mixture to 95°C for 5 min, and cool to 80°C.
4. Add 11 μl of a pre-heated (80°C) *Taq* mix, and carry out PCR for 25 cycles of 93°C for 1 min, 68°C for 1 min, 72°C for 2 min, followed by one cycle of 72°C for 10 min.
5. Extract the PCR products with 1 vol. chlorofom, and precipitate with 2 vol. ethanol. Analyse the DNA samples by non-denaturing acrylamide gel electrophoresis as in *Protocol 5*.

[a] It is not necessary to use MEGA-9 detergent in particular. Other cell lysates (e.g. S100, DEAE, or cell lysates prepared with CHAPS) can also be used in part B.
[b] Although the telomeric sequence in TAG-U is derived from *Tetrahymena*, it can be elongated by human telomerase. Similarly, other telomerases should be able to extend this G-rich primer.

2.4 Analysis of telomerase RNA

If you are following telomerase activity during purification or cellular fractionation studies, it may also be useful to analyse whether the telomerase RNA is also co-fractionating with your activity. The RNA is extracted and analysed by standard procedures, as outlined below.

Protocol 9. Detection of telomerase RNA by Northern analysis (14, 15, 37, 38)

Reagents

- Denaturing solution: 4 M guanidine thiocyanate in 25 mM sodium citrate pH 7, 0.1 M 2-mercaptoethanol, 0.5% (w/v) Sarkosyl
- Diethyl pyrocarbonate treated sterile water (DEPC water)
- 1.5 ml Eppendorf tube with a fitted pestle (Kimble)
- HYB solution: 0.5 M sodium phosphate pH 7.2, 1% (w/v) BSA, 1 mM EDTA, 7% (w/v) SDS
- Northern running buffer: 2.2 M formaldehyde, 10 mM morpholineproanesulfonic acid (MOPS) pH 7, 8 mM sodium acetate, 1 mM EDTA
- Northern loading dye: 1 × running buffer, 50% formamide
- 2 × SSC
- Radiolabelled hTR and 5S RNA probes (prepared by random hexamer labelling, e.g. Boehringer Mannheim)

Method

1. Isolate RNA from 10^7–10^8 cultured cells, or 50–500 mg tissue.[a]
2. Lyse cells in a minimal volume of denaturing solution. Lyse tissues using the Eppendorf tube and pestle.

Protocol 9. *Continued*

3. Extract with 1 vol. of 25:24:1 phenol:chloroform:isoamyl alcohol saturated with Na acetate.
4. Ethanol precipitate. Resuspend in DEPC water, and measure A_{260} (1 OD $\sim$ 40 μg RNA).
5. Load 10 μg RNA per sample onto a 1.5% (w/v) agarose gel in Northern running buffer, electrophorese for 4–8 h at 5 V/cm.[b]
6. Transfer the RNA to Nytran (Schleicher & Schuell) by gravity transfer (e.g. Southern transfer), and cross-link the RNA to the membrane using a Stratalinker (optimal cross-link setting).
7. Hybridize the membrane overnight in HYB solution, using 5×10^6 c.p.m./membrane for hTR, and 0.5×10^6 c.p.m./membrane for 5S RNA, at 65 °C.
8. Wash the membrane with 2 × SSC, 0.1% (w/v) SDS, five times for 5 min each, at room temperature, and two washes with 2 × SSC, 0.1% SDS, for 15 min each, at 65 °C.
9. Wrap the damp membrane in Saran Wrap, and expose it to XAR-5 film or a PhosphorImager screen.[c]

[a] If the amount of RNA in the sample is expected to be very low (i.e. less than 1 μg), reverse transcription and PCR analysis may be required to amplify the RNA. See ref. 14 for a description using mouse telomerase RNA-specific primers.
[b] Alternatively, replace steps 5 and 6 as follows: resolve the RNA samples on a 6% (w/v) non-denaturing acrylamide gel, and transfer to Nytran using a wet electroblotter (see refs 14 and 37 for example). The mouse and human telomerase RNAs are approx. 450 nucleotides in length (14, 15).

Recently, the *in situ* detection of human telomerase RNA in normal and tumour tissue samples has also been reported, using ^{35}S-labelled hTR-specific riboprobes (31, 39).

3. Telomeric restriction fragment analysis

Several studies have found that normal somatic cells that lack telomerase activity undergo telomere attrition *in vivo* and *in vitro* (reviewed in ref. 4). In yeast, several mutants have been characterized that lead to telomere shortening and 'senescence' (reviewed in ref. 40), most notably, the yeast telomerase RNA component (41), and the catalytic telomerase subunit, *EST2* (17, 42). Alterations in the DNA binding activity of the telomere repeat binding protein in yeast and humans can also lead to dramatic changes in telomere length (43, 44).

In mice, telomere length studies are more difficult, due to the extremely long telomeric restriction fragments in inbred laboratory mouse strains. For this reason, outbred mouse strains (e.g. *mus spretus*) have been employed for telomere length regulation analysis in mouse tissues and primary mouse fibroblasts (45).

Protocol 10. Telomeric restriction fragment analysis (21, 45, 46)

Reagents

- Proteinase K solution: 0.1 mg/ml proteinase K, 0.5% (w/v) SDS, 10 mM Tris–HCl pH 8, 100 mM NaCl, 10 mM EDTA
- TE: 10 mM Tris–HCl pH 8, 1 mM EDTA
- Restriction enzymes *Hin*fI and *Rsa*I and appropriate buffers (New England Biolabs)
- Whatman 3MM paper, or polyethylene film suitable for gel drying
- Denaturing solution: 1.5 M NaCl, 0.5 M NaOH
- Neutralizing solution: 1.5 M NaCl, 0.5 M Tris–HCl pH 8
- 1 × TAE
- 1 × PBS
- d(CCCTAA)$_3$ or d(TTAGGG)$_3$, 5′ end-labelled with [γ-^{32}P]ATP[a]
- HYB solution: 0.5 M sodium phosphate pH 7.2, 1% (w/v) BSA, 1 mM EDTA, 7% (w/v) SDS
- 2 × SSC

Method

1. Wash cells in 1 × PBS, and resuspend in TE at 5 × 10^7 cells/ml. Add 10 vol. (relative to cell pellet) proteinase K solution and digest at 48 °C overnight. Extract the DNA twice with 1 vol. phenol, once with 1 vol. chloroform, and precipitate with 2 vol. ethanol. Precipitate the DNA at 10 000 r.p.m. for 10 min, remove the ethanol, and dissolve the DNA in TE for several hours at 37 °C.
2. Digest the genomic DNA with *Hin*fI and *Rsa*I, extract and precipitate as above, and redissolve in TE. Measure the DNA concentration by A_{260} or fluorometry.
3. To 1 μg of each DNA sample, bring to 20 μl with TE and add agarose gel loading dye to 1 x. Load onto a 0.5% (w/v) agarose gel in 1 × TAE. Electrophorese the gel in 1 × TAE for approx. 13 h at 90 V.
4. Dry the gel onto Whatman 3MM paper or a removable polyethylene film at 60 °C for 30 min to 1 h. Remove the Whatman 3MM paper or film.
5. Wash the dried gel in denaturing solution for 15 min, followed by two washes in neutralizing solution for 10 min each.
6. Hybridize the gel to 5′-[γ-^{32}P]-d(CCCTAA)$_3$ (or [γ-^{32}P]-d(TTAGGG)$_3$) in HYB solution at 37 °C for 12 h. Wash the gel three times for 10 min each in 2 × SSC, followed by three washes for 7 min each in 0.1 × SSC at room temperature. Wrap the gel in Saran Wrap, and expose to film or PhosphorImager screen.

[a] The oligonucleotide is end-labelled with [γ-^{32}P]ATP and T4 polynucleotide kinase (Boehringer Mannheim). Longer oligonucleotides may also be used, however the washing and hybridization temperatures should be raised accordingly.

4. Telomere DNA binding activities

Both double-stranded telomere DNA binding activities and single-stranded telomere DNA binding activities have been identified in several organisms (reviewed in ref. 2). The specific incubation conditions are determined empirically for each protein. For example, the mammalian telomeric repeat binding factor, TRF1, specifically binds to double-strand telomeric DNA containing six or more telomeric repeats (47). To analyse this, 1 ng of labelled telomeric DNA probe (double-stranded d$(TTAGGG)_6$) is incubated with approx. 6 μg of cell lysate in a 20 mM Hepes buffer pH 8, containing 150 mM KCl, 1 mM $MgCl_2$, 0.5 mM DTT, 0.1 mM EDTA, and 5% (v/v) glycerol, in the presence of 1–3 μg of competitor *Hae*III digested *E. coli* DNA. After incubation for 30 min at 22°C, the mixture is resolved on a 5–6% (w/v) non-denaturing acrylamide gel, dried under vacuum, and exposed to film or PhosphorImager screen. Similar conditions were also developed to characterize the binding of purified ciliate telomerase to single-stranded telomeric primers (48).

As controls for the specificity of the binding reaction, various unlabelled DNA probes are tested for their ability to compete for the telomere binding protein–telomeric DNA complex. For example, mammalian TRF1 binds specifically to TTAGGG, and even TTGGGG will not compete for this binding (47). On the other hand, single-stranded telomere DNA binding proteins, such as *Tetrahymena* telomerase, and the *Saccharomyces* single-strand telomere DNA binding proteins Est1 and Cdc13, are able to bind several G-rich DNA oligonucleotides (48–51).

In vivo, the specificity of the telomere binding protein for telomeric DNA can be further demonstrated by their co-localization with telomeric DNA. For example, TRF1 has been demonstrated to co-localize with mammalian metaphase and interphase telomeric DNA (52).

An example of a protocol for the detection of telomere DNA using FISH is described in *Protocol 11* (52–58). The advantage of this particular method is that it the procedure is gentle, which minimizes perturbations in cell morphology (53). The method can also be combined with indirect immunofluorescence of telomere binding proteins, especially if one is interested in determining if a novel protein may be co-localized to the telomere.

FISH has also been used quantitatively to measure changes in mammalian telomere length (59–61).

Protocol 11. *In situ* analysis of mammalian telomeres and telomere binding proteins[a] (52–58)

Equipment and reagents

- Zeiss microscope equipped with a × 100 objective and FITC and rhodamine filters (or comparable model)
- LabTek Chamber Slides (Nunc)
- Plastic sealable box
- 22 × 50 mm coverslips (Sigma)
- Human telomeric probe, digoxigenin labelled (Oncor Inc.)

- Colcemid (Sigma)
- DAPI, or propidium iodide
- PBS
- Triton X-100
- Antifade mounting media
- Sheep anti-digoxigenin antibody, rhodamine-conjugated (Boehringer Mannheim)
- Rubber cement
- LRSC-conjugated anti-sheep antibody (Jackson Laboratories)

A. *Indirect immunofluorescence of telomere binding or telomerase proteins*[a]

1. Grow cells overnight in LabTek chamber slides. Cool the chamber on ice, aspirate off culture media, and wash in PBS.
2. Aspirate off PBS, and incubate in 2% (v/v) formaldehyde in PBS for 10 min at 4°C.
3. Aspirate off fixative, and wash cells three times with PBS (5 min/wash, last wash 10 min) at 4°C.
4. Layer primary antibody of interest[a] (diluted into 1 × PBS/0.3% (v/v) Triton X-100) so that it covers cells in the chamber. Incubate overnight at 4°C. Wash four times with 4°C PBS (5 min/wash).
5. Layer secondary antibody of interest (diluted into 1 × PBS/0.3% (v/v) Triton X-100) so that it covers cells in the chamber. Incubate overnight at 4°C. Wash as in step 3.
6. If FISH is also being performed, proceed to part B, step 4. If visualizing immediately, peel off chamber walls and gasket, mount in antifade medium, and place coverslip on the slide.[b]

B. *Fluorescence in situ hybridization of telomere DNA*

1. For metaphase arrested cells, treat cultured cells with colcemid (0.1 μg/ml) for 90 min,[c] and harvest by trypsinization. For interphase cells, begin at step 3 below.
2. Swell the cells in hypotonic solution for 20 min at 37°C, and using a pipette, drop the cells onto LabTek chamber slides.
3. If desired, at this point indirect immunofluorescence can be performed as outlined in part A, step 2. After completion, proceed with step 4 below.
4. Fix cells in 2% (v/v) formaldehyde in 1 × PBS for 10 min at 4°C. Peel off the chamber walls and gasket, and carry out all subsequent washes with the slide immersed in an appropriate container (e.g. a conical 50 ml polypropylene tube; Falcon).
5. Wash the chamber with 2 × SSC at room temperature. Incubate the slide in 70% deionized formamide in 2 × SSC at 80°C for 10 min. Wash once immediately with ice-cold 70% deionized formamide in 2 × SSC, and wash once in 2 × SSC at room temperature.
6. The labelled telomeric probe is pre-warmed at 37°C for 5 min,

Protocol 11. *Continued*

vortexed briefly, and centrifuged at 10 000 r.p.m. for 4 sec. Approx. 100 ng (in a volume of approx. 30 μl) is layered onto the chamber slide, covered with a 22 × 50 mm coverslip, and sealed with rubber cement. Incubate overnight at 37 °C in a humidified chamber (e.g. a sealed plastic box).

7. Wash the slide twice in 50% formamide in 2 × SSC at 37 °C for 30 min each, and wash twice in 2 × SSC at room temperature for 15 min each.
8. Incubate the slide overnight at 4 °C in 2 × SSC (or 0.3% Triton X-100 in 1 × PBS) plus 1% (w/v) BSA and rhodamine-conjugated anti-digoxigenin antibodies (20 μg/ml, Boehringer Mannheim). Wash as in part A, step 3.
9. Incubate slides at 37 °C for 1 h in 1 × PBS containing 5 μg/ml LRSC-conjugated anti-sheep antibody.
10. Wash the slides as in part A, step 3.
11. Mount slides in antifade medium containing either 0.1 μg/ml DAPI or 0.8 μg/ml propidium iodide. Visualize the telomeric signal using a Zeiss microscope, or comparable model, equipped with a 100 X/1.4 NA oil immersion lens and UV filter set.

[a] As an illustration of this method, to detect an HA tagged version of the mammalian telomere binding protein (52), TRF, fixed cells were incubated with the mouse HA monoclonal antibody 12CA5 as the primary antibody, followed by FITC labelled goat antibody to mouse IgG (Jackson ImmunoResearch Labs) as the secondary antibody. Indirect immunofluorescence of the telomerase catalytic subunit was also performed similar to that described here (58).

[b] Cy2 has similar fluorescent properties as FITC, but is more stable (Cy2 antibodies are commercially available; e.g. Amersham). If the signal is not sufficient for visualization, it may be necessary to amplify by following the primary antibody incubation with a secondary antibody conjugated to biotin, and then a tertiary Cy2 antibody conjugated to streptavidin (e.g. see ref. 58).

[c] To optimize this step, the colcemid arrest time will vary between cell types and it is best to test varying treatment times.

Much has been gleaned about the proteins involved in the process and regulation of DNA replication. It is increasingly apparent that the cell has evolved multiple checks and balances to ensure the ordered and precise replication and segregation of genetic material. With the continued identification of proteins important for telomere length regulation and telomerase activity, the next few years should herald exciting advances into the biochemistry and regulation of telomere length, and how telomere replication is coupled to the processes controlling DNA replication and genome stability.

Acknowledgements

I would like to thank all the authors who contributed to the development of the ideas and protocols outlined in this chapter (including those whose work was cited in reviews due to space constraints). I would particularly like to thank Rena Oulton for providing details regarding fluorescence *in situ* hybridization and indirect immunofluorescence.

References

1. Autexier, C. and Greider, C. W. (1996). *Trends Biochem. Sci.*, **21**, 387.
2. Fang, G. and Cech, T. R. (1996). In *Telomeres* (ed. E. H. Blackburn and C. W. Greider), p. 69. Cold Spring Harbor Laboratory Press, Cold Spring Harbor, NY.
3. Shore, D. (1997). *Trends Biochem. Sci.*, **22**, 233.
4. Greider, C. W. (1996). *Annu. Rev. Biochem.*, **65**, 337.
5. Greider, C. W. and Blackburn, E. H. (1985). *Cell*, **43**, 405.
6. Kim, N., Piatyszek, M. A., Prowse, K. R., Harley, C. B., West, M. D., Ho, P. L. C., *et al.* (1994). *Science*, **266**, 2011.
7. Piatyszek, M. A., Kim, N. S., Weinrich, S. L., Hiyama, E., Wright, W. E., and Shay, J. W. (1995). *Methods Cell Sci.*, **17**, 1.
8. Greider, C. W. and Blackburn, E. H. (1987). *Cell*, **51**, 887.
9. Greider, C. W. and Blackburn, E. H. (1989). *Nature*, **337**, 331.
10. Morin, G. B. (1989). *Cell*, **59**, 521.
11. Prowse, K. R., Avilion, A. A., and Greider, C. W. (1993). *Proc. Natl. Acad. Sci. USA*, **90**, 1493.
12. Blackburn, E. H. and Gall, J. G. (1979). *J. Mol. Biol.*, **120**, 33.
13. Greider, C. W. (1991). *Mol. Cell. Biol.*, **11**, 4572.
14. Blasco, M. A., Funk, W. A., Villeponteau, B., and Greider, C. W. (1995). *Science*, **269**, 1267.
15. Feng, J., Funk, W. D., Wang, S.-S., Weinrich, S. L., Avilion, A. A., Chiu, C.-P., *et al.* (1995). *Science*, **269**, 1236.
16. Collins, K., Kobayashi, R., and Greider, C. W. (1995). *Cell*, **81**, 677.
17. Lingner, J., Hughes, T. R., Schevchenko, A., Mann, M., Lundblad, V., and Cech, T. R. (1997). *Science*, **276**, 561.
18. Lingner, J. and Cech, T. R. (1996). *Proc. Natl. Acad. Sci. USA*, **93**, 10712.
19. Autexier, C. and Greider, C. W. (1994). *Genes Dev.*, **8**, 563.
20. Autexier, A., Pruzan, R., Funk, W. D., and Greider, C. W. (1996). *EMBO J.*, **15**, 5928.
21. Counter, C. M., Avilion, A. A., LeFeuvre, C. E., Stewart, N. G., Greider, C. W., Harley, C. B., *et al.* (1992). *EMBO J.*, **11**, 1921.
22. Steiner, B. R., Hidaka, K., and Futcher, B. (1996). *Proc. Natl. Acad. Sci. USA*, **93**, 2817.
23. Lue, N. F. and Wang, J. C. (1995). *J. Biol. Chem.*, **270**, 21453.
24. Lin, J.-J. and Zakian, V. A. (1995). *Cell*, **81**, 1127.
25. Cohn, M. and Blackburn, E. H. (1995). *Science*, **269**, 396.
26. Mantell, L. L. and Greider, C. W. (1994). *EMBO J.*, **13**, 3211.
27. Harrington, L. A. and Greider, C. W. (1991). *Nature*, **353**, 451.

28. Morin, G. B. (1991). *Nature*, **353**, 454.
29. Wright, W. E., Shay, J. W., and Piatyszek, M. A. (1995). *Nucleic Acids Res.*, **23**, 3794.
30. Ohyashiki, J. H., Ohyashiki, K., Toyama, K., and Shay, J. W. (1996). *Trends Genet.*, **12**, 395.
31. Yashima, K., Piatyszek, M. A., Saboorian, H. M., Virmani, A. K., Brown, D., Shay, J. W., *et al.* (1997). *J. Clin. Pathol.*, **50**, 110.
32. Savoysky, E., Akamatsu, K., Tsuchiya, M., and Yamazaki, T. (1996). *Nucleic Acids Res.*, **24**, 1175.
33. Heller, K., Kilian, A., Piatyszek, M. A., and Kleinhofs, A. (1996). *Mol. Gen. Genet.*, **252**, 342.
34. Fitzgerald, M. S., McKnight, T. D., and Shippen, D. E. (1996). *Proc. Natl. Acad. Sci. USA*, **93**, 14422.
35. Fajkus, J., Kovarik, A., and Kralovics, R. (1996). *FEBS Lett.*, **391**, 307.
36. Tatematsu, K., Nakayama, J., Danbara, M., Shionoya, S., Sato, H., Omine, M., *et al.* (1996). *Oncogene*, **13**, 2265.
37. Blasco, M. A., Rizen, M., Greider, C. W., and Hanahan, D. (1996). *Nature Genet.*, **2**, 200.
38. Avilion, A. A., Piatyszek, M. A., Gupta, J., Shay, J. W., Bacchetti, S., and Greider, C. W. (1996). *Cancer Res.*, **56**, 645.
39. Soder, A. I., Hoare, S. F., Muir, S., Going, J. J., Parkinson, E. K., and Keith, W. N. (1997). *Oncogene*, **14**, 1013.
40. Zakian, V. (1996). *Annu. Rev. Genet.*, **30**, 141.
41. Singer, M. and Gottschling, D. E. (1994). *Science*, **266**, 404.
42. Lendvay, T. S., Morris, D. K., Sah, J., Balasubramanian, B., and Lundblad, V. (1996). *Genetics*, **144**, 1399.
43. Cooper, J. P., Nimmo, E. R., Allshire, R. C., and Cech, T. R. (1997). *Nature*, **385**, 744.
44. van Steensel, B. and de Lange, T. (1997). *Nature*, **385**, 740.
45. Prowse, K. R. and Greider, C. W. (1995). *Proc. Natl. Acad. Sci. USA*, **92**, 4818.
46. Harley, C. B., Futcher, A. B., and Greider, C. W. (1990). *Nature*, **345**, 458.
47. Zhong, Z., Shiue, L., Kaplan, S., and De Lange, T. (1992). *Mol. Cell. Biol.*, **12**, 4834.
48. Harrington, L., Hull, C., Crittenden, J., and Greider, C. (1995). *J. Biol. Chem.*, **270**, 8893.
49. Lin, J.-J. and Zakian, V. A. (1996). *Proc. Natl. Acad. Sci. USA*, **93**, 13760.
50. Nugent, C. I., Hughes, T. R., Lue, N. F., and Lundblad, V. (1996). *Science*, **274**, 249.
51. Virta-Pearlman, V., Morris, D. K., and Lundblad, V. (1996). *Genes Dev.*, **10**, 3094.
52. Chong, L., van Steensel, B., Broccoli, D., Erdjument-Bromage, H., Hanish, J., Tempst, P., *et al.* (1995). *Science*, **270**, 1663.
53. Henderson, S., Allsopp, R., Spector, D., Wang, S.-S., and Harley, C. (1996). *J. Cell Biol.*, **134**, 1.
54. Luderus, M. E. E., van Steensel, B., Chong, L., Sigon, O. C. M., Cremers, F. F. M., and de Lange, T. (1996). *J. Cell Biol.*, **135**, 867.
55. Multani, A. S., Hopwood, V. L., and Pathak, S. (1996). *Anticancer Res.*, **16**, 3435.

56. Ausubel, F., *et al.* (ed.) (1992). In *Short protocols in molecular biology*, 2nd edn, pp. 14–25. John Wiley & Sons, Toronto, Ontario, Canada.
57. Gotta, M., Laroche, T., Formenton, A., Maillet, L., Scherthan, H., and Gasser, S. M. (1996). *J. Cell Biol.*, **134**, 1349.
58. Harrington, L., Zhou, W., McPhail, T., Oulton, R., Yeung, D. S. K., Mar, V., *et al.* (1997). *Genes Dev.*, **23**, 3109.
59. Lansdorp, P. M., Verwoerd, N. P., van de Rijke, F. M., Dragowska, V., Little, M. T., Dirks, R. W., *et al.* (1996). *Hum. Mol. Genet.*, **5**, 685.
60. Zijlmans, J. M., Martens, U. M., Poon, S. S., Raap, A. K., Tanke, H. J., Ward, R. K., *et al.* (1997). *Proc. Natl. Acad. Sci. USA*, **94**, 7423.
61. Blasco, M. A., Lee, H. W., Hande, M. P., Samper, E., Lansdorp, P. M., DePinho, R. A., *et al.* (1997). *Cell*, **97**, 25.

8

Saccharomyces cerevisiae as a model system to study DNA replication

MARCO FOIANI, GIORDANO LIBERI, SIMONETTA PIATTI, and PAOLO PLEVANI

1. Introduction

The budding yeast *Saccharomyces cerevisiae* represents a powerful tool to study DNA replication and the cell cycle regulatory pathways controlling entry into, and progression through S phase. In fact, the availability of the complete sequence of the *S. cerevisiae* genome, the identification of several DNA replication genes, together with the possibility of combining molecular genetic tools with classical techniques of genetics, biochemistry, and cell biology, make this system particularly useful to improve our knowledge of the DNA replication process at the molecular level. Moreover, by considering that the structure and function of most of the replication proteins and the regulatory circuits connecting DNA replication to other cell cycle events have been highly conserved during evolution, the information obtained using the *S. cerevisiae* or the *S. pombe* systems can be generally extended to other eukaryotic cells.

This chapter illustrates some of the experimental protocols which are routinely used in budding yeast to study DNA replication proteins, to characterize mutations in the corresponding genes, and to integrate this kind of analysis with the techniques used for monitoring cell cycle progression.

2. The *Saccharomyces cerevisiae* cell cycle

Progression through the mitotic cell cycle in *S. cerevisiae* is characterized by morphological changes which can be easily monitored microscopically (*Figure 1*). The most characteristic event is the emergence of a bud which grows continuously during the cell cycle, although, normally, it does not reach the size of the mother cell even at the time of cell division. In laboratory conditions, most of the *S. cerevisiae* strains complete the entire mitotic cell cycle in 90–120

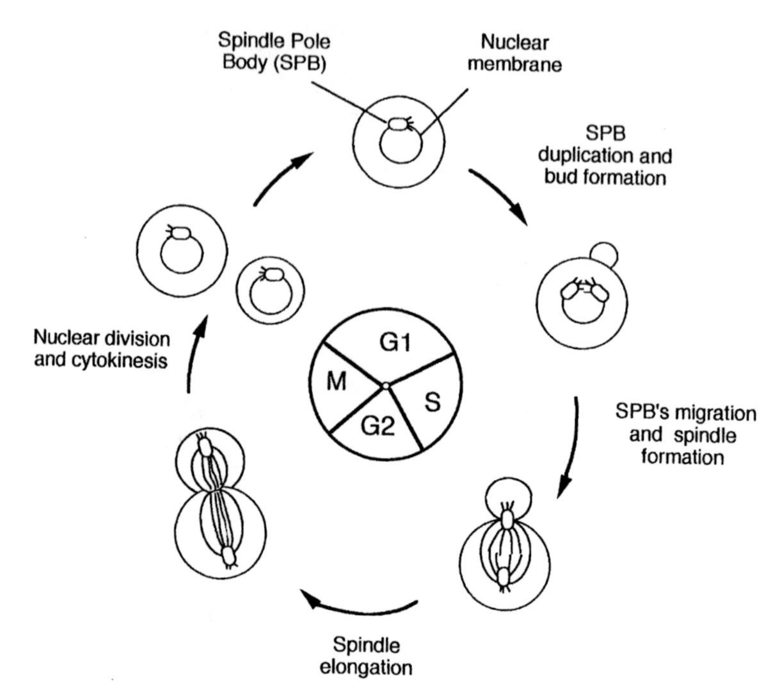

Figure 1. *Saccharomyces cerevisiae* mitotic cell cycle.

minutes. The yeast chromosomes are replicated in S phase which lasts, approximately, 20–30 minutes and, usually, initiation of DNA synthesis coincides with the time of bud emergence and spindle pole body (SPB) duplication. Although budding, DNA synthesis and SPB duplication occur in a short window of the cell cycle, these events can be genetically uncoupled (1) and, therefore, are considered as parallel pathways.

Due to the short length of the *S. cerevisiae* cell cycle, it is important to correlate S phase progression with other cell cycle events, such as bud emergence, nuclear division, and spindle formation. This kind of analysis relies on the ability to efficiently synchronize cells with a well-defined genetic background and the following protocols are routinely used with *S. cerevisiae*.

2.1 Synchronization by elutriation

At the time of cell division *S. cerevisiae* generates a small daughter cell which has to spend more time in G1 than the mother cell, in order to reach the critical size required to pass a restriction point in G1, called START (1), after which cells are committed to cell division. Due to the size difference between newborn daughters and either unbudded mothers or budded cells, it is relatively easy to separate the small cell population from the others by the use of gradients. The more homogeneous is the size of the small daughters selected, the higher is the degree of synchrony achieved.

The protocol described below allows this goal to be achieved, avoiding the use of drugs or mutations which might perturb the cell cycle and is, therefore, considered to be the most physiological system, although it requires a special and expensive piece of equipment.

Protocol 1. Elutriation of daughter cells

Equipment and reagents

- Centrifugal elutriator (Beckman)
- Sonicator
- Spectrophotometer
- Light microscope
- YEP or minimal medium (2)

Method

1. Grow 2–6 litres of culture to $OD_{600} = 2$ (which corresponds to 2×10^7 cells/ml). In order to enrich for small daughter cells raffinose is preferred to glucose as carbon source, especially when rich medium is used.
2. Spin down cells using a refrigerated centrifuge, resuspend in 100–500 ml of medium, and keep on ice until release.
3. Sonicate cells for 1–2 min, until unbudded cells are well separated.
4. Before loading cells into the elutriator, wash the chamber extensively by pumping in ethanol and sterile water and, finally, fill it with medium, eliminating any bubbles in the tubing and in the chamber.
5. Spin at 4000 r.p.m. in the elutriator and keep pumping ice-cold medium into the chamber.
6. Load the cells, avoiding bubbles which would perturb the gradient.
7. When almost all the cells are loaded, replace cell suspension with ice-cold medium. In order to avoid bubbles at this step, switch off the pump briefly while rapidly moving the tube from the cell suspension into the medium.
8. Increase the pump speed very slowly in order to allow the gradient to move to the top of the chamber: as the first daughter cells come out (to be checked at the microscope), start collecting 200 ml fractions.
9. Spin down only the fractions which are not contaminated by budded cells, and resuspend daughter cells in a small volume (2–10 ml) of ice-cold medium. In order to obtain better synchrony, it is advisable to use daughter cells with a homogeneous size, unless large volumes of culture are needed. Release the cells in the desired conditions. Typically, from 4–6 litres of wild-type culture grown in YEPRaf it is possible to recover up to 500–1000 OD_{600} of unbudded cells.

2.2 Synchronization by α-factor treatment

Haploid *S. cerevisiae* strains can exist as two different mating types, called **a** and α (3). Yeast strains of the **a** mating type (*MAT***a**) in the presence of the pheromone (α-factor) produced by cells of the opposite mating type accumulate G1 unbudded cells. This short peptide is commercially available and widely used to synchronize *MAT***a** strains. This method is easy to apply and it does not rely on any particular equipment, although, certain genetic backgrounds cannot be efficiently synchronized. However, in some cases, this problem can be bypassed by deleting the *BAR1* gene, which results in a higher sensitivity of cells to the action of the pheromone (4).

Protocol 2. Synchronization by α-factor treatment

Equipment and reagents

- Sonicator
- Filter units
- Light microscope
- Spectrophotometer
- *MAT***a** strains
- α-factor (Sigma) 1 mg/ml
- YEP or minimal medium (2)

Method

1. Grow cells to $OD_{600} = 0.5$.
2. Add 2 μg/ml α-factor to the cultures and shake until more than 95% of the cells are unbudded (usually a wild-type strain reaches these conditions in about 90 min at 30 °C or 150 min at 25 °C). It is important to sonicate cells briefly before checking their morphology at the microscope.
3. Filter and wash the cells with at least an equal volume of medium to remove the α-factor and release from the α-factor block by re-suspending the cells in pre-warmed medium at $OD_{600} = 0.5–1$.

2.3 Synchronization by nocodazole treatment

Nocodazole treatment results in a rapid and complete disassembly of cytoplasmic and intranuclear microtubules which causes a reversible block in G2/M, that is characterized by the accumulation of large budded yeast cells with a single nucleus.This protocol is particularly useful to study cell cycle events linking DNA synthesis to the completion of the preceding mitosis (5).

Protocol 3. Synchronization by nocodazole treatment

Equipment and reagents

- Light microscope
- Filter units
- Spectrophotometer
- Dimethyl sulfoxide (DMSO)
- Nocodazole (Sigma) 0.5 mg/ml in DMSO
- YEP medium (2)

Method

1. Grow cells in rich medium to OD_{600} = 0.5 (in minimal medium cells do not arrest properly).
2. Add 5 μg/ml nocodazole and leave the cultures shaking until more than 90% of the cells are large budded (usually after 150 min in YEPD at 25 °C).
3. Filter and wash the cells using at least three volumes of YEP medium containing 1% DMSO. Release the cells by resuspending in medium without DMSO at OD_{600} = 0.5–1.

2.4 Synchronization using *cdc* mutants

The *CDC15* gene product is required for metaphase to anaphase transition (1), while *CDC28* encodes the master regulator of the budding yeast cell cycle (1). *cdc15–2* and *cdc28–13* are temperature-sensitive mutations causing, respectively, a reversible arrest in mitosis or G1 at restrictive temperature. However, it is important to remember that the use of mutations altering the function of important cell cycle regulators might cause potential artefacts.

Protocol 4. Synchronization using *cdc15* or *cdc28* mutants

Equipment and reagents

- Filter units
- Light microscope
- Spectrophotometer
- *cdc15–2* (K. Nasmyth, I. M. P., Vienna, Austria) or *cdc28–13* mutants (S. Reed, La Jolla, California)

Method

1. Grow cells at 25 °C to OD_{600} = 0.3–0.5.
2. Filter and resuspend cells in the same volume of pre-warmed medium at 37 °C.
3. Incubate the cultures at 37 °C with shaking, until 90% of the cells display either the *cdc15* large budded or the *cdc28* large unbudded phenotype (usually after 150–180 min).
4. Release the cells from the temperature block by filtering and re-suspend in medium at 25 °C.

The function of the *CDC28* gene product in G1 requires its association with at least one G1 cyclin encoded by the *CLN1, 2, 3* genes (6). Therefore, synchronization in G1 can be obtained by using a *S. cerevisiae* strain carrying a triple deletion of the *CLN1, 2, 3* genes rescued by a copy of the *CLN2* gene, under the control of the repressible *MET3* promoter (7). This strain can grow in minimal medium without methionine, while G1 arrest by Cln depletion is obtained by methionine addition to the medium.

Protocol 5. Synchronization by Cln depletion

Equipment and reagents

- Spectrophotometer
- Light microscope
- Filter units
- Minimal medium
- *cln1Δ cln2Δ cln3Δ MET3-CLN2* strain (K. Nasmyth, I. M. P., Vienna, Austria)
- 50 mM methionine

Method

1. Grow cells in (–)Met medium at 25–30 °C to OD = 0.5.
2. Add 2 mM methionine and incubate the cultures by shaking until more than 90% of the cells are large unbudded.
3. Filter and wash cells with three volumes of (–)Met medium. Release from the G1 arrest by resuspending the cells in (–)Met medium.

2.5 Use of fluorescence activated cell sorter (FACS) to analyse cell cycle progression of *S. cerevisiae* cells

FACS analysis is one of the most useful tools to study cell cycle progression in budding yeast. In this section we will describe some applications of this technique based on the staining of DNA with propidium iodide. This procedure allows:

(a) Determination of the genome size of yeast strains and, therefore, their ploidy.
(b) Analysis of the degree of synchrony of a cell culture (*Figure 2*).
(c) Characterization of yeast mutants defective in DNA synthesis and/or proper cell cycle progression.

In applying this technique to cell cycle analysis of budding yeast cultures, several aspects must be considered:

(a) The coefficients of variations (CV) of DNA histograms for the peaks in G1 or G2 + M are higher than those obtained with mammalian cells.
(b) *S. cerevisiae* cells have a cell wall which, by limiting enzymatic diffusion, makes RNA digestion by ribonuclease treatment less efficient.
(c) Budding yeast has a very large ratio of mitochondrial to nuclear DNA.

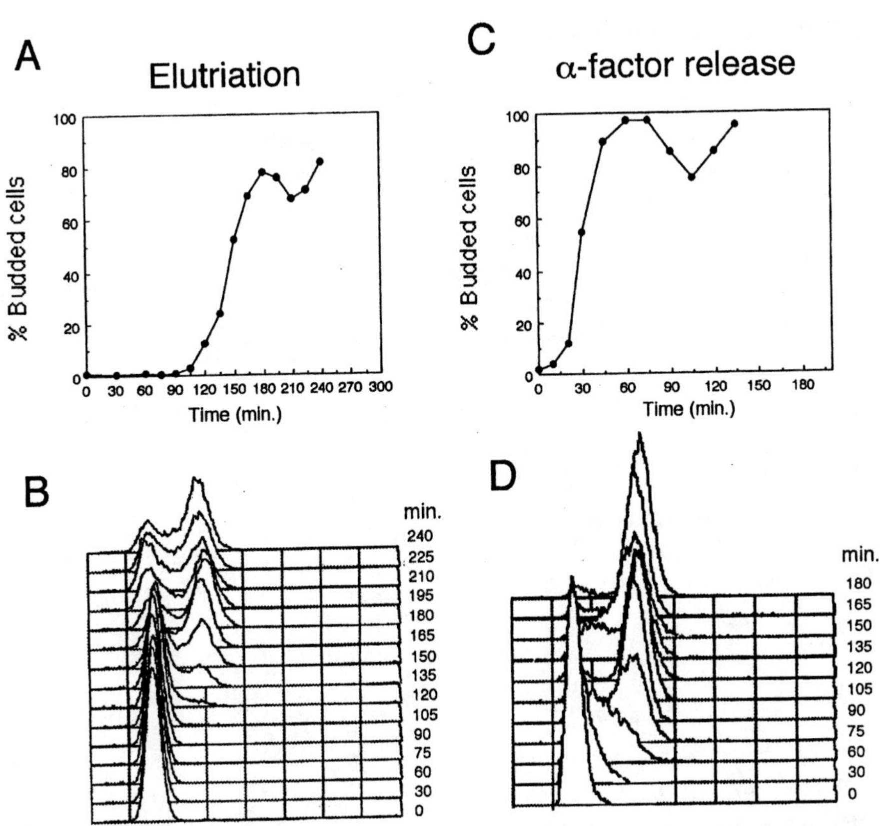

Figure 2. Synchronization procedures. (A, B) Budding and FACS profiles of yeast cells synchronized using the elutriation protocol (*Protocol 1*). (C, D) Budding and FACS profiles of yeast cells synchronized using α-factor treatment.

In fact, mitochondrial DNA represents between 5–20% of total yeast DNA and, since mitochondrial DNA is synthesized continuously during the cell cycle, it might interfere with the interpretation of the data. This problem can be overcome by using ρ° strains lacking mitochondrial DNA. However, these generally grow more slowly.

Protocol 6. FACS analysis

Equipment and reagents

- FACScan (Becton Dickinson)
- Sonicator
- 50 mM Tris–HCl pH 7.5
- 0.5 mg/ml propidium iodide
- 10 mg/ml RNase A
- Pepsin: 5 mg/ml freshly dissolved in 55 mM HCl
- FACS buffer: 0.2 M Tris–HCl pH 7.5, 0.21 M NaCl, 0.078 M $MgCl_2$

Protocol 6. *Continued*

Method

1. Spin down at least 10^7 cells, resuspend them in 1 ml 70% ethanol, and leave the sample in ethanol for at least 30 min at room temperature.
2. Spin down the cells, wash once with 1 ml Tris–HCl. Resuspend in 1 ml Tris–HCl and add 20 μl RNase A. Incubate overnight at 37 °C.
3. Spin down the cells, resuspend in 0.5 ml pepsin, and incubate for 30 min at 37 °C.
4. Spin down the cells, wash once with 1 ml FACS buffer, and resuspend in 0.5 ml FACS buffer containing 50 μl propidium iodide.
5. Sonicate the samples 10–20 sec, transfer 50–100 μl to 1 ml Tris–HCl, and analyse them by FACS, reading at least 10 000 cells/sample. If necessary samples can be stored at –20 °C.

2.6 Budding analysis

A simple way to monitor the degree of synchrony of a wild-type *S. cerevisiae* strain is to analyse the kinetics of bud emergence (*Figure 2*). In fact, three morphologically distinguishable populations can be easily scored microscopically in budding yeast cell cultures: unbudded, small budded, and large budded cells which, roughly, correspond to cells in G1, early S, and late S/G2/M phases, respectively.

Protocol 7. Budding analysis

Equipment and reagents

- Light microscope
- Fixing solution: 3.7% formaldehyde, 0.9% NaCl

Method

1. Collect 1 ml of culture $OD_{600} = 0.5$, sonicate, and add an equal volume of fixing solution. If necessary, fixed cells can be kept at 4 °C until the day after.
2. Score the fraction of unbudded, small budded, and large budded cells.
3. The kinetics of budding profile can be monitored on samples taken at different times on cells pre-synchronized by elutriation or α-factor treatment (*Protocols 1* and *2*).

3. Characterization of *S. cerevisiae* DNA replication mutants

A large number of *cdc* mutations specifically affect the yeast DNA replication process (1), and mutants in DNA replication genes can be easily produced and characterized in *S. cerevisiae* by reverse genetics techniques (8). Many of the methodologies described in this chapter can be used to physiologically characterize these mutants. For example, some mutants defective in DNA synthesis exhibit a longer S phase which can be detected by FACS analysis on synchronized cultures. Due to this cell cycle delay, exponentially growing mutant cells are characterized by an increase in large budded cells and a decrease in unbudded cells compared to the isogenic wild-type strain. This phenotype can be monitored by FACS and by microscopic examination.

The use of specific genetic and biochemical assays can also be helpful to gain some insight on the step altered in a mutant. For example, search for second-site mutations conferring a synthetic lethal phenotype (9) may help in identifying the pathway affected in the mutant. Moreover, it has been shown that mutations in DNA replication genes affect plasmid stability (see *Protocol 8*) and lead to the accumulation of replication intermediates which can be detected by sedimentation analysis (*Protocol 9*) or gel electrophoresis. If temperature-sensitive alleles of replication genes are available, FACS analysis may help to define the altered step of the DNA replication process after shift to the restrictive temperature. In fact, it is expected that mutants defective in an early step of DNA synthesis will accumulate cells with unreplicated DNA, while most of the replication mutants defective in elongation usually arrest with a G2/M DNA content. Alternative approaches are available to detect specific defects in initiating DNA synthesis at an origin and will be discussed more in detail in other chapters of this book. This analysis can be carried out in combination with other methods monitoring cell morphology, nuclear division, and spindle formation. Finally, a role in initiation or elongation of DNA synthesis can be assessed by performing reciprocal shift experiments (10), which allow establishment of the order of function of cell cycle genes.

3.1 Minichromosome stability assay

A subset of cell division cycle (*CDC*) yeast mutants exhibit elevated mitotic loss of endogenous chromosomes and artificial minichromosomes (10). An increase in minichromosome loss rate is associated to mutations in some genes required for initiation of DNA synthesis, such as *MCM*, *ORC*, *CDC6*, and this defect can be rescued by increasing the number of origins of replication (ARS) in the plasmid molecule (10). Mutations affecting elongation of DNA replication may also show increased rate of plasmid loss, but this phenotype is not rescued by the addition of multiple ARSs.

Protocol 8. Minichromosome stability assay

Equipment and reagents

- YPD plates (2)
- Plasmid pDK243 (single ARS) (11)
- Plasmid pDK368-7 (multiple ARSs) (11)

Protocol 8. *Continued*

Method

1. Transform mutant and isogenic wild-type strains with plasmids pDK243 and pDK368-7.
2. Resuspend, for each transformed strain, ten independent *ADE3 LEU2* colonies in 1 ml water and spread on three YPD plates (200 cells per colony per plate).
3. Incubate the plates at 25 °C, estimate the frequency of *ade3 leu2* colonies after five to six days, and calculate the rates of mitotic plasmid loss (12).

3.2 Alkaline sucrose gradient analysis of newly synthesized DNA

Mutations affecting certain DNA replication proteins (such DNA polymerases) cause a failure to synthesize high molecular weight DNA products and lead to the concomitant accumulation of DNA chains of heterogeneous length, gaps, and single-stranded DNA regions. These defects, which can cause other phenotypes such as hyper-recombination and checkpoint activation (13, 14) can be monitored by sizing of newly synthesized DNA chains by alkaline sucrose sedimentation.

Protocol 9. Alkaline sucrose gradient analysis of newly synthesized DNA

Equipment and reagents

- SW55 centrifuge rotor (Beckman) and polyallomer centrifuge tubes (Beckman 326819)
- Sucrose gradient fraction collector
- GF/C glass fibre filters (Whatman)
- YPD medium (2)
- Buffer A: 0.1 M Tris–HCl pH 9, 0.01 M EDTA
- Buffer B: 0.01 M $KiPO_4$ pH 7, 0.01 M EDTA
- 10 mg/ml zymolase
- 10% NP-40 (Sigma)
- Buffer C: 0.5 M NaCl, 0.25 M NaOH, 1 mM EDTA
- 5% sucrose and 20% sucrose in buffer C
- 70% sucrose
- 4 M NaOH
- 2 mg/ml salmon sperm DNA (Sigma)
- [5,6-^{3}H]uracil

Method

1. Grow 50 ml of wild-type and mutant cultures in YPD medium at 25 °C to a concentration of 10^7 cells/ml.
2. Add [5,6-^{3}H]uracil to a concentration of 27 μCi/ml and incubate the cultures for 3 h at the temperature requested.
3. Collect and wash the cells in buffer A, and incubate for 10 min at room temperature.
4. Spin down and resuspend the cells in buffer B.

5. Spin down and resuspend the cells in 200 μl buffer B.
6. Add 20 μl zymolase and 30 μl NP-40, and incubate for 20 min at 37 °C.
7. Load the extract onto 4.8 ml 5–20% (w/v) sucrose gradients cushioned with 0.2 ml 70% sucrose.
8. Centrifuge at 16 000 r.p.m. using a Beckman SW55 for 16 h at 20 °C.
9. Collect 160 μl fractions, add 10 μl NaOH, and heat for 30 min at 80 °C to hydrolyse RNA.
10. Add 10 μl salmon sperm DNA, 10 μl 1 M Tris–HCl pH 7, and 6 μl HCl.
11. Process 150 μl of each fraction on GF/C glass fibre to measure trichloroacetic acid insoluble radioactive material (15).

3.3 DNA replication mutants and checkpoints

Eukaryotic cells have developed a network of highly conserved surveillance mechanisms (checkpoints), ensuring that damaged chromosomes are repaired before being replicated or segregated. These mechanisms are essential for maintaining genome integrity and cell viability by delaying cell cycle progression in response to DNA damage. DNA replication mutants often cause the accumulation of replication intermediates which are sensed by the checkpoint mechanism(s). The checkpoint activation can be monitored by analysing the phosphorylation state of the master checkpoint regulator Rad53p, which is specifically phosphorylated in response to DNA damage or replication blocks (15).

Many temperature-sensitive DNA replication mutants are still able to apparently complete DNA replication after shift to the restrictive temperature (at least as judged by FACS analysis), but arrest with a G2/M DNA content. This arrest is usually dependent on the *RAD9* gene product which is required to prevent entry into mitosis in the presence of damaged or unproperly replicated DNA molecules (16). However, genetic approaches are available to test whether the arrest of replication mutants at the restrictive temperature is dependent on *RAD9* (17).

Moreover, a subset of DNA replication mutants exhibit some checkpoint defects (14, 18, 19), suggesting that certain replication proteins may be directly involved in these surveillance mechanisms. These mutants are generally sensitive to genotoxic compounds and specific tests are used to monitor checkpoint defects (14, 18, 20).

4. Analysis of the level and post-translational modifications of DNA replication proteins during the cell cycle

Most of the *S. cerevisiae* DNA replication genes are periodically transcribed at the G1/S boundary (21). However, at least for most of the genes analysed so far, this cell cycle regulated transcriptional programme is not required for S phase entry due to the stability of the corresponding gene products. To measure protein levels during cell cycle by Western blot analysis it is essential to reduce proteolytic artefacts during extract preparation and to use highly specific immunological reagents. Several protocols are available for yeast extract preparation, but the one described below gives, in our hands, the most reproducible results. The following protocol (22) prevents the action of proteases and phosphatases by pre-treating yeast cells with trichloroacetic acid (TCA) before protein extraction. If good antibodies against the protein of interest are not available, it is possible to produce tagged versions of the corresponding genes and to use commercially available monoclonal antibodies (23).

Protocol 10. Preparation of denatured yeast crude extracts

Equipment and reagents

- Microcentrifuge
- Glass beads 425–600 μm diameter (Sigma)
- Trichloroacetic acid (TCA)
- Laemmli loading buffer
- 2 M Tris

Method

1. Spin down 10^8 cells, wash once with 10 ml H_2O, and once with 10 ml 20% TCA.
2. Spin down the cells and resuspend them in 100 μl 20% TCA. At this stage the cells can be stored at –20°C if necessary.
3. Extracts should be prepared from thawed cells at room temperature. Add an equal volume of glass beads and vortex for 4 min.
4. Transfer the extract to a fresh tube without centrifuging (use a Pasteur pipette).
5. Wash the beads twice with 100 μl 5% TCA without centrifuging the sample using a Pasteur pipette. Combine the washing to the crude extract. At this stage the final volume should be 300 μl of crude extract in 10% TCA.
6. Pellet the proteins by centrifuging the sample at 3000 r.p.m. in a microcentrifuge at room temperature. Discard the supernatant.
7. Resuspend the pellet in 100 μl Laemmli loading buffer. Sample should turn yellow. Neutralize by adding 50 μl 2 M Tris. Sample should turn blue.
8. Boil the sample for 3 min, centrifuge at 3000 r.p.m. for 10 min, and discard the pellet.
9. Use 10 μl of crude extract for Western blot analysis (using these conditions, most of the replicative proteins can be visualized).

4.1 *In vivo* labelling of DNA replication proteins and immunoprecipitation

Since the stability of certain proteins varies throughout the cell cycle, it may be necessary to carefully measure the half-life of the protein by pulse-chase analysis.

Protocol 11. Pulse-chase analysis to analyse yeast protein stability

Equipment and reagents

- GF/C fibre glass disks (Whatman)
- Glass beads
- Gamma Bind Plus suspension (Pharmacia)
- [^{35}S]methionine (Amersham)
- 50 mM methionine
- 10% and 50% TCA
- Acetone
- Buffer A: 50 mM Tris–HCl pH 7.5, 1 mM EDTA, 1% SDS
- IP buffer: 50 mM Tris–HCl pH 7.5, 1 mM EDTA, 150 mM NaCl, 0.5% Tween 20
- RIPA buffer: 150 mM NaCl, 1% NP-40, 0.5% Na deoxycholate, 0.1% SDS, 50 mM Tris–HCl pH 8
- 1% SDS

Method

1. Grow cells in synthetic medium until they reach the concentration of 5×10^6 cells/ml.
2. Add 250 μCi [^{35}S]methionine/ml of culture for 3–10 min.
3. Chase by adding cold methionine at a final concentration of 2 mM.
4. At each time point after the chase withdraw 1 ml of culture, add 100 μl 50% cold TCA, and keep in ice for at least 10 min.
5. Spin the cells down, wash them twice with 1 ml cold acetone, resuspend them in 50 μl buffer A, and an equal volume of glass beads.
6. Vortex for 4 min and boil for 5 min.
7. Add 1 ml IP buffer, vortex for 30 sec, and recover crude protein extract by centrifugation for 2 min in a microcentrifuge.
8. Spot 3 μl of extract on GF/C filters and keep them for 3 min in 10% boiling TCA.
9. Wash the filters three times with H_2O, twice with ethanol, and count them in a liquid scintillation apparatus.
10. Adjust aliquots of extracts containing 5×10^7 total counts for each time point to 1 ml with IP buffer.
11. Pre-clear the extracts overnight at 4°C by absorption with 60 μl Gamma Bind Plus suspension.
12. Pre-bind 5 μg antibody to 60 μl Gamma Bind Plus suspension and incubate with the pre-cleared extracts for 1 h at 4°C.

Protocol 11. *Continued*

13. After centrifugation, wash the immunoprecipitate four times with RIPA buffer, and resuspend the final pellet in 100 μl 1% SDS before boiling for 4 min.
14. Spin the samples for 2 min in a microcentrifuge and add 900 μl IP buffer to the supernatant.
15. Add another aliquot of antibody pre-bound to 60 μl Gamma Bind Plus suspension and repeat steps 12–15. This second immunoprecipitation greatly reduces the background.
16. Analyse the samples by SDS–PAGE and autoradiography.

Metabolic labelling with [^{32}P]orthophosphate is required to firmly establish whether a protein is phosphorylated, although this *in vivo* labelling technique is quite inefficient in *S. cerevisiae* cells. An alternative, but indirect method is based on the detection of isoforms with different electrophoretic mobility which are sensitive to phosphatase treatments (24).

Protocol 12. *In vivo* labelling of yeast proteins with [^{32}P]orthophosphate

Equipment and reagents

- Glass beads
- [^{32}P]orthophosphate
- Wickerham's medium (2)
- Sodium phosphate
- TE: 10 mM Tris–HCl pH 8, 1 mM EDTA
- 5% TCA

Method

1. Grow cells in 50 ml Wickerham's medium (2) supplemented with 100 μM sodium phosphate until they reach the concentration of 5×10^6 cells/ml.
2. Spin down the cells, wash, and resuspend them in an equal volume of the same medium without sodium phosphate. Incubate the cells until they reach the concentration of 10^7 cells/ml.
3. Spin down the cells, wash them twice with 10 mM Tris–HCl pH 8, 1 mM EDTA, and resuspend them in 300 μl Wickerham's salts, 0.5% glucose.
4. Add 0.5–1 μCi [^{32}P]orthophosphate and incubate the sample for 2 h at 30 °C.
5. Spin down the cells, resuspend them in 1 ml 5% cold TCA, and leave the sample in ice for at least 10 min.
6. Extract preparation and immunoprecipitation are performed as in *Protocol 11*, steps 5–16. Process at least 10^9 total c.p.m. for immunoprecipitation.

A common problem in molecular biology is to establish whether two proteins physically interact or whether they belong to the same pathway. *S. cerevisiae* offers several advantages to address these problems. The epitope tagging approach, which allows the surveillance of the protein of interest using specific monoclonal antibodies (8), is routinely used in yeast to study cellular localization of the tagged polypeptide, its post-translational modifications, and interactions with other proteins. Moreover, *S. cerevisiae* offers the possibility to test rapidly whether the tagged proteins mantain their normal function *in vivo*. Protein–protein interactions can be studied using the two-hybrid technique (25) which allows one to identify and characterize in great detail the protein domains involved in the physical interaction. Finally, genetic approaches such as synthetic lethality screening, in which one mutation exacerbates the severity of another (26), can be very helpful in defining the components of specific pathways or subunits of the same complex.

References

1. Pringle, J.R. and Hartwell, L.H. (1981). In *The molecular biology of the yeast Saccharomyces*, Vol. 1, p. 97. Cold Spring Harbor Laboratory Press, NY.
2. Sherman, F. (1991). In *Methods in enzymology* (ed. C. Guthrie and G.R. Fink), Vol. 194, p. 3. Academic Press, London.
3. Thorner, J. (1981). In *The molecular biology of the yeast Saccharomyces*, Vol. 1, p. 143. Cold Spring Harbor Laboratory Press, NY.
4. MacKay, V.L., Welch, S.K., Insley, M.Y., Manney, T.R., Holly, J., Saari, G.C., *et al.* (1988). *Proc. Natl. Acad. Sci. USA*, **85**, 55.
5. Piatti, S., Böhm, T., Cocker, J.H., Diffley, J.F.X., and Nasmyth, K. (1996). *Genes Dev.*, **10**, 1516.
6. Richardson, H.E., Wittenberg, C., Cross, R., and Reed, S.I. (1989). *Cell*, **59**, 1127.
7. Amon, A., Irniger, S., and Nasmyth, K. (1994). *Cell*, **77**, 1037.
8. Guthrie, C. and Fink, G.R. (ed.) (1991). *Methods in enzymology*, Vol. 194. Academic Press, London.
9. Longhese, M.P., Fraschini, R., Plevani, P., and Lucchini, G. (1996). *Mol. Cell. Biol.*, **16**, 3235.
10. Hartwell, L.H. (1976). *J. Mol. Biol.*, **104**, 803.
11. Hogan, E. and Koshland, D. (1992). *Proc. Natl. Acad. Sci. USA*, **89**, 3098.
12. Lea, D.E. and Coulson, C.A. (1948). *J. Genet.*, **49**, 264.
13. Longhese, M.P., Jovine, L., Plevani, P., and Lucchini, G. (1993). *Genetics*, **133**, 183.
14. Marini, F., Pellicioli, A., Paciotti, V., Lucchini, G., Plevani, P., Stern, D.F., *et al.* (1997). *EMBO J.*, **16**, 639.
15. Chang, L.M.S. (1977). *J. Biol. Chem.*, **252**, 1873.
16. Sun, Z., Fay, D. S., Marini, F., Foiani, M., and Stern, D.F. (1996). *Genes Dev.*, **10**, 395.
17. Weinert, T.A. and Hartwell, L.H. (1988). *Science*, **241**, 317.
18. Navas, T.A., Zhou, Z., and Elledge, S. (1995). *Cell*, **80**, 29.
19. Longhese, M.P., Neecke, H., Paciotti, V., Lucchini, G., and Plevani, P. (1996). *Nucleic Acids Res.*, **24**, 3533.

20. Paulovich, A.G. and Hartwell, L.H. (1995). *Cell*, **82**, 847.
21. Johnston, L.H. and Lowndes, N.F. (1992). *Nucleic Acids Res.*, **20**, 2403.
22. Reid, G. and Shatz, G. (1982). *J. Biol. Chem.*, **257**, 13056.
23. Kolodziej, P.A. and Young, R.A. (1991). In *Methods in enzymology* (ed. C. Guthrie and G.R. Fink), Vol. 194, p. 508. Academic Press, London.
24. Foiani, M., Liberi, G., Lucchini, G., and Plevani, P. (1995). *Mol. Cell. Biol.*, **15**, 883.
25. Chien, C.T., Bartel, P.L., Sternglanz, R., and Fields, S. (1991). *Proc. Natl. Acad. Sci. USA*, **88**, 9578.
26. Guarente, L. (1993). *Trends Genet.*, **9**, 362.

9

The use of *Xenopus laevis* interphase egg extracts to study genomic DNA replication

JOHANNES WALTER and JOHN NEWPORT

1. Introduction

Cell-free extracts derived from the eggs of *Xenopus laevis* are a powerful experimental system to study genomic DNA replication in higher eukaryotes (1–3). DNA templates (either plasmid DNA or *Xenopus* sperm chromatin) added to these extracts are packaged into nuclei and then replicated under normal cell cycle control. This process occurs rapidly and efficiently and can be extensively manipulated in order to elucidate underlying biochemical mechanisms. This *in vitro* system has proven extremely useful for studying initiation and elongation of genomic DNA replication (4–10), the spatial organization and structural requirements for DNA replication (3, 11–16), the relationship of S phase to other phases of the cell cycle (17–20), and the regulatory system which limits DNA replication to once per cell cycle (6–8, 21, 22).

In this article, we will discuss the most common experimental approaches currently used to study mechanistic and regulatory aspects of S phase in *Xenopus* egg extracts. First, we will describe how to prepare interphase *Xenopus* egg extracts which are highly efficient for DNA replication. Next, we will present several methods to assay DNA replication in these extracts. Finally, protocols will be described which are aimed at understanding what role different proteins play in the replication process. Although it will not be discussed here, *Xenopus* can also be used to study DNA replication *in vivo* (23).

2. Egg collection and extract preparation

2.1 Induction of oocyte maturation and ovulation

Unfertilized eggs from *Xenopus* female frogs are the source of the interphase extract used to study DNA replication. *Xenopus laevis* females can either be purchased from a commercial distributor (NASCO, *Xenopus-1*) or maintained in the laboratory as a colony. For laboratory maintenance, female frogs are kept in water tanks containing 40 litres of tap-water at a density of one animal per litre. The water temperature should be between 17–19 °C, and it is changed twice a week. Frogs are sensitive to chlorine. In most areas, chlorine concentrations in tap-water are low enough to use the water untreated. However, if necessary, water can be dechlorinated by letting it stand overnight or by adding sodium thiosulfate to a final concentration of 0.1 mM. If the temperature of the water exceeds 21 °C, poor egg quality may result.

Frogs are primed for ovulation by injecting each frog with 100 units (0.5 ml) of pregnant mare serum gonadotropin (Calbiochem) two to eight days before the eggs are required. The gonadotropin is introduced into the dorsal lymph sac along the thigh of the hind leg by subcutaneous injection using a 27 gauge needle. Frogs calm down enough to be injected when their eyes are covered. This can be achieved either by wrapping them in a wet paper towel or by holding them in one hand (usually the left) so that their eyes are pressed into the palm of the hand. The latter technique is more convenient because it leaves the right hand free to perform the injection. To do this, grab the frog from the top with your index finger going between the hind legs and the thumb and other fingers wrapping around the torso with your smallest finger pressing the frog's head into the palm of your hand. To induce ovulation and egg laying, inject the frogs with 400 units (0.4 ml) of human chorionic gonadotropin (hcg; USB) between 16–20 hours before the eggs are needed. The injection technique is the same as for the primary injection of PMSG. After use, the frogs should be allowed to recover for four to six months to insure a good yield of eggs the next time they are used.

After inducing ovulation with the hcg, each frog is placed in a separate tank containing 2.5–5 litres of 100 mM NaCl at 17–19 °C. A lid with holes is placed on the tank to prevent frogs jumping out of the tank. It is important to maintain frogs in separate tanks so that poor quality eggs laid by a particular frog do not contaminate high quality eggs produced by another frog. Frogs generally start laying eggs 8–10 hours after injection and they may lay for up to 20 hours. Eggs are recovered by pouring off most of the excess water from each tank and decanting the eggs into a 200 ml glass beaker.

2.2 Preparation of low speed interphase extracts (LS extracts)

Mature unfertilized *Xenopus* eggs are arrested in metaphase of the second meiotic division. They are maintained in this state by high levels of maturation promoting factor (MPF) which consists of the cdc2 kinase and its regulatory subunit cyclin B (24, 25). When eggs are crushed in the absence of calcium chelators, the release of calcium from intracellular stores initiates a process that leads to the proteolysis of cyclin B. As a result, MPF activity drops, and the extract leaves the mitotic state and enters interphase. Eggs can also be

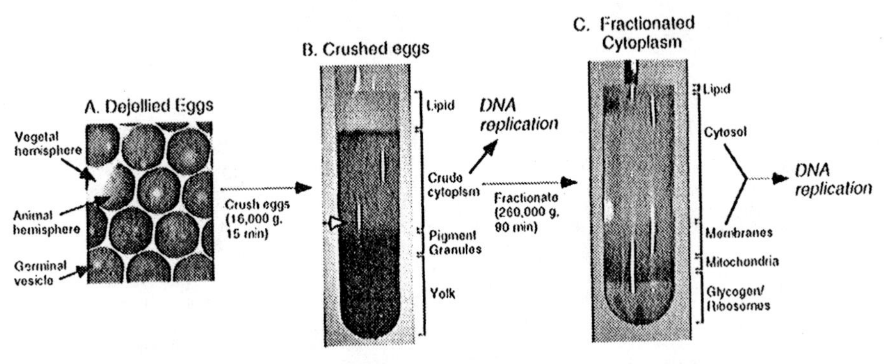

Figure 1. Steps in the preparation of unfractionated and fractionated interphase extracts from *Xenopus* eggs. (A) Dejellied eggs. (B) Eggs crushed by centrifugation; the white arrowhead indicates where the needle should be injected to withdraw the crude cytoplasm. (C) Crude cytoplasm after fractionation by ultracentrifugation.

activated prior to crushing either by adding calcium ionophore or by electrical activation (26).

When preparing the extract (*Protocol 1*), abnormal appearing eggs are avoided because they will result in an extract of poor quality. Healthy eggs have a uniformly pigmented brown or green animal hemisphere with a white spot in the centre called the germinal vesicle (*Figure 1A*). Abnormal eggs should be discarded. These include large, whitish eggs, eggs in which the pigment looks severely mottled or variegated, or eggs which lack a germinal vesicle (oocytes). If more than about 10% of the eggs in a batch from one frog are abnormal, the batch should be discarded. Small numbers of abnormal eggs can be removed with a Pasteur pipette after the eggs have been washed with ELB, but it is not necessary to remove every single abnormal egg. If a batch of eggs undergoes continual egg lysis or necrosis during the washing steps, it should be discarded.

Protocol 1. Preparation of low speed (LS) interphase egg extract

Equipment and reagents

- Clinical centrifuge
- Sorvall RC refrigerated centrifuge
- HB-4 swinging-bucket rotor
- 2% cysteine pH 7.7, freshly made
- MMR solution: 100 mM NaCl, 2 mM KCl, 1 mM $MgSO_4$, 2 mM $CaCl_2$, 0.1 mM EDTA, 5 mM Hepes pH 7.8
- 1 M DTT, stored at −20 °C
- 10 mg/ml cycloheximide, stored at 4 °C
- 10 mg/ml aprotinin (Boehringer Mannheim) dissolved in water, stored in small aliquots at −20 °C
- Egg lysis buffer (ELB): 250 mM sucrose, 2.5 mM $MgCl_2$, 50 mM KCl, 10 mM Hepes pH 7.7, 50 μg/ml cycloheximide, 1 mM DTT—ELB is made on the day of the experiment by combining water, 1 M sucrose, and a 10 × stock of $MgCl_2$, KCl, and Hepes in appropriate amounts, and then adding DTT and cycloheximide from stock solutions
- 10 mg/ml leupeptin (Boehringer Mannheim) dissolved in water, stored in small aliquots at −20 °C
- 5 mg/ml cytochalasin B (Sigma) dissolved in DMSO, stored in small aliquots at −20 °C

Protocol 1. *Continued*

Method

1. Decant as much of the 100 mM NaCl solution from each batch of eggs as possible, and dejelly each batch separately by adding 100 ml 2% cysteine at room temperature. Incubate with periodic swirling for as long as it takes for eggs to dejelly (usually 4–6 min). Eggs are dejellied once they pack closely against each other (*Figure 1A*) and jelly coats are seen floating above the eggs.
2. Decant the cysteine and wash the eggs three times with about 50 ml of one-quarter strength MMR (0.25 × MMR) at room temperature.
3. Wash the eggs twice with ELB at room temperature. At this point, remove abnormal eggs using a long Pasteur pipette. After sorting, combine all good batches of eggs and wash once more in ELB.
4. Transfer eggs to a 15 ml polypropylene Falcon 2059 tube. Allow eggs to settle and remove the supernatant with a Pasteur pipette. Pack the eggs by spinning at 170 *g* (1000 r.p.m.) in a clinical centrifuge for 30 sec at room temperature. Remove all the supernatant with a Pasteur pipette. Add aprotinin, leupeptin, and cytochalasin B (1/2000 the volume of the packed eggs). Do not cool the eggs before the centrifugation in step 5 as this will impede egg activation.
5. Lyse the packed eggs in a Sorvall RC centrifuge by spinning at 10 000 r.p.m. (16 000 *g*) for 15 min at 4°C in an HB-4 swinging-bucket rotor. From now on, keep the lysate on ice.
6. Insert a 21 gauge needle 4 mm above the dark pigment layer (*Figure 1B*, white arrowhead). Remove this needle and immediately insert another 21 gauge needle (bevel up) attached to a syringe of the appropriate size and withdraw the crude cytoplasmic extract until the lipids begin to contaminate the extract. If significant contamination by pigment or lipids occurs, recentrifuge the extract for 10 min as in step 5 and harvest the extract as before.
7. To 1 ml of crude extract, add 10 μl cycloheximide, and 1 μl each of DTT, aprotinin, leupeptin, and cytochalasin B. The extract is stable for up to 3 h when stored on ice, and it can be diluted by up to 30% with ELB without any loss in the efficiency of nucleus formation or DNA replication.

2.3 Freezing low speed egg extracts

It is desirable to be able to freeze replication competent extracts so that it is not necessary to prepare a fresh extract every day and so that an extended series of experiments can be performed on the same extract. There are two ways to freeze interphase extracts. One way is to fractionate the extract into cytosolic and membrane fractions which are frozen separately (see below). Alternatively, the unfractionated extract can be frozen directly (2, 27), but it has been reported that it takes four to seven hours for all the nuclei assembled in these extracts to complete replication. Below, we describe a similar procedure for freezing unfractionated S phase extract. We find that these extracts complete replication almost as fast as fresh extracts, which is usually in 60–90 minutes (J. Fang and J. Newport, unpublished results). However, in our experience, there is considerable variability between extracts prepared by this procedure; for example, some extracts become apoptotic after incubation at room temperature. Apoptosis in these extracts, which may occur immediately or after several hours, is marked by fragmentation of the chromatin and disintegration of the nucleus (28).

Protocol 2. Freezing LS egg extracts

Equipment and reagents

- See *Protocol 1* (but no ELB is required)
- Extraction buffer: 50 mM Hepes pH 7.6, 50 mM KCl, 5 mM $MgCl_2$, 2 mM 2-mercaptoethanol
- 2 mg/ml calcium ionophore A23187 (Sigma) dissolved in DMSO
- 50% glycerol (filter sterilized)
- Liquid nitrogen

Method

1. Dejelly the eggs and wash them three times in 0.25 × MMR as described in *Protocol 1*.
2. Activate the washed eggs in 0.25 × MMR using the Ca ionophore at a concentration of 0.2 μg/ml. As soon as the pigment on the animal hemisphere begins to contract, wash once with 0.25 × MMR.
3. Wash the eggs three times with ice-cold extraction buffer and place the eggs on ice.
4. Follow *Protocol 1*, steps 4–6.
5. Recentrifuge and isolate the cytosol as in *Protocol 1*, steps 5 and 6.
6. Add glycerol to a final concentration of 3%, and quick-freeze 50 μl aliquots of extract in 0.5 ml Eppendorf tubes by immersion in liquid nitrogen; store extract at –70°C. To use, quickly thaw between fingers.

2.4 Preparation of high speed fractionated egg extracts (HS + M extracts)

The LS extract described above can be fractionated into a cytosolic fraction and a membrane fraction which are frozen separately (3). The cytosolic fraction alone supports decondensation of sperm chromatin (*Figure 2*, 5 min time point) and the assembly of many replication factors onto the chromatin (5–9). In addition, single-stranded DNA can be replicated in this fraction

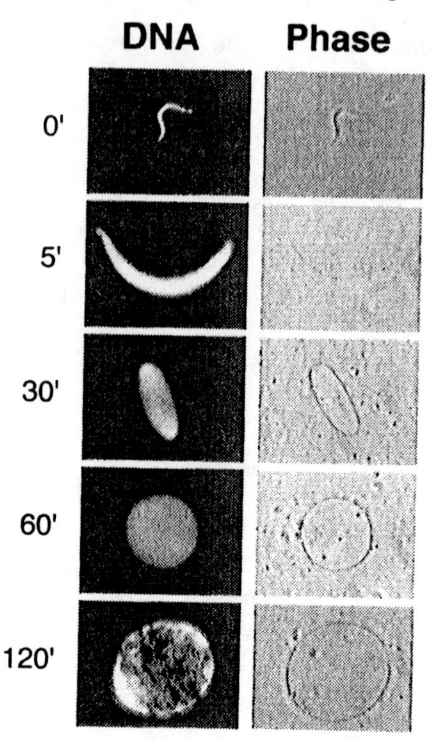

Figure 2. Nuclear assembly in egg extracts. Demembranated sperm chromatin (2000/μl) was added to an unfractionated extract containing nocodazole and an ATP regeneration system (see *Protocol 5*). Aliquots were removed at different times after mixing the sperm with the extract and examined using the DAPI channel of a fluorescence microscope to visualize the DNA (left column; see *Protocol 5*). Each sample was also examined with phase-contrast optics to visualize the nuclear envelope (right column). A robust nuclear envelope first becomes apparent after 30 min.

(*Protocol 5*). However, sperm incubated in the cytosol does not have a functional nuclear envelope and chromosomal DNA replication does not occur (3). When membrane and cytosolic fractions are present, a fully functional nucleus which can carry out chromosomal DNA replication is formed. In our experience, nuclei formed in this type of extract are almost as active for DNA replication as nuclei formed in LS extracts.

Protocol 3. Preparation of fractionated interphase egg extract

Equipment and reagents

- See *Protocol 1*
- Beckman TL-100 table-top ultracentrifuge, TL-55 swinging-bucket rotor, and appropriate tubes

Method

1. Centrifuge the fresh LS extract containing cycloheximide, leupeptin, aprotinin, and DTT (*Protocol 1*) at 55 000 r.p.m. (260 000 *g*) for 90 min at 2 °C in a TL-55 swinging-bucket rotor using a TL-100 centrifuge. This separates the LS extract into a number of distinct fractions shown in *Figure 1C*.
2. Remove any lipid from the top of the gradient by aspiration using a pulled Pasteur pipette. Then remove the cytoplasmic layer using a p200 Pipettman and a cut-off tip, and be careful to avoid the underlying membrane layer.
3. Recentrifuge the cytoplasmic fraction for 30 min as in step 1 and harvest as before. Freeze the cytosolic fraction in liquid nitrogen in 0.5 ml Eppendorf tubes as 20–75 μl aliquots.
4. While carrying out the second centrifugation of the cytoplasmic fraction, remove the golden layer of membranes which was located just beneath the cytoplasmic layer after the first spin (*Figure 1C*). Use p200 Pipettman and a cut-off tip to remove the membrane layer, being careful to avoid the tan underlying layer of mitochondria (*Figure 1C*).
5. Wash the membranes with five volumes of ELB containing 10 μg/ml aprotinin and leupeptin and 5 μg/ml cytochalasin B. Mix well by inverting and incubate on ice for 10–15 min.
6. Layer the membranes over 200 μl ELB containing 0.5 M (instead of 0.25 M) sucrose, and centrifuge at 22 000 r.p.m. (34 000 *g*) for 20 min at 2 °C in the TL-55 swinging-bucket rotor using a TL-100 centrifuge.
7. Aspirate the supernatant. Stir the membranes thoroughly with a p200 Pipettman and a cut-off pipette tip, and then very slowly pipette them up and down while avoiding air bubbles. Then freeze the membranes as aliquots which are one-tenth the size of the cytoplasmic aliquots. If the membranes are too viscous to be pipetted, add 0.2 vol. ELB containing 0.5 M sucrose, mix, and freeze as above. Membrane and cytoplasmic fractions are stable for at least a year when stored at −70 °C.

2.5 Mitotic extracts and cycling extracts

In some cases, it is desirable to examine the fate of replication factors in mitosis and during the transition from mitosis to interphase. In this case, one can prepare a cytostatic factor (CSF) arrested extract which is arrested in mitosis and can be induced to enter interphase. As described in Section 2.3, mature unfertilized *Xenopus* eggs are arrested in metaphase of the second meiotic division by the activity of CSF which maintains high levels of MPF in the egg. If eggs are crushed in the presence of EGTA, the release of calcium which normally leads to the inactivation of CSF and MPF is prevented, and the extract is arrested in mitosis. When demembranated sperm chromatin is added to this CSF arrested extract it will form mitotic chromosomes. If calcium is added to 0.5 mM, CSF and MPF are inactivated, and the extract enters interphase. As a result, the mitotic chromosomes are assembled into a nucleus and the DNA is replicated. If no cycloheximide is present, MPF activity reaccumulates, and the extract enters mitosis anew, where it usually arrests (26). If one wants to carry out multiple cell cycles *in vitro*, a so-called cycling extract is prepared (17, 26, 29). In this case, eggs are activated electrically or by the addition of calcium ionophore prior to crushing to insure complete release from the mitotic state. The extract is then prepared in the absence of cycloheximide so that cyclins can be synthesized repeatedly in every cell cycle. Demembranated sperm chromatin added to a cycling extract will be assembled into nuclei which will repeatedly undergo DNA replication and division. The preparation of the CSF arrested and cycling extracts is described elsewhere in detail (26, 29).

2.6 Preparation of demembranated sperm chromatin

Demembranated *Xenopus* sperm chromatin is the best source of DNA for replication in egg extracts as it supports highly efficient nuclear assembly and is usually replicated completely.

Protocol 4. Preparation of demembranated *Xenopus* sperm chromatin

Equipment and reagents

- Clinical centrifuge
- Beckman TL-100 table-top ultracentrifuge, TL-55 swinging-bucket rotor
- Beckman ultra clear 11 × 34 mm tubes
- 10 × buffer X: 100 mM Hepes pH 7.4, 800 mM KCl, 150 mM NaCl, 50 mM $MgCl_2$, 10 mM EDTA; mix with solid sucrose and BSA where appropriate

Method

1. All procedures are carried out at room temperature unless stated otherwise. Anaesthetize four to six male frogs by immersion in ice water until they no longer right themselves when turned on their backs (about 10–15 min). Kill the animals by cranial dislocation and pithing. Open the peritoneal cavity through a mid line incision through the abdominal wall using dissection scissors and move the organs of the intestinal tract to one side. The almond shaped testes are greyish in colour and about 5–8 mm long, and they are located in the mid body, on either side of the mid line. Remove them by cutting them at the base using dissection scissors. Blot off blood and place them in a small Petri dish containing buffer X plus 200 mM sucrose.

2. Mince the testes into 1–2 mm pieces using two pairs of forceps; this releases the sperm.
3. Transfer the minced testes to a 15 ml screw-cap polypropylene tube. Rinse out the Petri dish with buffer X, 200 mM sucrose and combine with the testes. Vortex the testes vigorously (1 min), and pellet the larger pieces by a mild centrifugation in a clinical table-top centrifuge (1000 *g* for 1 min).
4. Transfer the supernatant to a new 15 ml screw-cap polypropylene tube. Add 2–3 ml buffer X, 200 mM sucrose to the pellet, vortex 1 min, and recentrifuge as above. Combine the supernatants and repeat the extraction of the pellet two or three times until the supernatant is not very cloudy.
5. Centrifuge the combined supernatants in the clinical centrifuge (1700 *g*, 40 sec) to pellet the larger pieces of tissue. Transfer the supernatant to a 15 ml Falcon tube and pellet the sperm by centrifugation in an HB-4 swinging-bucket rotor at 4000 r.p.m. (2500 *g*) and 4°C for 10 min in a Sorvall centrifuge.
6. Prepare sucrose step gradients in four 2.5 ml tubes for the TL-100 rotor. To make each gradient, overlay 0.25 ml buffer X, 2.5 M sucrose with 1.7 ml buffer X, 2.3 M sucrose.
7. Completely resuspend the sperm pellet in 0.8 ml buffer X, 2 M sucrose by pipetting up and down. Overlay the sucrose gradient with the sperm (0.2–0.4 ml/tube). Completely disrupt the interface between the sperm and the 2.3 M sucrose by stirring with a Pasteur pipette tip. Centrifuge at 33 000 r.p.m. and 2°C for 25 min in a TL-100 ultracentrifuge. The red blood cells should band on top of the 2.3 M sucrose layer. Part of the sperm bands on top of the 2.5 M sucrose cushion but most of the sperm goes to the bottom of the tube. If there is still a cloudy layer of sperm on top of the 2.3 M sucrose layer, the interface between this and the above layer should be disrupted again and the sample respun for 15–20 min to collect the rest of the sperm.
8. Aspirate off the top-half of the gradient containing the red blood cells. Resuspend the sperm in the lower-half of the gradient by pipetting up and down, then transfer to a 15 ml Falcon tube. Be careful not to take any red blood cells from the top-half of the gradient.
9. Dilute the sperm to 12 ml with buffer X, 0.2 M sucrose. Pellet the sperm by centrifugation in an HB-4 rotor at 5000 r.p.m. (4080 *g*) and 4°C for 10 min.
10. Resuspend the sperm pellet in 1 ml buffer X, 0.2 M sucrose, 0.4% Triton X-100, containing aprotinin, leupeptin, and DTT. Incubate on ice for 30 min.

Protocol 4. *Continued*

11. Prepare two 1.5 ml microcentrifuge tubes containing 0.5 ml buffer X, 0.5 M sucrose, 3% BSA (plus aprotinin, leupeptin, and DTT). Overlay each sucrose cushion with half of the sperm preparation. Centrifuge in an IEC clinical table-top centrifuge at setting 5 (4000 *g*) for 10 min at room temperature.
12. Remove the supernatant and resuspend the pellet in 0.1 ml buffer X, 0.2 M sucrose, 3% BSA (plus aprotinin, leupeptin, and DTT). Avoid the sides of the tubes containing residual Triton X-100. Transfer the sperm chromatin to a new 1.5 ml microcentrifuge tube, dilute to 1 ml with the same buffer mix, and centrifuge as in step 11.
13. Resuspend the sperm chromatin pellet in 0.5 ml buffer X, 0.2 M sucrose, 3% BSA (plus aprotinin, leupeptin, and DTT). Count the sperm in a haemocytometer using 8 μg/ml Hoechst dye and a fluorescence microscope (DAPI channel) to visualize the sperm. Dilute the sperm to 50 000/μl and freeze as 5 μl aliquots in liquid N_2. Store at −70 °C. Typical yields are $2–4 \times 10^7$ sperm per frog.

3. DNA replication assays

3.1 Monitoring nuclear formation and DNA replication

A complete or nearly complete round of genomic DNA replication can be achieved by adding demembranated sperm DNA to a LS egg extract or to a HS egg extract which has been mixed with the purified membrane fraction (HS + M). DNA replication can also be carried out using purified DNA such as lambda phage or plasmid DNA, but it is significantly less efficient than using sperm DNA (2, 3).

In order to carry out efficient DNA replication in egg extracts, the extract is supplemented with several components (see *Protocol 5*):

(a) An ATP regeneration system is added to supply a continual source of ATP.

(b) Nocodazole is added to depolymerize microtubules and to reduce viscosity in the reaction.

(c) Deoxynucleotide triphosphates are added in case a large number of nuclei are to be replicated in the extract.

(d) A radioactively labelled deoxynucleotide is added to detect DNA replication.

All of these components, including the DNA to be replicated, should not exceed 30% of the volume of the extract to insure efficient and rapid DNA replication. Reactions are carried out at 20–22 °C.

Following the addition of sperm DNA to the extract, an ordered series of events leads to formation of nuclei and DNA replication (2, 3) (see *Figure 2*). Briefly, after five minutes in the extract, the sperm increase their volume approximately 30-fold as histones and other chromatin proteins are added. After 15–30 minutes, the sperm thicken and become cigar-shaped as the nuclear envelope is assembled around the sperm and transport across the nuclear envelope begins. DNA replication begins at this stage. As replication proceeds, the nuclei continue to grow, becoming spherical. At low concentrations of sperm (less than 2000/μl), individual nuclei take only about 15 minutes to replicate although the synthetic period appears longer when measured in an ensemble of nuclei due to asynchrony between individual nuclei (30, 31). Replicon size in these nuclei is about 7 kb and the rate of polymerization is about 500 nucleotides/minute (31, 32). At higher concentrations of sperm, nuclei take longer to replicate due to larger replicon sizes which can reach 60 kb at 10 000 nuclei/μl (31). Thus, the concentration of nuclei can be altered to adjust both replicon size and the length of S phase. It should be noted that only extracts made from the highest quality eggs will completely replicate 10 000 nuclei/μl.

In many instances, it is useful to carry out DNA replication using M13 single-stranded DNA as a template. For example, if the origin recognition complex (ORC) has been immunodepleted from the extract (see *Protocol 10*) there is very little replication. In order to demonstrate that the block to replication does not result from non-specific inactivation of the entire replication machinery, replication with M13 single-stranded DNA is carried out. Since the template is single-stranded, it should be replicated in the absence of initiation factors such as ORC. Indeed, this is the case (5). To perform this control, follow *Protocol 5*, but substitute 0.25 ng/μl M13 DNA for the demembranated sperm chromatin. It is important to note that replication of M13 DNA does not require the presence of nuclei and therefore can be carried out in HS extract.

Protocol 5. Basic DNA replication assay using agarose gel electrophoresis

Reagents

- 0.2 M ATP (Sigma): dissolved in water, pH adjusted to 7, stored at –20°C in 50 μl aliquots
- 1 M phosphocreatine (Sigma): dissolved in 10 mM KPO_4 pH 7, stored at –20°C as 50 μl aliquots
- 5 mg/ml creatine phosphokinase (Sigma): dissolved in 10 mM Hepes pH 7.5, 50% glycerol, stored at –20°C
- 0.5 mg/ml nocodazole (Sigma): dissolved in DMSO, stored at –20°C
- 6 mM dNTPS (Pharmacia)
- 3000 Ci/mmole [α-^{32}P]dATP
- HS egg extract and membrane fraction (see *Protocol 3*) or LS extract (*Protocol 1*)
- Demembranated sperm chromatin (*Protocol 4*), purified plasmid DNA, or lambda DNA
- Stop solution: 8 mM EDTA, 0.13% phosphoric acid, 10% Ficoll, 5% SDS, 0.2% bromphenol blue, 80 mM Tris pH 8
- 20 mg/ml proteinase K: dissolved in water, stored at –20°C as 50 μl aliquots
- Hoechst fix: 200 mM sucrose, 10 mM Hepes pH 7.6, 7.4% formaldehyde, 8 μg/ml Hoechst dye

Protocol 5. *Continued*

Method

1. Remove aliquots of purified membranes and HS extract from the freezer and immediately begin thawing the membrane fraction by holding it between fingers; then store it on ice. Combine HS extract and membranes in a 10:1 ratio and mix thoroughly by pipetting up and down. Store on ice while adding the components listed below.
2. Add phosphocreatine to a final concentration of 20 mM, ATP to a final concentration of 2 mM, and creatine phosphokinase to a final concentration of 5 μg/ml. This is most easily done by mixing these three components together first before adding them to the extract (combine 10 μl phosphocreatine, 5 μl ATP, and 0.5 μl creatine phosphokinase; this is the amount needed for 500 μl of extract). Add nocodazole to a final concentration of 3 μg/ml.[a]
3. Add [α-^{32}P]dATP to 0.1 μCi/μl of extract.
4. Add demembranated sperm to the desired concentration and mix by pipetting up and down ten times. For most purposes, 1000 sperm/μl gives the best results since at this concentration S phase is not extended. Transfer the tube to room temperature (22 °C) to start the reaction. To monitor the progress of nuclear formation, remove 1 μl of the reaction and add it to 1 μl Hoechst fix on a microscope slide, and cover the mixture with a 12 mm diameter coverslip. Examine the nuclei in a fluorescence microscope (*Figure 2*).
5. To monitor the extent of replication, place 3 μl aliquots of the reaction in 0.5 ml Eppendorf tubes once all the components have been mixed together. Then add 5 μl stop solution at the desired time to terminate the replication reaction.
6. When all time points have been collected, add 1 μl proteinase K solution to each time point and incubate for 60 min at 37 °C. After digestion, vortex vigorously to reduce viscosity.
7. Load each time point on a 0.8% TBE agarose gel. Run the gel until the bromphenol blue has travelled at least 1.5 inches. Cut the gel just above the dye front and discard the bottom part of the gel. Dry the rest of the gel, first by placing it between filter paper and paper towels for 15 min with a weight; then place it on a gel dryer supported by filter paper and dry at 80 °C. To quantify the amount of DNA replication at each time point, expose the gel to a PhosphorImager.

[a] The amount of DMSO in the reaction should not exceed 2% as this will inhibit DNA replication. Also, if replicating more than 8000 sperm/μl extract, add 40 μM of each dNTP.

3.2 Denaturing gel electrophoresis

In many cases it is necessary to examine daughter DNA strands directly during DNA replication, and this is done using a denaturing gel (*Protocol 6*). This procedure is useful for following the early stages of DNA replication, and it is most powerfully applied in conjunction with a synchronization protocol which causes replication to initiate at the same time at all origins of replication in the reaction (Section 3.4).

Protocol 6. Alkaline gel analysis of replication products

Reagents

- See *Protocol 5*
- 1 M spermine, 1 M spermidine (dissolved in water, stored at −20°C)
- Buffer A: 5 mM EDTA, 20 mM Hepes pH 7.6, 50 mM NaCl
- Gel buffer: 50 mM NaCl, 1 mM EDTA
- 5 × alkaline loading buffer: 300 mM NaOH, 6 mM EDTA, 18% Ficoll, 0.15% bromcresol blue, 0.25% xylene cyanol FF
- Running buffer: 1 mM EDTA, 30 mM NaOH
- 100% trichloroacetic acid
- DEAE paper

Method

1. First prepare the alkaline gel: cast a 0.8–1.5% gel in gel buffer. Once the agarose has polymerized, cover the gel in running buffer, and incubate for at least 1 h before use.
2. Carry out the replication reaction as described in *Protocol 5*, except that three to five times more [α-^{32}P]dATP is used so that short replication products can be detected.
3. Remove 5 μl aliquots of the reaction at appropriate times and mix with 500 μl ice-cold buffer A containing 0.5 mM spermine and 0.5 mM spermidine.
4. When all samples have been collected, isolate nuclei by centrifugation for 5 min at 14000 r.p.m. (16000 *g*), in a microcentrifuge at 4°C. Discard the supernatant.
5. Resuspend the pellet in 50 μl buffer A containing 0.5% SDS and 0.5 mg/ml proteinase K. Incubate for 60 min at 37°C.
6. Phenol:chloroform extract once, chloroform extract once, and ethanol precipitate using 0.3 M NaAc as a salt and 40 μg of carrier RNA.
7. Resuspend pellets in an appropriate volume of alkaline loading buffer and load onto the alkaline denaturing gel. Run the gel at 3 V/cm for the desired distance. Recirculate the running buffer from the back to the front reservoir using a peristaltic pump.
8. Fix the gel for 30 min in 7% trichloroacetic acid. Place the gel between

Protocol 6. *Continued*

two sheets of DEAE paper and two sheets of Whatman paper. Dry the gel for 15 min by placing it between paper towels, then dry it under vacuum supported by two pieces of Whatman filter paper.[a] Expose the gel to film or to a PhosphorImager.

[a] The vacuum pump should be cleaned immediately after use to remove the trichloroacetic acid.

3.3 BrdU substitution

An important technique for the study of DNA replication is equilibrium density substitution (*Protocol 7*). DNA replication is carried out in the presence of a deoxynucleotide analogue, such as bromo-dUTP (BrdU), which is more dense than its counterpart, dTTP and radioactive dATP. The replicated DNA is analysed on a CsCl gradient which separates the DNA according to density. DNA which has not incorporated BrdU will migrate at the normal density of DNA which is about 1.7 g/ml, DNA in which one strand has most or all T residues substituted with BrdU migrates at about 1.75 g/ml, and fully substituted DNA migrates at about 1.8 g/ml. This technique addresses two important questions. First, was DNA replication semi-conservative? If so, DNA should migrate at the intermediate density. Secondly, how many rounds of DNA replication were there? If there was one complete round of DNA replication, the DNA should migrate as a single symmetrcial peak centred around the intermediate density. If there was more than one round of DNA replication, some or all of the DNA should migrate at 1.8 g/ml. Under normal conditions, sperm DNA added to egg extracts undergoes one complete round of replication (2), but a second round of DNA replication can be induced in egg extracts by various treatments (2).

Protocol 7. BrdU substitution

Equipment and reagents

- Beckman 16 × 76 mm Quick-Seal tubes, Ti70.1 rotor, and ultracentrifuge
- Bromo-deoxyuridine 5′ triposphate (Sigma): 50 mM stock in TE stored at –20 °C
- Buffer A (see *Protocol 6*)
- 10% SDS solution (w/v)
- 20 mg/ml proteinase K
- 1.75 g/ml CsCl dissolved in TE: to make about 100 ml, mix 100 g CsCl with 75 ml TE; check the refractive index before use

Method

1. Carry out DNA replication as described in *Protocol 5*. There should be about 100 000 sperm to get a good signal. Add [α-^{32}P]dATP to ~ 0.2 μCi/μl of extract and add BrdUTP to a concentration of 0.5 mM.
2. To stop the reaction and to remove unincorporated [α-^{32}P]dATP and

BrdUTP, add 1 ml cold buffer A, incubate 5 min on ice, and centrifuge for 5 min in a microcentrifuge at top speed (16000 *g*).

3. Resuspend the pellet in 100 μl buffer A. Add SDS to 0.5% and proteinase K to 0.4 mg/ml. Incubate for 60 min at 37 °C.
4. Phenol:chloroform extract, chloroform extract, and ethanol precipitate using 0.3 M NaAc and 20 μg carrier RNA. Resuspend the pellet in 100 μl TE.
5. Mix the resuspended DNA with 12.7 ml 1.75 g/ml CsCl. Fill into Beckman 16 × 76 mm Quick-Seal tubes, and centrifuge for 45 h at 30 000 r.p.m. and 20 °C in a Beckman Ti70.1 rotor.
6. Collect 50 × 0.25 ml fractions by inserting a 21 gauge needle at the bottom of the tube; also insert a 27 gauge needle in the top of the gradient to avoid a vacuum.
7. Measure the refractive index and the amount of radioactivity in each fraction and plot c.p.m. versus density in g/ml.

3.4 Immunofluorescence with BiodUTP

The above replication assays which are based on the incorporation of radioactive nucleotides measure replication averaged over all the nuclei in the reaction. In some cases it is important to observe replication in individual nuclei. For this purpose, biotinylated dUTP is used during DNA replication, and once replication is complete, nuclei are isolated, fixed, and reacted with fluorescently labelled streptavidin (*Protocol 8*). Thus, using a fluorescence microscope, the amount and spatial localization of replication can be analysed within individual nuclei. A similar technique uses BrdUTP and fluorescently labelled anti-BrdU antibodies, but the background in this technique is reported to be higher. Drawbacks of fluorescence microscopy are that it is not highly quantitative and that nuclear structure is difficult to preserve during the fixation process used to prepare samples for microscopy. In addition, there is sometimes a non-specific background in the fluorescence technique, so a negative control using 20–50 μg/ml aphidicolin to block replication should always be used.

Protocol 8. DNA replication immunofluorescence

Equipment and reagents

- Poly-L-lysine treated coverslips: 12 mm circular coverslips are marked with a VWR marker to keep track of the treated side and inverted for 10 min on an 80 μl drop of 0.1 mg/ml poly-L-lysine for 10 min; the liquid is blotted off and the coverslip is allowed to air dry
- 10 ml plastic syringes (Becton Dickinson)
- ELB (see *Protocol 1*)
- ELB–M (ELB lacking Mg)
- 1 mM biotinylated dUTP (Boehringer Mannheim)
- Texas red-conjugated avidin (Pierce)

Protocol 8. *Continued*

- 200 mM EGS (Pierce): dissolved in DMSO, stored at –20 °C as one-use aliquots
- 10 mg/ml poly-L-lysine (Sigma) stored at –20 °C
- Fetal calf serum
- 20% Triton X-100
- 10 × PBS
- 1 M glycine pH 7
- Mounting solution: 50% glycerol, 1 × PBS, 4 μg/ml Hoechst dye

Method

1. Carry out DNA replication as in *Protocol 5* in the presence of 10 μM BiodUTP in a volume of 20 μl with ~ 1000 nuclei/μl.
2. Stop the reaction by diluting it with 0.8 ml cold ELB–M and incubate for 2–10 min on ice. Warm the reaction to room temperature, add EGS to 2 mM, and invert immediately to cross-link the nuclei. Incubate at room temperature for 40 min.
3. Place the coverslip with the poly-L-lysine coated side facing up in an appropriate sized receptacle (see below) and add 1 ml ELB containing a final concentration of 0.5 M sucrose. Layer the nuclei on top of the cushion and spin for 10 min at room temperature in an IEC clinical centrifuge at setting No. 3 (1700 *g*). To fashion a receptacle for the coverslip, we cut 3 cm off the top of the plunger of a 10 ml plastic syringe as well as the conical part of the rubber seal and we place the shortened plunger backwards inside the syringe. This generates a well of the right size to hold the coverslip, the sucrose cushion, and the nuclei.
4. Aspirate the supernatant, remove the coverslip with forceps, and place it in a small beaker or a 6-well tissue culture plate containing 1 × PBS for 3 min to rinse off the sucrose. Repeat the wash two more times.
5. Place the coverslip (nuclei facing down) onto a 70 μl drop of 4% formaldehyde in 1 × PBS for 10 min at room temperature.
6. Wash the coverslip as in step 4. Invert the coverslip onto a 70 μl drop of 10% FCS (fetal calf serum), 0.1% Triton X-100, 0.1 M glycine, and 1 × PBS, for 10 min at room temperature.
7. Wash the coverslip as in step 4. Invert the coverslip onto a 70 μl drop of 10% FCS in 1 × PBS containing a 1:100 dilution of avidin conjugated to the fluorescent dye Texas Red. Incubate for 30 min in an enclosed chamber containing wet Kimwipes to prevent dehydration of the sample.
8. Wash the coverslip as in step 4 except incubate for 5 min instead of 3 min.
9. Mount the coverslip on a microscope slide for viewing. Put a 1.5 μl drop of mounting solution on a microscope slide and place the coverslip (from which most of the liquid has been removed gently by

blotting the edge onto a Kimwipe) nuclei-down on the drop. Using the rolled up corner of a Kimwipe, carefully remove excess liquid from around the edges of the coverslip without disturbing the coverslip. To seal the coverslip with nail polish, place small drops of nail polish at three points around the edge of the coverslip to attach it to the slide. Once the drops have hardened, seal the remainder of the coverslip using more nail polish. View the sample with a fluorescent microscope using the DAPI channel to view the Hoechst dye, and the FITC channel to view the Texas Red.

3.5 Synchronization experiments using AraC, aphidicolin, or Cip1

In many situations it is critical to synchronize all the replication forks in a replication reaction. Although initiations appear to be quite synchronous within a single nucleus replicating in the egg extract system, there is considerable asynchrony between nuclei, ranging from about 20 minutes to several hours (30, 31). To synchronize all the replication forks in a reaction, an inhibitor of DNA replication is added long enough for all origins of replication in all the nuclei in the reaction to fire or to be competent to fire, and then the inhibition is reversed.

Several different inhibitors can be used. A deoxycytidine triphosphate analogue, β-D-arabinofuranoside 5′-triphosphate (AraC; Sigma) inhibits DNA polymerase but has no effect on initiation (33). When added at 200 μM, the rate of the replication fork slows from 500 nucleotides/min (32) to about 20 nucleotides/min (31). This inhibition can be reversed almost completely by adding 1–2 mM dCTP to the reaction, which avoids having to pellet the nuclei to remove the inhibitor. A slight drawback of this approach is that replication forks move different distances during the inhibition so synchrony is not perfect (31). Another easily reversible inhibitor is the cyclinE/cdk2 kinase inhibitor Cip1 (34). When used at a concentration of 100 nM, it completely blocks an early event in the initiation process. The inhibition can be reversed by diluting the extract fivefold with fresh extract (J. Walter and J. Newport, unpublished results) or by adding excess cyclinE/cdk2 (34). In this case, although there is no replication at all in the presence of Cip, substantial asynchrony is induced when the inhibition is reversed.

For the best synchrony, aphidicolin is used at a concentration of 20–50 μg/ml. This leads to much more efficient inhibition of the replication fork than AraC (50–100 nucleotides are synthesized in 60 minutes; Z. Murthy and J. Newport, unpublished results). To reverse the inhibition by aphidicolin, nuclei are diluted tenfold in NIBS buffer (50 mM KCl, 50 mM Hepes pH 7.6, 5 mM $MgCl_2$, 0.5 mM spermidine, 0.5 mM spermine, 1 mg/ml leupeptin, pepstatin, and aprotinin, and 2 mM 2-mercaptoethanol), and then centrifuged for 5 min through a 15% sucrose cushion in NIBS buffer at 4°C in a Sorvall centrifuge using an HB-4 rotor at 6000 r.p.m. (5860 *g*). After removing the supernatant, the nuclei are resuspended in 100 μl of fresh, cold extract by pipetting them up and down and allowed to recover on ice for 10 min before transferring back to 21 °C.

4. Assaying the behaviour of proteins involved in DNA replication

There are several assays which are commonly used to investigate the properties of proteins involved in DNA replication. One class of assays is designed to determine the spatial localization of replication factors before, during, and after DNA replication. A different class of experiments use immunodepletion of specific proteins from the extract to elucidate their function.

4.1 Chromatin spin down experiments

A chromatin spin dowm assay is used to determine how much of a specific replication factor is bound to chromatin in decondensed sperm or in nuclei (*Protocol 9*). Sperm is allowed to decondense in a high speed cytosolic extract or it is allowed to form nuclei in an egg extract containing membranes. The chromatin is then isolated by centrifugation and analysed by immunoblotting to determine the amount of a specific replication factor found in the chromatin fraction. The procedure is slightly different depending on whether binding to chromatin on sperm or in whole nuclei is examined. In the former case the sperm are pelleted and extracted for analysis by Western blotting; in the latter case, the nuclei are pelleted and then resuspended in detergent to dissolve the nuclear envelope and release nucleoplasmic proteins. These lysed nuclei are then pelleted again to yield the chromatin bound fraction.

The advantage of this approach is that it provides a quantitative measure of how much of a specific protein is bound to the chromatin fraction. A disadvantage of this approach is that it does not reveal which component of the chromatin the factor binds. Still, by measuring how much of a certain protein is associated with chromatin at different times during DNA replication or under different conditions can reveal much about its function (6, 7, 9).

Protocol 9. Chromatin binding of DNA replication factors

Equipment and reagents

- Horizontal microcentrifuge (Beckman E type)
- 5 × 44 mm microcentrifuge tubes (Beckman)
- ELB (see *Protocol 1*)

Method

1. Set-up a replication reaction as described in *Protocol 5*, but leave out the [α-^{32}P]dATP, and add sperm to a concentration of 2000/μl. If binding to chromatin within nuclei is to be examined, use an LS

extract or a HS + M extract. If binding to decondensed chromatin is to be examined, use the HS extract without membranes. In both cases, set-up a negative control lacking sperm. Incubate reactions at 21–23°C long enough for nuclei to form (~ 30–45 min) or for the sperm to decondense completely (~ 15 min).

2. Remove 15 μl of the reaction and mix with 60 μl cold ELB.
3. Layer the mixture over 100 μl ELB containing 0.5 M sucrose in Beckman 5 × 44 mm microcentrifuge tubes. Spin for 15 sec (HS + M or HS) or 25 sec (LS) at 4°C and 16000 *g* to pellet the nuclei/sperm. If longer spins are used at this step, a sperm-independent background results.
4. Aspirate the supernatant, leaving behind 3–5 μl. If binding to decondensed sperm in a HS extract is being examined, resuspend the pellet in 10 μl SDS sample buffer and analyse by Western blotting. If binding to chromatin within nuclei is being examined, resuspend the pellet in 150 μl ELB containing 0.6% Triton X-100 by pipetting up and down ten times using thin pipette tips.
5. Spin the nuclei through a sucrose cushion as in step 3, but spin for 1 min.
6. Remove the supernatant by aspiration, leaving about 3 μl of buffer. Resuspend the pellet in 10 μl SDS sample buffer and analyse by Western blotting.

4.2 Protein immunofluorescence

Nuclei formed in egg extracts are very large, frequently reaching a diameter of 25 μm. This makes them ideal for immunolocalization of proteins which can be very informative as to their function. The protocol for immunofluorescence of proteins in nuclei and decondensed sperm is essentially the same as for detection of biotinylated dUTP (*Protocol 8*). Instead of using streptavidin, however, a primary antibody is added at step 7, and after step 8 an appropriate fluorescently labelled secondary antibody is added in 10% FCS, 1 × PBS for 60 min. The rest of the procedure is the same.

4.3 Immunodepletion

Immunodepletion is a widely used technique to understand what role a specific protein plays in DNA replication in *Xenopus* egg extracts. Antibodies (either affinity purified or a high titre serum) are used in conjunction with protein A Sepharose beads to selectively remove a protein of interest, and the depleted extracts are then subjected to appropriate functional assays. Below, we describe the immunodepletion of Xorc2 (5), a component of the origin recognition complex (5, 10, 35) but the protocol can be adapted for use with other proteins. Xorc2 is easiest to deplete from a crude interphase extract, resulting in a severe reduction in the rate of replication. While immunodepletion of Xorc2 from fractionated egg extract was also efficient, it was more difficult to observe a reproducible phenotype. We have observed this difference between crude and fractionated extracts with other proteins. To immunodeplete Xorc2, it is necessary to deplete the extract three times.

Protocol 10. Immunodepletion of Xorc2 from egg extracts

Reagents

- Protein A Sepharose (Pharmacia): 20 mg/ml IgG binding capacity
- ELB (See *Protocol 1*)

Method

1. To deplete 100 μl of extract, prepare 60 μl antibody resin. Wash 60 μl swollen protein A Sepharose resin twice with ELB. Pellet the resin in a microcentrifuge at 5000 r.p.m. (2040 *g*) for 1 min after each wash.
2. Combine the resin with 30 μl ELB and 15 μl of high titre Xorc2 antiserum in a 0.5 ml Eppendorf tube.[a] Incubate for 45 min on a rotating wheel at room temperature. Although the liquid stays in the bottom of the tube, the resin moves freely through the solution. For low titre sera, up to 200 μl of serum can be used at this step. Control beads are prepared by the same procedure using an equivalent amount of pre-immune serum.
3. Wash the beads once in ELB, twice in ELB/0.5 M KCl, twice in ELB, and split into three 20 μl aliquots in 0.5 ml Eppendorf tubes.
4. Using a 27 gauge needle connected to an aspirator, remove the void volume from one 20 μl aliquot of antibody resin (the resin should appear 'dry' after this step). Immediately add 100 μl of freshly prepared crude interphase egg extract containing 3 μg/ml nocodazole. Mix by pippetting and rotate end-over-end at 4°C for 45 min in a 0.5 ml Eppendorf tube.
5. Spin the extract for 2 min at 10 000 r.p.m. and 4°C in a microcentrifuge to pellet the resin.[b] Remove the extract (leaving behind the beads).
6. Repeat steps 4 and 5 twice.
7. To assess the extent of immunodepletion, analyse 1 μl of depleted extract and 1 μl of mock-depleted extract alongside serially diluted undepleted extract using immunoblotting with the anti-Xorc2 antibody. Depletion is typically between 99% and 99.8%.
8. The depleted extract can be used in any of the above protocols to measure DNA replication or binding of proteins to the chromatin.

[a] In the case of affinity purified antibody, 10–20 μg of specific antibody or an equivalent amount of an affinity purified control antibody should be used. Monoclonal antibodies have also been successfully used in immunodepletions (12).

[b] When depleting a HS + M extract, the centrifugation is carried out at 2000 r.p.m. to avoid pelleting of membranes.

References

1. Lohka, M.J. and Masui, Y. (1983). *Science*, **220**, 719.
2. Blow, J.J. and Laskey, R.A. (1986). *Cell*, **47**, 577.
3. Newport, J. (1987). *Cell*, **48**, 205.
4. Fang, F. and Newport, J.W. (1991). *Cell*, **66**, 731.
5. Carpenter, P.B., Mueller, P.R., and Dunphy, W.G. (1996). *Nature*, **379**, 357.
6. Kubota, Y., Mimura, S., Nishimoto, S., Takisawa, H., and Nojima, H. (1995). *Cell*, **81**, 601.
7. Chong, J.P.J., Mahbubani, H.M., Khoo, C-Y., and Blow, J.J. (1995). *Nature*, **375**, 418.
8. Madine, M.A., Khoo, C.-Y., Mills, A.D., and Laskey, R.A. (1995). *Nature*, **375**, 421.
9. Coleman, T.R., Carpenter, P.B., and Dunphy, W.G. (1996). *Cell*, **87**, 53.
10. Rowels, A., Chong, J.P.J., Brown, L., Howell, M., Evan, G.I., and Blow, J.J. (1996). *Cell*, **87**, 287.
11. Mills, A.D., Blow, J.J., White, J.G., Amos, W.B., Wilcock, D., and Laskey, R.A. (1989). *J. Cell Sci.*, **94**, 471.
12. Newport, J.W., Wilson, K.L., and Dunphy, W.G. (1990). *J. Cell Biol.*, **111**, 2247.
13. Adachi, Y. and Laemmli, U.K. (1992). *J. Cell Biol.*, **119**, 1.
14. Jenkins, H., Holman, T., Lyon, C., Lane, B., Stick, R., and Hutchison, C. (1993). *J. Cell Sci.*, **106**, 275.
15. Yan, H. and Newport, J. (1995). *Science*, **269**, 1883.
16. Lawlis, S.J., Keezer, S.M., Wu, J.R., and Gilbert, D.M. (1996). *J.Cell Biol.*, **135**, 1207.
17. Hutchison, C.J., Cox, R., Drepaul, R.S., Gomperts, M., and Ford, C.C. (1987). *EMBO J.*, **6**, 2003.
18. Dasso, M. and Newport, J.W. (1990). *Cell*, **61**, 811.
19. Smythe, C. and Newport, J. (1992). *Cell*, **68**, 787.
20. Kumagai, A. and Dunphy, W.G. (1995). *Mol. Biol. Cell*, **6**, 199.
21. Blow, J.J. and Laskey, R.A. (1988). *Nature*, **332**, 546.
22. Hua, X.H., Yan, H., and Newport, J. (1997). *J. Cell Biol.*, **137**, 183.
23. Hyrien, O. and Mechali, M. (1993). *EMBO J.*, **12**, 4511.
24. Dunphy, W.G., Brizuela, L., Beach, D., and Newport, J.W. (1988). *Cell*, **54**, 423.
25. Draetta, G., Luca, F., Westendorf, J., Ruderman, J., and Beach, D. (1989). *Cell*, **56**, 829.
26. Murray, A. (1991). *Methods Cell Biol.*, **36**, 581.
27. Hutchison, C.J. (1990). In *The cell cycle: a practical approach* (ed. P.A. Fantes and R. Brooks). Oxford University Press.
28. Neumeyer, D.D., Farschon, D.M., and Reed, J.C. (1994). *Cell*, **79**, 353.
29. Smythe, C. and Newport, J.W. (1991). *Methods Cell. Biol.*, **35**, 449.
30. Blow, J.J. and Watson, J.V. (1987). *EMBO J.*, **6**, 1997.
31. Walter, J. and Newport, J.W. (1997). *Science*, **275**, 993.
32. Mahbubani, H.M., Pauli, T., Elder, J.K., and Blow, J.J. (1992). *Nucleic Acids Res.*, **20**, 1457.
33. Cozzarelli, N.R. (1977). *Annu. Rev. Biochem.*, **46**, 641.
34. Strausfeld, U.P., Howell, M., Rempel, R., Maller, J.L., Hunt, T., and Blow, J.J. (1994). *Curr. Biol.*, **4**, 876.
35. Bell, S.P. and Stillman, B. (1992). *Nature*, **357**, 128.

Note added in proof

We have recently developed a variation of the egg extract system in which DNA replication does not require the formation of an intact nucleus (Walter, J., Sun, L., and Newport, J. (1998). *Mol. Cell*, **1**, 519).

10

Viral *in vitro* replication systems

PETER A. BULLOCK

1. Introduction

It is possible to study the duplication of DNA using a number of *in vitro* replication systems, including those employing prokaryotic (1) as well as eukaryotic replication factors (2). In many instances, the eukaryotic replication factors were identified by fractionation of cellular extracts that support the *in vitro* replication of viral DNA (3–5 for reviews). Prominent among the *in vitro* viral DNA replication systems are those that support adenovirus, papillomavirus, and papovavirus DNA replication. Additional systems for studying viral DNA replication (e.g. herpesvirus, poxvirus, and parvovirus) have been described (2).

Perhaps the best characterized eukaryotic *in vitro* replication systems is that based on the replication of simian virus 40 (SV40) DNA in primate cells (6–8). The recognition of the SV40 origin by the virally-encoded T-antigen has been analysed in detail (reviewed in ref. 9). Related studies have provided considerable insight into the mechanisms that operate during subsequent replication events. For example, the SV40 replication system was used to provide basic insights into origin unwinding (10) and DNA synthesis events at replication forks (9, 11). Procedures used to study SV40 DNA replication, particularly initiation events, are described herein. These techniques may be useful for analyses of initiation events in less well characterized eukaryotic replication systems (e.g. insects, mammals, plants, yeast) (2).

2. The SV40 *in vitro* replication reaction

2.1 Preparation of cellular extracts

Extracts prepared from a number of primate cell lines support SV40 DNA replication *in vitro* (12), provided they are supplemented with adequate amounts of virally-encoded T-antigen (T-ag), a protein whose isolation and functions in replication are described below.

Methods for preparing extracts from human HeLa cells, adapted from Wobbe *et al.* (8), are described in the following section. HeLa cell extracts are frequently used for replication assays since HeLa cells are relatively easy to acquire and to grow.

Protocol 1. Preparation of HeLa cell crude extracts[a]

Equipment and reagents

- Low speed centrifugation equipment (e.g. Beckman GS-6R centrifuge with a Beckman GH3.7 rotor)
- High speed centrifugation equipment (e.g. Sorvall RC-5B centrifuge with a Sorvall SS-34 rotor and Beckman L8-70 M centrifuge with a Beckman 70Ti rotor)
- Hypotonic buffer: 20 mM Hepes pH 7.5, 5 mM KCl, 1.5 mM $MgCl_2$,1 mM DTT, and 1 × protease inhibitors (0.2 μg/ml leupeptin, 0.1 μg/ml antipain, 50 μM EGTA)
- Extract storage buffer: 20 mM Hepes pH 7.5, 0.1 mM EDTA, 50 mM NaCl, 1 mM DTT, 0.1 mM PMSF, 10% (v/v) glycerol
- HeLa cells: can be obtained from the American Type Culture Collection and grown in media provided by a number of different companies (e.g. Mediatech's 'cellgro RPMI 1640' supplemented with 10% fetal bovine serum). Grow ~ 10 litres of cells until they reach a density of ~ 5×10^5 cells/ml. Alternatively, HeLa cells can be ordered from the National Cell Culture Center. After washing the cells in PBS buffer, the cell pellets (~ 15 ml/10 litres of cells) are shipped overnight on ice.
- PBS: 137 mM NaCl, 2.7 mM KCl, 10.6 mM $Na_2HPO_4.7H_2O$, 1.4 mM NaH_2PO_4

Method

1. Pellet HeLa cells via centrifugation at ~ 1100 g[b,c] for 10 min at 4°C. Resuspend the cells in ice-cold PBS (15 ml/litre of culture) and repellet via centrifugation at 1100 g.
2. After removing the supernatant, resuspend the HeLa cell pellet in hypotonic buffer (10 ml/litre of cells). Centrifuge at 1100 g for 10 min, 4°C.
3. In order to swell the cells, gently resuspend the cell pellet in ice-cold hypotonic buffer (3.3 ml/litre of cells), and incubate at 4°C for 10 min.
4. Disrupt the cells by Dounce homogenization using a tissue grinder (Kontes) and a B pestle (~ 10 strokes). The intact nuclei should be visible under a light microscope.
5. Determine the volume of lysate; then slowly add 5 M NaCl to a final concentration of 0.2 M.
6. Spin the sample at 47 936 g[d] for 30 min at 4°C. Remove the supernatant and centrifuge at 100 000 g for 60 min at 4°C.[e]
7. Using dialysis tubing having a M_r cut-off of 6000–8000, dialyse the supernatant overnight at 4°C against 4 litres of extract storage buffer.
8. To clarify the sample, centrifuge the dialysed extract at 100 000 g for 60 min at 4°C. Remove the supernatant and determine the protein concentration of the extract. Store the samples at –70°C. When prepared according to this procedure, the protein concentrations of the crude cytoplasmic extracts are ~ 8–12 mg/ml.

[a] Related protocols for the preparation of HeLa cell crude extracts have been published (13, 14).
[b] In all instances, relative centrifugal forces are given as g_{max} values.
[c] Beckman GH3.7 rotor run at 2200 r.p.m.
[d] Sorvall RC-5B centrifuge and a Sorvall SS-34 rotor run at 20 000 r.p.m.
[e] Beckman L8-70 M centrifuge and a Beckman 70Ti rotor run at 31 170 r.p.m.

2.2 Obtaining purified T-antigen

T-ag has several important roles during the initiation of replication, including recognition of the SV40 origin, recruitment of cellular initiation factors, and functioning as a DNA helicase. T-ag can be purchased from CHIMERx; however, for many reasons including cost, it may be necessary to purify T-ag using standard techniques.

T-ag is most commonly isolated from recombinant adenovirus-infected human cells or from recombinant baculovirus-infected insect cells (the T-ag sold by CHIMERx is isolated from a baculovirus vector). Detailed protocols for viral infection and subsequent production of a cell lysate have been published (14–17). Owing to space limitations, it will be assumed that one such T-ag containing lysate is available. The procedures required to purify T-ag from infected cell lysates, using standard immunoaffinity techniques (18, 19), are described in *Protocol 2*.

Protocol 2. Isolation of T-antigen[a]

Reagents

- Baculovirus recombinant-infected insect cell lysate: procedures for preparing lysates from baculovirus-infected cells have been reported (14)
- DEAE Sepharose Fast Flow column (~ 25 ml bed volume) (Pharmacia)
- Sepharose CL-4B column (~ 5 ml bed volume) (Sigma)
- Immunoaffinity column containing PAb 419 antibodies cross-linked to protein A Sepharose CL-4B beads (Sigma): detailed protocols for preparing a PAb 419 immunoaffinity column are available (13, 14, 19, 20)
- T-ag storage buffer: 20 mM Tris–HCl pH 8, 50 mM NaCl, 10% (v/v) glycerol, 1 mM EDTA, 1 mM DTT, 0.1 mM PMSF, 1 × PI
- Buffer 1: 20 mM Tris–HCl pH 8, 0.3 M NaCl, 10% (v/v) glycerol, 1 mM EDTA, 0.1 mM PMSF, 1% NP-40, 1 mM DTT, and 1 × protease inhibitors (PI; 0.2 μg/ml leupeptin, 0.1 μg/ml antipain, 50 μM EGTA)
- Buffer 2: 50 mM Tris–HCl pH 8, 1 M NaCl, 10% (v/v) glycerol, 1 mM EDTA, 0.1 mM PMSF, 1 mM DTT, 1 × PI
- Buffer 3: 50 mM Tris–HCl pH 8.5, 0.5 M NaCl, 10% (v/v) glycerol, 1 mM EDTA, 10% (v/v) ethylene glycol, 0.1 mM PMSF, 1 mM DTT, 1 × PI
- Buffer 4: 20 mM Tris–HCl pH 8.5, 1 M NaCl, 10% (v/v) glycerol, 1 mM EDTA, 55% (v/v) ethylene glycol, 0.1 mM PMSF, 1 mM DTT, 1 × PI

Method

1. To remove compounds such as nucleic acids, pass the lysate over a DEAE Sepharose column pre-equilibrated in buffer 1 (~ 10 × bed volume). After loading the lysate, wash the column with buffer 1 until very little additional protein is eluted; combine the wash and the original flow-through.
2. Run the Sepharose CL-4B and PAb 419 immunoaffinity columns in series. Pre-equilibrate the columns in buffer 1 (~ 10 × bed volume) and load the DEAE flow-through (and wash) at ~ 10–12 ml/h. To insure complete binding of T-ag, recycle the flow-through over the CL-4B and PAb 419 columns.

Protocol 2. *Continued*

3. Wash the columns with 500 ml buffer 1.
4. Separate the PAb 419 immunoaffinity column from the Sepharose CL-4B column, wash the PAb 419 column with 100 ml buffer 2, and then with 50 ml buffer 3.
5. Elute the T-ag with ~ 30 ml buffer 4. To insure complete elution, several millilitres of buffer 4 should be allowed to enter the column prior to temporarily stopping the column for ~ 20–30 min. Collect 25 ~ 1 ml fractions.
6. Fractions containing active T-ag molecules can be identified by *in vitro* replication reactions (*Protocol 3*) or by a filter binding assay (8). T-ag containing fractions can also be identified by SDS–PAGE.
7. Pool aliquots containing T-ag and dialyse overnight against 2 litres of T-ag storage buffer. If the T-ag concentration is low (< 200 μg/ml), microconcentrators (e.g. Millipore 'Centricon 30') can be used to increase the protein concentration. Once purified, T-ag is stored at –70 °C.

[a] Similar protocols for the purification of T-ag from infected cell lysates were previously reported (13, 14).

2.3 Assembling an SV40 *in vitro* replication reaction

There are many reasons for conducting an SV40 *in vitro* replication reaction. For example one may wish to test the effect of potentially inhibitory compounds or to determine the relative activities of newly isolated preparations of T-ag or HeLa cell crude extracts. The steps required to assemble a basic *in vitro* replication assay are described in *Protocol 3*.

Protocol 3. Performing an SV40 *in vitro* replication reaction[a]

Equipment and reagents

- Glass fibre filters ('ENZOFILTERS', Enzo Biochem, Inc.)
- Scintillation counter (Beckman LS 3801)
- HeLa cell crude extract and T-ag: see *Protocols 1* and *2* respectively
- DNA: a useful construct for replication assays is plasmid pSV01ΔEP (available on request), stored at 150 μg/ml in TE buffer (10 mM Tris–HCl pH 8, 1 mM EDTA). This ~ 2800 bp plasmid has the origin-bearing *Eco*RII G fragment of SV40 inserted into the *Eco*RI site of a pBR322 derivative (8, 21). Additional SV40 origin-containing plasmids have been described (14).
- [α-^{32}P]dCTP labelling mix (appropriate for ~ ten 30 μl reactions): combine 4 μl H_2O, 1 μl 10 mM dCTP, and 5 μl [α-^{32}P]dCTP (3000 Ci/mmol) (DuPont, NEN, Life Science Products)
- Replication mix (suitable for ten 30 μl reactions): assemble on ice 2.1 μl 1 M $MgCl_2$, 0.15 μl 1 M DTT, 12 μl 0.1 M ATP, 24 μl 0.5 M creatine phosphate (di-Tris salt pH 7.6), 1.4 μl 5 mg/ml creatine phosphate kinase[b] (Boehringer Mannheim Biochemicals), 25 μl pSV01ΔEP (150 μg/ml), 3.6 μl 16.7 mM rNTP–A mix, 9 μl 3.3 mM dNTP–C mix, and 6 μl [α-^{32}P]dCTP labelling mix

- 16.7 mM rNTP–A mix: add 50 μl CTP, GTP, and UTP (all at 100 mM; Boehringer Mannheim) to 150 μl dH_2O
- 3.3 mM dNTP–C mix: add 16.5 μl dATP, dGTP, and dTTP (all at 100 mM; Boehringer Mannheim) to 450.5 μl dH_2O
- Pyrophosphate stop mix: combine equal volumes of 100 mM tetra sodium pyrophosphate and a solution containing 1 mg/ml salmon testis DNA (Sigma)

Method

1. For each reaction, combine in an Eppendorf tube on ice, 8.3 μl replication mix with ~ 15 μl HeLa cell crude extract (~ 150 μg). Calculate the volume of T-ag needed to introduce 0.5 μg into the reaction; however, T-ag should be added last. As a contol, assemble one reaction in the absence of T-ag, to this tube add an equal volume of T-ag storage buffer. Adjust the final volume of the reactions to 30 μl with H_2O. Upon addition of T-ag, mix the tubes gently and then place them in a 37 °C water-bath.[c]
2. To stop the reactions, add 100 μl of the pyrophosphate stop mix to each sample along with 1 ml 5% (w/v) TCA. Following a ~ 5 min incubation on ice, pass the contents of the tube over a glass fibre filter, then wash the filter three times with 1% (w/v) TCA solution, and once with 95% ethanol. The glass filters can be either air dried or dried with the aid of a heat lamp. Determine the total radioactivity in the samples using a scintillation counter.
3. The pmol of dCTP incorporated into nascent DNA can be determined by dividing the total number of counts (c.p.m.) present on a given filter by the specific activity of the [α-^{32}P]dCTP labelling mix. To determine the specific activity, measure the c.p.m. in 0.6 μl of the labelling mix and divide this number by 600, the total number of pmols of dCTP in a single reaction.

[a] Similar protocols for conducting SV40 *in vitro* replication reactions have been previously reported (13, 14).
[b] Resuspended in 50 mM imidazole buffer pH 6.6.
[c] The final concentrations of the reagents are 7 mM $MgCl_2$, 0.5 mM DTT, 4 mM ATP, 40 mM creatine phosphate (di-Tris salt pH 7.7), 1.4 μg creatine phosphokinase, 12.5 μg/ml pSV01ΔEP, dATP, dGTP, and dTTP (100 μM each), CTP, GTP, and UTP (200 μM each), and [α-^{32}P]dCTP (20 μM, ~ 5 c.p.m./fmol).

In vitro replication reactions reconstituted from purified proteins (22–25) are not discussed in this chapter. However, protocols for conducting replication reactions with purifed proteins have been described (14).

DNA synthesized in any replication reaction is the product of many discrete steps. Methods for analysing individual steps during SV40 replication, particularly those operating during initiation, are described in the following sections.

3. Detecting origin-specific DNA unwinding

A critical event during initiation of SV40 DNA replication is T-ag-dependent unwinding of the core origin (26–28). The importance of this process is suggested by studies indicating that DNA unwinding is a major point at which DNA replication is regulated (27, 29, 30). Unwinding creates single-stranded regions of DNA that serve as templates for polymerases, such as the pol α-primase complex.

In addition to T-ag, a DNA helicase (31), origin-specific DNA unwinding requires a single-strand DNA binding protein. With covalently closed DNA molecules, a topoisomerase that can remove positive supercoils is also necessary (26). It is presumed that the requirement for a single-strand DNA binding protein is to prevent reannealing of unwound single-stranded regions (26). Techniques for detecting T-ag-dependent unwinding of covalently closed DNA molecules, using either purified proteins (26, 27) or extracts prepared from HeLa cells (32), are described in *Protocol 4*. Additional procedures for detecting DNA unwinding on linear molecules have been described (33), and are also discussed in Chapter 11.

Protocol 4. Detection of T-ag catalysed DNA unwinding

Reagents

- HeLa cell crude extract and T-ag: see *Protocols 1* and *2* respectively
- DNA: an SV40 origin-containing plasmid, such as pSV01ΔEP (stored in TE buffer, at 150 μg/ml)
- *E. coli* SSB (Pharmacia)
- Human single-strand binding protein (hSSB; also termed RP-A) (34) can be purified from HeLa cells (35); alternatively, recombinant RP-A can be isolated from *E. coli* (36)
- A topoisomerase that can remove positive supercoils (e.g. calf thymus topoisomerase I and DNA gyrase, both from Gibco BRL)
- Unwinding mix: unwinding assays are conducted under conditions suitable for SV40 DNA replication; however, the reaction volumes are doubled in order to aid in the detection of the unwound product, form U (26). For ten 60 μl reactions combine: 4.2 μl 1 M $MgCl_2$, 0.64 μl 0.5 M DTT, 24 μl 0.1 M ATP, 48 μl 0.5 M creatine phosphate (di-Tris salt pH 7.6), 2.9 μl 5 μg/μl creatine phosphate kinase (CPK), 10 μl 1 mg/ml bovine serum albumin (BSA), and 52.5 μl 150 μg/ml pSV01ΔEP.
- Stop mix: for stopping ten 60 μl reactions combine 18 μl 0.5 M EDTA, 2 μl 10 mg/ml tRNA, 18 μl 10% *N*-lauroylsarcosine pH 7.7, and 30 μl 10 mg/ml proteinase K
- ~ 6 × gel loading buffer: combine 20% (v/v) Ficoll, 0.1 M EDTA, 0.25% (w/v) bromphenol blue, and 0.25% (w/v) xylene cyanol (37)
- 1.8% agarose gel containing chloroquine: combine 3.6 g agarose (e.g. FMC 'SeaKem ME'), 4 ml 50 × Tris, acetate, EDTA (TAE) buffer (final concentration of 40 mM Tris–acetate, 2 mM EDTA) (37), and H_2O to 200 ml. Boil the gel mix in a microwave oven; when the temperature of the molten agarose falls below 75°C, add 30 μl 10 mg/ml chloroquine. Cast the 1.8% agarose gel containing chloroquine in a suitable gel system (Hoefer SE600). Although vertical gel systems have been traditionally used for unwinding assays, it is likely that horizontal gels will also work.
- Gel running buffer: combine 20 ml 50 × TAE buffer (37) with 150 μl 10 mg/ml chloroquine, and adjust the volume to 1 litre with H_2O

A. *Unwinding using purified proteins*

1. To assemble a reaction, combine 14.2 μl of the unwinding mix[a] with 1 μg SSB, 10 U topoisomerase I (or gyrase), 1 μg T-ag, and add H_2O to

60 µl. As a control, one reaction should be conducted in the absence of T-ag. Incubate for > 15 min at 37 °C.

2. To stop the reactions, add 6.8 µl of the stop mix[b] and continue the incubation for ~ 30 min. Adjust the volume to ~ 200 µl with TE and extract the samples with phenol:chloroform. Add 200 µl 5 M ammonium acetate and precipitate the DNA by adding ~ 1 ml 100% ethanol. Wash the pellets with 80% ethanol and dry. Resuspend the samples in 10 µl TE and ~ 4 µl gel loading buffer.
3. Load the samples on the chloroquine-containing agarose gel: convenient markers include an aliquot of form I pSV01ΔEP DNA and a DNA ladder (e.g. phage λ DNA cut with *Hin*dIII). Run the gel at ~ 2.8 V/cm for approx. 14 h. The 1 × TAE running buffer must also contain 1.5 µg/ml chloroquine.
4. Soak the gel in 50 mM NaCl for 1 h (to remove the chloroquine), stain for ~ 30 min in ethidium bromide (0.5 µg/ml), and destain for 30 min in H_2O. Visualize the DNA under ultraviolet illumination.

B. *Unwinding using HeLa cell crude extracts*

Using similar techniques, it is possible to detect DNA unwinding in HeLa, and presumably other, crude cell extracts (32). The crude extracts contain the hSSB and topoisomerases required for unwinding.

1. To assemble a reaction, combine 14.2 µl of the unwinding mix, ~ 30 µl HeLa cell (or equivalent) crude extract, 1 µg T-ag, and add H_2O to 60 µl. As with the purified system, a control reaction should be assembled that lacks T-ag.
2. Incubate for > 15 min at 37 °C.
3. Stop the reactions and resolve the topoisomerase on a chloroquine-containing agarose gel that is processed for photography as described (part A, steps 2–4).

[a] The final concentrations of these reagents in the unwinding reactions are 7 mM $MgCl_2$, 0.5 mM DTT, 4 mM ATP, 40 mM creatine phosphate, 23.3 µg/ml CPK, 16.7 µg/ml BSA, and 13 µg/ml pSV01ΔEP.

[b] The final concentrations of these reagents are 13.5 mM EDTA, 30 µg/ml tRNA, 0.3% *N*-lauroylsarcosine pH 7.7, and 0.45 mg/ml proteinase K.

Upon addition of T-ag to an otherwise complete SV40 replication reaction, it takes ~ 10 min to detect DNA unwinding (32). During this period of time, multiple protein:protein and protein:DNA interactions take place, culminating in the formation of a poorly defined pre-initiation complex (reviewed in refs 9 and 38). In the presence of adequate levels of nucleotides, this complex can initiate DNA synthesis. Techniques useful for studying these initial DNA synthesis events, especially in the vicinity of the SV40 origin, are described in the following sections.

4. The formation of nascent DNA

For studies of initiation of DNA replication in the vicinity of the SV40 origin, it is often convenient to use 'pulse assays' to limit the extent of DNA synthesis. Pulse assays are essentially SV40 replication assays in which the dNTPs and rNTPs (with the exception of ATP) are initially withheld from the reactions (32). At a subsequent point, the nucleotides required for DNA synthesis are added for a brief period of time (including a radioactive tracer such as [α-^{32}P]dCTP), and the nascent DNA formed is then purified and analysed.

The assembly of a typical 'pulse reaction' is described in *Protocol 5*. For many experiments, such as those designed to map initiation sites for the pol α-primase complex (see below), it is necessary to pool the nascent DNA products formed in ~ eight separate 120 μl reactions prior to analyses. Moreover, to measure background incorporation, one reaction lacking T-ag is always conducted. Furthermore, an additional reaction is always planned in order to insure an adequate supply of reagents. Therefore, the reagents required for ten reactions are described in the following section. However, the number of reactions that one assembles is determined by the experimental objective. For example, a single reaction is sufficient for examining the topological requirements for initiation (32).

Protocol 5. Assembling a 'pulse' replication reaction

Equipment and reagents

- Disposable glass culture tube (Fisherbrand 13 × 100 mm disposable culture tubes)
- HeLa cell crude extract and T-ag: see *Protocols 1* and *2* respectively
- Replication mix: for ten 120 μl reactions combine 8.4 μl 1 M $MgCl_2$, 0.64 μl 1 M DTT, 48 μl 0.1 M ATP, 96 μl 0.5 M creatine phosphate (di-Tris salt pH 7.6), 5.8 μl 5 mg/ml creatine phosphate kinase, and 100 μl of an SV40 origin-containing plasmid, such as pSV01ΔEP (150 μg/ml)
- T-ag storage buffer: see *Protocol 4*
- Pulse mix: for ten reactions combine 14.4 μl rNTP–ATP mix (containing 16.7 mM CTP, GTP, and UTP), 36 μl dNTP–C mix (containing 3.3 mM dATP, dGTP, and dTTP), 2.7 μl 1.3 mM dCTP, and 30 μl [α-^{32}P]dCTP (~ 3000 Ci/mmol) (DuPont, NEN, or Life Science Products)
- Stop mix: for ten reactions combine 36 μl 0.5 M EDTA, 4 μl 10 mg/ml *E. coli* tRNA, 36 μl 10% *N*-lauroylsarcosine pH 7.7, and 54 μl 10 mg/ml proteinase K
- Pyrophosphate stop mix: see *Protocol 3*

Method

1. Set-up nine Eppendorf tubes on ice. To each tube add the following: 25.9 μl reaction mix[a] and ~ 60 μl HeLa cell extract (~ 600 μg). Calculate the volume of T-ag needed to introduce 2 μg into the reaction, but do not add T-ag at this point. Add water to bring the final volume to 113 μl.
2. At 2 min intervals, place the Eppendorf tubes in a 37 °C water-bath.
3.[b] Incubate the reactions at 37 °C for 45 min in the absence of T-ag. This

incubation period lowers the T-ag-independent labelling of form II DNA (circular duplex DNA with at least one single-stranded break) (32).

4. Following the 45 min pre-incubation, add 2 μg T-ag to all tubes at 2 min intervals (except to the one containing the control reaction; to this reaction add an equal volume of T-ag storage buffer).
5. Upon addition of T-ag, either at step 1 or step 4, incubate the reactions for 15 min to allow formation of the poorly defined 'pre-initiation complex'.
6. Upon completion of the 15 min 'T-ag-dependent' incubation period, add 8.3 μl of the pulse mix[c] and gently vortex the tube.
7. Following the desired length of synthesis (e.g. 5 sec), stop the reactions by the addition of 13 μl stop mix,[d] followed by immediate vortexing.
8. If desired, the efficiency of labelling can be monitored by a trichloroacetic acid precipitation (step 10). To initiate this procedure, remove a 10 μl aliquot and add it to a disposable glass culture tube containing 200 μl pyrophosphate stop mix.
9. After addition of the stop mix, incubate the reactions for 30 min at 37 °C. The trichloroacetic acid precipitations (step 10) can be completed during this incubation period.
10. TCA precipitations were described in *Protocol 3*. Fill the glass tubes, containing pyrophosphate stop mix and aliquots from the pulse reactions with a 5% (w/v) TCA solution. Incubate for ~ 5 min on ice, then pass the contents of the tube over a glass fibre filter that is subsequently washed three times with 1% (w/v) TCA solution, and once with 95% ethanol. Determine the total radioactivity in the samples using a scintillation counter.

    The pmol of dCTP incorporated into nascent DNA can be determined by dividing the c.p.m. of a given sample by the specific activity of the pulse mix. To determine the specific activity of the pulse mix, establish the total number of c.p.m. in 8.3 μl of the pulse mix and divide by 364, the total number of pmols of dCTP in a single reaction.
11. Following the 30 min incubation in stop mix, dilute the reactions to 200 μl with TE buffer and then extract them with an equal volume of phenol:chloroform. To precipitate the DNA, add an equal volume of 5 M ammonium acetate along with 1 ml 100% ethanol. After spinning in a microcentrifuge (~ 20 min) remove the ethanol, and collect the pellets into a single 50 μl aliquot of TE. Use an additional 50 μl aliquot to rinse the Eppendorf tubes, and combine this aliquot with the initial aliquot. If desired, a 1 μl aliquot can be removed for analyses of product lengths via alkaline agarose gel electrophoresis (*Protocol 6*).

Protocol 5. *Continued*

To remove additional unincorporated nucleotides, precipitate the samples a second time (add a 100 μl aliquot of 5 M ammonium acetate to the Eppendorf tube containing the pooled samples, followed by 2.5 vol. 100% EtOH). After mixing, collect the sample via centrifugation (~ 20 min) in an Eppendorf microcentrifuge. Rinse the pellet in 80% EtOH and dry. (This is a convenient point in the procedure to break for the day; store the samples overnight at −20°C in 80% EtOH.)

12. If desired, the maturation of DNA molecules formed during the initial pulse (step 6) can be followed by chasing the reactions with an excess of unlabelled dCTP. After conducting the pulse, add 6 μl 100 mM dCTP (final concentration of ~ 4.7 mM dCTP). Allow synthesis to continue for the desired period of time and then process as described in steps 7–11.

[a] The final concentrations of the reagents are 7 mM $MgCl_2$, 0.5 mM DTT, 4 mM ATP, 40 mM creatine phosphate (di-Tris salt pH 7.7), 2.8 μg creatine phosphokinase, 1.5 μg SV40 origin-containing pSV01ΔEP.
[b] If background labelling of form II DNA is not a concern step 3 can be omitted. If step 3 is omitted, add 2 μg T-ag to the reactions and an equal volume of T-ag storage buffer to the control reactions. However, T-ag is usually added at step 4.
[c] After addition of the pulse mix, the reactions contain dATP, dGTP, and dTTP (final concentration of each 100 μM), CTP, GTP, and UTP (final concentrations of each 200 μM), and [α-^{32}P]dCTP (final concentration ~ 3 μM; ~ 225 c.p.m./fmol).
[d] The final concentrations of the stop mix components are 13.5 mM EDTA, 30 μg/ml tRNA, 0.3% *N*-lauroylsarcosine pH 7.7, and 0.45 mg/ml proteinase K.

The reactant concentrations suggested in *Protocol 5* reflect those used in standard SV40 replication reactions (3–5). However, reaction conditions can be varied to test reaction parameters of interest. For example, the concentration of the ribonucleotides could be changed to reflect intracellular concentrations in particular cells (39, 40). Furthermore, while *Protocol 5* suggests a 5 sec pulse, DNA synthesis can be conducted for longer periods of time.

5. Gel-based assays of newly synthesized DNA

Figure 1B (lane 4) presents the entire population of nascent DNA formed during a 5 sec pulse, resolved on an alkaline agarose gel. It is apparent that the T-ag-dependent nascent DNA population formed during a brief pulse (*Protocol 5*) is bimodal; the smaller distribution was termed primer: RNA:DNA (41, 42) and the second distribution contains larger 'Okazaki sized' DNA molecules. *Figure 1A* presents the same nascent DNA population displayed on a 10% denaturing polyacrylamide gel. These experiments

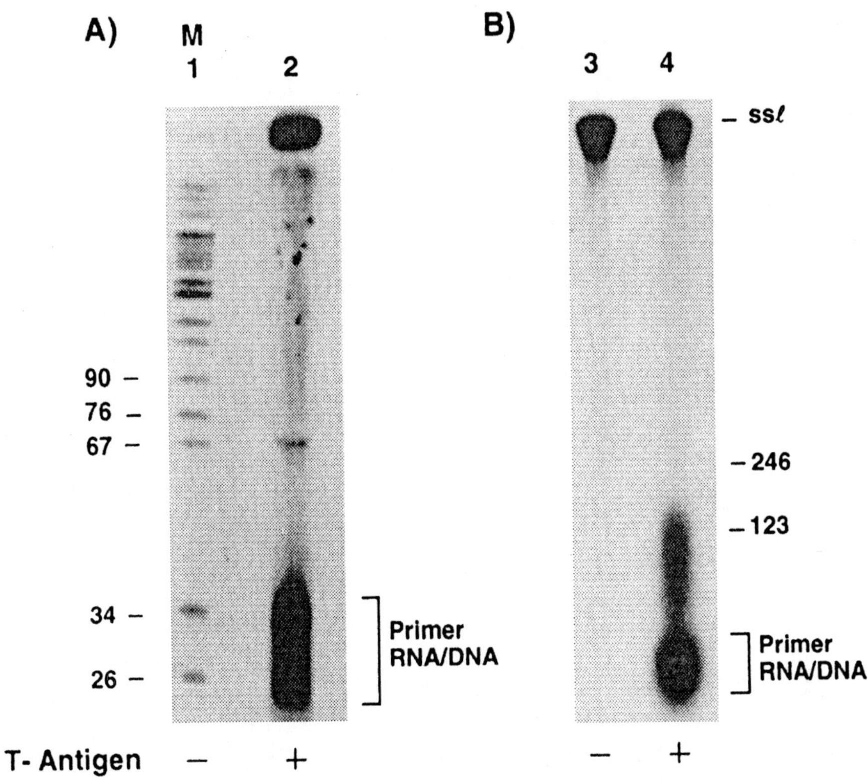

Figure 1. Autoradiograms showing the nascent DNA population formed during a 5 sec pulse displayed on polyacrylamide and alkaline agarose gels. (A) Nascent DNA formed during a 5 sec pulse was applied to a denaturing 10% polyacrylamide gel (lane 2). The position of primer:RNA:DNA is indicated. The nature of the species migrating at ~ 67 bp is not known. Primer:RNA:DNA can be removed as described in *Protocol 8* and used for a variety of purposes, such as the primer extension experiments described in *Protocol 10*. Size markers were from an *Msp*I digest of pBR322 labelled with kinase by standard methods (37) (lane 1). (B) An aliquot of the nascent DNA population was applied to a 1.8% alkaline agarose gel (lane 4). On this gel system, the nascent DNA population was resolved into two populations: primer:RNA:DNA and larger Okazaki sized DNA molecules. The products of a pulse reaction, conducted in the absence of T-ag, are presented in lane 3. Kinased size markers (data not shown) were derived from a 123 bp DNA ladder (Gibco BRL). Reprinted from ref. 43.

demonstrated that primer:RNA:DNA molecules extend between ~ 24–34 nucleotides (41–43).

Gel-based techniques for analysing the products of replication reactions are described in *Protocol 6*. It is noted that primer:RNA:DNA molecules have been observed, to date, only during SV40 DNA replication in primate cells (41, 42, 44). The techniques described in this section may be useful for detecting primer:RNA:DNA in other replication systems.

Protocol 6. Analysing the T-ag-dependent products of replication reactions

Reagents

- 1.8% alkaline agarose gels: combine 3.6 g agarose, 2 ml 5 M NaCl, 0.4 ml 0.5 M EDTA, and add H_2O to 200 ml (37). Microwave the sample, pour the molten agarose into a horizontal mould, and allow to harden. Incubate the gel in running buffer for at least 30 min prior to use.
- Alkaline agarose gel running buffer: combine 7.5 ml 10 M NaOH, 5 ml 0.5 M EDTA, and add dH_2O to 2.5 litres (final concentration of 30 mM NaOH, 1 mM EDTA)
- 1 × Tris–borate running buffer: 0.089 M Tris–borate, 0.089 M boric acid (37)
- Formamide loading buffer: 90% formamide, 1 × TBE, 0.02% bromphenol blue, and 0.02% xylene cyanol FF (37)
- TE: 10 mM Tris–HCl pH 8, 1 mM EDTA
- 10% polyacrylamide gel: combine 15 ml 40% (w/v) acrylamide (19:1 acrylamide to *bis*acrylamide) (Intermountain Scientific Corporation) with 6 ml 10 × TBE (37), 26 g RNase-free urea, and 0.8 ml 10% (w/v) APS. Add H_2O to 60 ml. Add TEMED and allow the gel to polymerize in a suitable mould (gels used to isolate primer:RNA:DNA were cast in a mould that was 20 cm long, 16 cm wide, and 0.8 mm thick).
- Size markers: an appropriate size marker for alkaline agarose gels is the 123 bp DNA ladder available from Gibco BRL. A convenient size marker for the 10% acrylamide gel is an *Msp*I digest of pBR322 (New England Biolabs). Markers are labelled using standard methods (37).

A. *Alkaline agarose gel*[a]

1. To establish the overall size distribution of a nascent DNA population, load a small aliquot of the reaction products on an alkaline agarose gel. Sample(s), e.g. 1 μl (see *Protocol 5*, step 11; ~ 2000 c.p.m.) are added to ~ 15 μl alkaline loading buffer (50 mM NaOH, 1 mM EDTA, 2.5% (w/v) Ficoll (type 400; Pharmacia), 0.025% bromocresol green, and 0.04% xylene cyanol FF) (37).
2. After loading the sample(s), run the alkaline agarose gel at 60 V (~ 200 mA) for 14 h.
3. Following electrophoresis, place the gel on Whatman DE81 paper, cover with plastic wrap, and dry at room temperature prior to autoradiography for ~ two days.

B. *10% acrylamide gel*

1. Resuspend the sample(s) containing the newly synthesized DNA (*Protocol 5*, step 11; ~ 750,000 c.p.m.) in 10 μl TE and 15 μl formamide loading buffer.
2. To dissociate the nascent DNA from the parental template, boil the sample(s) for 4 min and load a 10% polyacrylamide gel; the gel should be pre-electrophoresed in 1 × TBE buffer (37) for 30 min (~ 900 V, 46 mA, and 40 W). Gels are run in 1 × TBE buffer, at ~ 900 V, 46 mA, and 40 W, until the bromphenol blue dye migrates 15 cm.
3. Techniques for autoradiography and isolation of primer:RNA:DNA from the acrylamide gel are described in *Protocol 8*.

[a] An alternative alkaline gel procedure is described in Chapter 1, Section 3.3.

It is possible to modify the procedures in *Protocols 5* and *6* to investigate the functions of particular proteins in the initiation process. For example, by including antibodies that neutralize PCNA in the crude extracts (45), it was demonstrated that PCNA was required for the synthesis of nascent DNA molecules larger than primer:RNA:DNA (41). These experiments could be repeated with neutralizing antibodies directed against other initiation factors. Moreover, similar experiments could be used to establish the efficacy of drugs targeted against particular initiation factors.

6. Isolation of nascent DNA

6.1 Isolation of the total nascent DNA population

Nascent DNA molecules formed during a brief pulse can be used for various purposes. For example, it can be used as a probe in hybridization experiments (41, 42). Alternatively, it can be cloned for subsequent analyses. A procedure useful for the isolation of the entire population of nascent DNA molecules formed during a brief pulse is described in *Protocol 7*.

Protocol 7. Purification of the entire nascent DNA population

Equipment and reagents

- Beckman L5-65 ultracentrifuge (or equivalent)
- Beckman SW55Ti rotor (or equivalent) plus suitable polyallomer centrifuge tubes (e.g. Beckman No. 326819)
- Beckman SW41 rotor (or equivalent) plus suitable tubes
- Gradient maker (e.g. Hoefer SG Series)
- 100 ml stock of 5% (w/v) sucrose mix:[a] combine 2 ml 1 M Tris pH 8, 200 μl 0.5 M EDTA, 20 ml 5 M NaCl, 5 g sucrose; adjust volume to 100 ml with H_2O
- 100 ml stock of 20% (w/v) sucrose mix:[b] 2 ml 1 M Tris pH 8, 200 μl 0.5 M EDTA, 20 ml 5 M NaCl, 20 g sucrose; adjust volume to 100 ml with H_2O

Method

1. Add 2.2 ml 20% sucrose mix to the rear chamber, and 2.4 ml 5% sucrose mix to the front chamber of the gradient maker. Form the gradient in a 5 ml sterile polyallomer centrifuge tube.
2. Resuspend the sample from the pulse (or pulse-chase experiment) (e.g. *Protocol 5*, step 11) in 50 μl TE pH 8. Boil for 4 min and load onto the sucrose gradient.
3. Spin the sample(s) at 304 000 *g* for 2 h at 4°C.[c]
4. Drip the gradient and collect 20–25 fractions; use a scintillation counter to determine the location of the nascent DNA peak. Pool the peak fractions and dilute them with ~ 2 × the vol. of H_2O. Precipitate the samples by adding 100% ethanol (2.5 × the sample volume) and centrifuge at 234 745 *g* for 1 h.[d] Wash the pellet in 80% ethanol and air

Protocol 7. *Continued*

dry. Resuspend the samples in TE. The volume used depends on the further analysis to be carried out.

5. If desired, an aliquot can be removed and applied to an alkaline agarose gel (*Protocol 6*) in order to establish the population of nascent DNA molecules present in the purified sample.

[a] Final concentration of 20 mM Tris pH 8, 1 mM EDTA, 1 M NaCl, and 5% sucrose.
[b] Final concentration of 20 mM Tris pH 8, 1 mM EDTA, 1 M NaCl, and 20% sucrose.
[c] 50 000 r.p.m. in a Beckman SW55Ti.
[d] 37 000 r.p.m. in a Beckman SW41 rotor.

6.2 Isolation of primer:RNA:DNA molecules

Primer:RNA:DNA is approximately 30 nt long and contains either eight or nine nucleotides of RNA covalently linked to a small stretch of DNA (43, 46) (*Figures 1* and *3*). It is formed exclusively on lagging strand templates (42, 44). Synthesis of primer:RNA:DNA does not require proliferating cell nuclear antigen (PCNA) or PCNA-dependent polymerases (41). These and related studies indicate that primer:RNA:DNA is the product of the pol α-primase complex. Therefore, by characterizing primer:RNA:DNA, it is possible to gain insights into the pol α-primase complex; for instance, the template sequences utilized by this enzyme complex during initiation of DNA synthesis. Methods used to purify primer:RNA:DNA are described in *Protocol 8*.

Protocol 8. Purification of primer:RNA:DNA

Equipment and reagents

- Beckman L5-65 ultracentrifuge (or equivalent)
- SW55Ti rotor and suitable tubes, e.g. 0.5 × 2 inch (13 × 51 mm) polyallomer centrifuge tube (Beckman)
- Rotating platform (e.g. a Clay Adams 'Nutator')
- Millex- HV 0.45 μm filter unit (Millipore)
- Nascent DNA separated on a 10% acrylamide gel (see *Protocol 6*, part B)
- Elution buffer: 0.1% (w/v) SDS, 0.5 M ammonium acetate, 10 mM magnesium acetate (37)
- Yeast tRNA

Method

1. After separating the nascent DNA products on a 10% polyacrylamide gel, remove one glass plate and place a piece of Whatman 3MM paper over the gel. Transfer the gel to the 3MM paper (the 3MM paper should be cut so that it fits neatly into an autoradiography cassette). Following transfer, tape phosphorescent markers to the four corners of the 3MM paper. Wrap the gel and paper support in plastic wrap and subject to autoradiography at –80 °C using a screen and two pieces of Kodak X-OMAT AR film. If the gel contains the pool of nascent DNA

formed in several reactions (e.g. ~ 400 000 c.p.m.), the exposure time is ~ 1 h.

2. After developing the films, remove the region corresponding to primer: RNA:DNA from one autoradiogram and align the film with the gel based on the positions of the phosphorescent markers. (The second autoradiogram is saved as a record of the experiment.)

   Once the autoradiogram is aligned, remove regions of the gel containing primer:RNA:DNA with a clean razor blade and place the gel slices in a 2 ml Eppendorf tube (~ 75 000 c.p.m.). To aid diffusion, crush the gel slices by briefly spinning the samples in a microcentrifuge. Add 1.5 ml elution buffer to the sample(s) and incubate overnight at 4°C on a rotating platform.

3. Pellet the acrylamide fragments by spinning (~ 1 min) the samples in a microcentrifuge. Pass the primer:RNA:DNA containing aqueous layer through the 0.45 μm filter unit into a polyallomer centrifuge tube. Add an additional 0.5 ml elution buffer to the acrylamide pellet and incubate the sample(s) at 4°C for 30 min. After a second brief spin, pass the 'elution buffer wash' over the 0.45 μm filter unit into the centrifuge tube. Add 40 μg yeast tRNA as carrier, fill the tubes with 100% ethanol, and centrifuge for 1 h at 304 000 *g*.[a] Following centrifugation, remove the ethanol with a pipette and air dry the tubes.

4. The type of experiment to be conducted determines the volume of H_2O, or TE, used to resuspend the primer:RNA:DNA pellet. For example, if it is to be used as a probe in Southern blots (42), the pellet can be resuspended in a convenient volume of TE (~ 100 μl). For the primer extension reactions (*Protocol 10*), it is useful to resuspend the samples in 40 μl H_2O.

[a] 50 000 r.p.m. in a Beckman SW55Ti.

7. Techniques for characterization of nascent DNA

Several procedures have been used to analyse the entire nascent DNA population formed in the vicinity of the SV40 origin. One straightforward method is to use these molecules as probes in Southern blots.

7.1 Southern blots conducted with newly synthesized DNA molecules

Southern blots can provide basic information about initiation, such as whether leading or lagging strand templates are used to synthesize origin proximal DNA molecules. Useful techniques for conducting these studies are described in *Protocol 9*.

Protocol 9. **Determining the template specificity of initiation events via Southern blots**

Reagents

- Pre-hybridization and hybridization mix: for 50 ml of mix, combine 25 ml formamide, 6 ml 1 M Na_2HPO_4 pH 7.2, 2.5 ml 5 M NaCl, 100 µl 0.5 M EDTA, and 3.5 g SDS (to dissolve the SDS, briefly warm the mix). Adjust the volume to 50 ml with H_2O. (Protocol adopted from the Bio-Rad 'Zeta-Probe Blotting Membranes' instruction manual.)
- Single-stranded DNA having the same sequence as the template used to form primer:RNA:DNA—subclone fragments of interest into M13mp19 and isolate single-stranded DNA via standard protocols (37). It is noted that PCR techniques afford an alternative method for generating single-stranded DNA) (47). Upon isolation, confirm the sequence of the purified single-stranded DNA (48, 49).
- Dot blots for hybridization: affix the single-stranded DNA to Zeta Probe filters (Bio-Rad) using a dot blot apparatus (e.g. Bio-Rad 'Bio-Dot'). Apply templates for leading strand DNA synthesis (~ 1 µg) to one side of a filter and templates for lagging strand synthesis to the opposite side (41, 42).

Method

1. Incubate the Zeta Probe filters in hybridization buffer[a] for 2–4 h at 42 °C.
2. Heat nascent DNA probes (e.g. the products of *Protocol 7*; ~ 50 000 c.p.m.), to 95 °C for 4 min and then add to the DNA-containing Zeta Probe membranes in 7 ml of hybridization buffer.
3. Following hybridization for 10–14 h at 42 °C, wash the filters twice in 2 × SSC, 0.1% SDS (37) for 15 min at 65 °C, and then twice in 0.1% SSC, 0.1% SDS for 30 min at 65 °C. Air dry the filters and either expose to Kodak X-OMAT AR film or to a PhosphorImager screen for quantitation.

[a] Final concentration of 50% formamide, 0.12 M Na_2HPO_4, 0.25 M NaCl, 7% SDS, and 1 mM EDTA. To lower background hybridization, salmon sperm DNA (50 µg/ml) can be added to the hybridization mix.

Using these techniques, it was demonstrated that nascent DNA initiates on SV40 templates for lagging strand DNA synthesis, at locations outside of the core origin (41, 42).

7.2 Utilizing primer:RNA:DNA to map initiation sites for lagging strand synthesis

The template positions at which primer:RNA:DNA is initiated can be established using primer extension reactions. To perform these experiments, ^{32}P-labelled primer:RNA:DNA, or primer:DNA molecules, are hybridized to complementary single-stranded DNA (*Figure 2*). These molecules are then elongated using a DNA polymerase, such as Sequenase 2.0 (United States

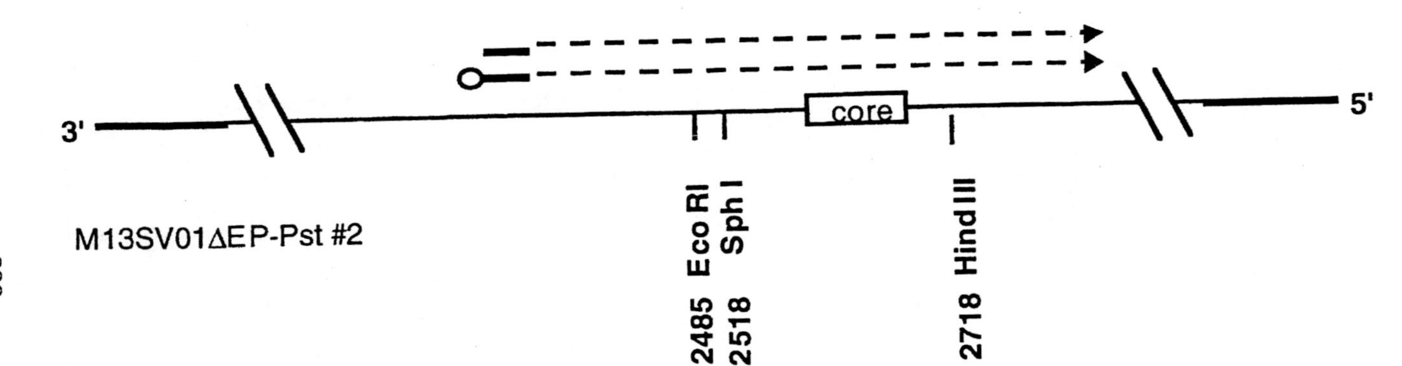

Figure 2. A depiction of a primer extension reaction using M13SV01ΔEP-*Pst*2 (43). Single-stranded DNA derived from plasmid pSV01ΔEP is symbolized by the thin line; M13 DNA is symbolized by thicker lines. ^{32}P-labelled primer DNA is symbolized by the small black rectangle, while the small oval represents primer RNA. Removal of the RNA primers from half of the samples with alkali (depicted by the rectangles not associated with a small circle) permits the determination of primer:DNA start sites. The dashed arrows symbolize Sequenase 2.0 catalysed primer extension products that can be cleaved at convenient restriction endonuclease sites (e.g. *Eco*RI, *Sph*I, or *Hin*dIII). The SV40 core origin is indicated.

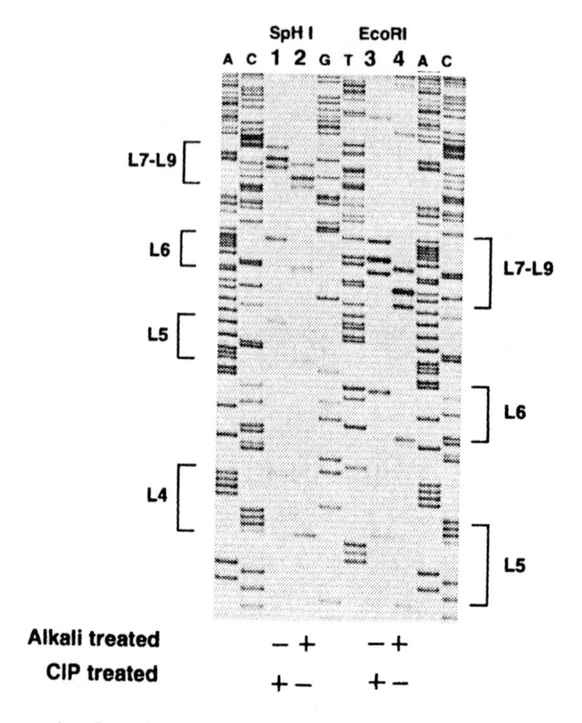

Figure 3. An example of a primer extension reactions used to map initiation sites on the late side of the SV40 origin. Aliquots of the primer extension products formed with M13SV01ΔEP-*Pst2* and either CIP treated (lanes 1 and 3) or alkali treated (lanes 2 and 4) primer:RNA:DNA. Reactions were cleaved with either *Sph*I at pSV01ΔEP position 2518 (lanes 1 and 2) or *Eco*RI at pSV01ΔEP position 2485 (lanes 3 and 4). Individual primer:RNA and primer:DNA start sites (L stands for late) are indicated. The sequencing ladders used as size markers are indicated by the letters A, C, G, and T. Adapted from ref. 43.

Biochemical). Upon cleavage at a given restriction endonuclease site, distinct primer extension products are generated whose size(s) can be determined on acrylamide gels containing suitable sequencing ladders. The procedures used to perform the primer extension techniques are described in detail in *Protocol 10* and an example of a primer extension reaction is reproduced in *Figure 3.*

Primer extension reactions revealed the existence of an initiation signal for the pol α-primase complex that is located within those template sequences encoding primer:RNA (43). More precisely, it is situated proximal to the nucleotide encoding the 5′ end of the RNA primer. It was also observed that initiation signals are present, on average, once every 19 nucleotides (43). These conclusions were based on results from many primer extension reactions (43). Experiments are being conducted to determine to what extent varying the reaction conditions, such as those discussed in *Protocol 5*, influence initiation site selection.

Protocol 10. Determining initiation sites for primer:RNA:DNA formation using primer extension reactions

Reagents

- Single-stranded DNA having the same sequence as the template used to form primer:RNA:DNA as described in *Protocol 9*
- Sequenase version 2.0 (USB): routinely diluted 1:5 in enzyme dilution buffer
- rRNasin ribonuclease inhibitor (Promega, No. N2511)
- Calf intestinal phosphatase (Boehringer Mannheim Biochemicals)
- A mix containing all four dNTPs, each at 1.25 mM (dITP in place of dGTP)
- 0.1 M DTT
- Purified primer:RNA:DNA (see *Protocol 8*), and resuspended in 40 μl dH_2O
- Carrier DNA, e.g. phage lambda cut with *Hin*dIII
- 8% polyacrylamide gel containing 8 M urea
- Sequencing markers prepared by the dideoxy sequencing method employing a kit from USB (for details regarding the sequencing ladders used as markers in the primer extension reactions presented in *Figure 3*, see ref. 43)

Method

1. Remove RNA primers from one 20 μl aliquot of primer:RNA:DNA by adding NaOH to 0.3 M and incubating the sample at 65°C for 45 min. After completing this incubation, neutralize the sample with 0.5 μl acetic acid and add 1 μl 10 mg/ml tRNA. Ethanol precipitate the primer: DNA containing sample (no additional salt is required), wash with 80% ethanol, and dry.
2. To remove the phosphate residues on the 5′ termini of primer: RNA:DNA, treat the second 20 μl aliquot with CIP. To perform the reaction, combine 5 μl 10 × CIP, 1 μl 50 mM DTT, 1 μl rRNasin, 2 μl CIP, and an additional 20 μl dH_2O with the original 20 μl aliquot. Incubate the sample for 30 min at 55°C. To stop the reaction, add 0.5 μl 0.5 M EDTA, 1 μl 10 mg/ml tRNA, and heat the sample for 10 min at 75°C. Add 50 μl TE; to remove the CIP, extract the sample twice with phenol:chloroform. Following the extraction, add 10 μl 3 M Na acetate and precipitate the sample with 100% ethanol. Wash the pellet with 80% ethanol and dry.
3. The resulting two pellets contain RNA-free primer:DNA and CIP treated primer:RNA:DNA. To conduct the primer extension reactions, it is necessary to hybridize these molecules to single-stranded DNA. To a given pellet, add 1 μg of a suitable single-stranded template (e.g. either M13SV01ΔEP-*Pst*1 or M13SV01ΔEP-*Pst*2) (~ 8 μl) along with 4 μl 5 × Sequenase reaction buffer, and H_2O to 14 μl. After assembling the reactions, place the Eppendorf tubes at 65°C for 2 min and then allow the samples to cool to 37°C in a ~ 200 ml beaker.
4. Upon completion of the hybridization step, initiate the primer extension reactions by adding 3 μl of the deoxynucleotide mix containing dATP, dCTP, dITP, and TTP (1.25 mM of each), 1 μl 0.1 M DTT,

Protocol 10. *Continued*

and 2 μl of diluted Sequenase. Allow to polymerize for ~ 10 min at 37 °C.

Inactivate the Sequenase by placing the tubes at 75 °C for 20 min. Then adjust the volume of the reaction mixture to ~ 100 μl with TE. To ensure complete removal of the Sequenase, extract the reactions twice with phenol:chloroform. To each reaction add carrier DNA (~ 3 μg) and 10 μl 3 M Na acetate, and precipitate with ethanol. Wash the pellets with 80% ethanol and dry.

5. For every digest to be conducted (~ 4000 c.p.m. of primer extension product per digest), resuspend the pellet in 5 μl TE along with an additional 5 μl for an undigested control. After resuspension, digest the samples with the desired restriction endonuclease, according to the manufacturer's recommendations. To prevent degradation of the RNA primer, add RNasin (1 μl) to each reaction. Allow restriction endonuclease digestions to proceed for 2 h in 40 μl reactions with 20 U of a given enzyme. Stop the digestions by adding TE to 100 μl and conduct a single phenol:chloroform extraction. Then ethanol precipitate, using 10 μl 3 M Na acetate and 100% ethanol, and a subsequent 80% ethanol wash. (The samples may be stored overnight in the 80% ethanol wash.) The non-restriction endonuclease treated control should be processed in exactly the same manner.
6. After drying, resuspend the samples in 2.5 μl of the stop mix (supplied with the USB sequencing kit). Boil the samples for 4 min and then apply to a denaturing 8% polyacrylamide gel, along with sequencing ladders that serve as size markers.

Finally, when the products of the primer extension reactions were cleaved at origin distal restriction endonuclease sites, such as the *Hin*dIII site at 2718 (*Figure 2*), it was noted that initiation sites for primer:RNA:DNA were greatly suppressed over the SV40 core origin *in vitro* (50, 51). One interpretation of these results is that the unwound region must attain a particular size before DNA synthesis can initiate. Alternatively, the pol α-primase complex may be initially positioned beyond the limits of the core origin. Future experiments, perhaps utilizing some of the techniques described in this chapter, may help to distinguish between these and related possibilities.

Acknowledgements

The protocols in this chapter reflect the efforts of many individuals, particularly those associated with the Hurwitz, Kelly, and Stillman laboratories. Supported by a grant from the National Institutes of Health (9RO1GM55397).

References

1. Kornberg, A. and Baker, T. A. (1992). *DNA replication*. W. H. Freeman and Co., New York.
2. DePamphilis, M. L. (1996). In *DNA replication in eukaryotic cells*. Cold Spring Harbor Laboratory Press, Cold Spring Harbor, New York.
3. Challberg, M. D. and Kelly, T. J. (1989). *Annu. Rev. Biochem.*, **58**, 671.
4. Hurwitz, J., Dean, F. B., Kwong, A. D., and Lee, S.-H. (1990). *J. Biol. Chem.*, **265**, 18043.
5. Stillman, B. W. (1989). *Annu. Rev. Cell Biol.*, **5**, 197.
6. Li, J. and Kelly, T. (1984). *Proc. Natl. Acad. Sci. USA*, **81**, 6973.
7. Stillman, B. W. and Gluzman, Y. (1985). *Mol. Cell. Biol.*, **5**, 2051.
8. Wobbe, C. R., Dean, F., Weissbach, L., and Hurwitz, J. (1985). *Proc. Natl. Acad. Sci. USA*, **82**, 5710.
9. Bullock, P. A. (1997). *Crit. Rev. Biochem. Mol. Biol.*, **32**, 503.
10. Borowiec, J. A., Dean, F. B., Bullock, P. A., and Hurwitz, J. (1990). *Cell*, **60**, 181.
11. Brush, G. S. and Kelly, T. J. (1996). In *Mechanisms for replicating DNA* (ed. M. L. DePamphilis), p. 1. Cold Spring Harbor Laboratory Press, Cold Spring Harbor, NY.
12. Li, J. J. and Kelly, T. J. (1985). *Mol. Cell. Biol.*, **5**, 1238.
13. Kenny, M. K. (1993). In *In vitro SV40 DNA replication* (ed. P. Fantes and R. Brooks), p. 197. IRL Press, Oxford.
14. Brush, G. S., Kelly, T. J., and Stillman, B. (1995). In *Identification of eukaryotic DNA replication proteins using simian virus 40 in vitro replication system* (ed. J. L. Campbell), Vol. 262. Academic Press, San Diego.
15. O'Reilly, D. R. and Miller, L. K. (1988). *J. Virol.*, **62**, 3109.
16. Lanford, R. E. (1988). *Virology*, **167**, 72.
17. Murphy, C. I., Weiner, B., Bikel, I., Piwnica-Worms, H., Bradley, M. K., and Livingston, D. M. (1988). *J. Virol.*, **62**, 2951.
18. Dixon, R. A. F. and Nathans, D. (1985). *J. Virol.*, **53**, 1001.
19. Simanis, V. and Lane, D. P. (1985). *Virology*, **144**, 88.
20. Harlow, E. and Lane, D. (1988). *Antibodies: a laboratory manual*. Cold Spring Harbor Laboratory Press, Cold Spring Harbor Laboratory, NY.
21. Myers, R. M. and Tijan, R. (1980). *Proc. Natl. Acad. Sci. USA*, **77**, 6491.
22. Weinberg, D. H., Collins, K. L., Simancek, P., Russo, A., Wold, M. S., Virshup, D. M., *et al.* (1990). *Proc. Natl. Acad. Sci. USA*, **87**, 8692.
23. Ishimi, Y., Claude, A., Bullock, P., and Hurwitz, J. (1988). *J. Biol. Chem.*, **263**, 19723.
24. Tsurimoto, T., Melendy, T., and Stillman, B. (1990). *Nature*, **346**, 534.
25. Lee, S.-H., Eki, T., and Hurwitz, J. (1989). *Proc. Natl. Acad. Sci. USA*, **86**, 7361.
26. Dean, F. B., Bullock, P., Murakami, Y., Wobbe, C. R., Weissbach, L., and Hurwitz, J. (1987). *Proc. Natl. Acad. Sci. USA*, **84**, 16.
27. Dodson, M., Dean, F. B., Bullock, P., Echols, H., and Hurwitz, J. (1987). *Science*, **238**, 964.
28. Wold, M. S., Li, J. J., and Kelly, T. J. (1987). *Proc. Natl. Acad. Sci. USA*, **84**, 3643.
29. Roberts, J. M. and D'Urso, G. (1989). *J. Cell Sci. Suppl.*, **12**, 171.
30. Murakami, Y. and Hurwitz, J. (1993). *J. Biol. Chem.*, **268**, 11018.
31. Stahl, H., Droge, P., and Knippers, R. (1986). *EMBO J.*, **5**, 1939.

32. Bullock, P. A., Seo, Y. S., and Hurwitz, J. (1989). *Proc. Natl. Acad. Sci. USA*, **86**, 3944.
33. Goetz, G. S., Dean, F. B., Hurwitz, J., and Matson, S. W. (1988). *J. Biol. Chem.*, **263**, 383.
34. Wold, M. S. (1997). *Annu. Rev. Biochem.*, **66**, 61.
35. Wobbe, C. R., Weissbach, L., Borowiec, J. A., Dean, F. B., Murakami, Y., Bullock, P., *et al.* (1987). *Proc. Natl. Acad. Sci. USA*, **84**, 1834.
36. Henricksen, L. A., Umbricht, C. B., and Wold, M. S. (1994). *J. Biol. Chem.*, **269**, 11121.
37. Sambrook, J., Fritsch, E. F., and Maniatis, T. (ed.) (1989). *Molecular cloning: a laboratory manual.* Cold Spring Harbor Laboratory, Cold Spring Harbor, NY.
38. Fanning, E. and Knippers, R. (1992). *Annu. Rev. Biochem.*, **61**, 55.
39. Hauschka, P. V. (1973). *Methods Cell Biol.*, **3**, 362.
40. Traut, T. W. (1994). *Mol. Cell. Biochem.*, **140**, 1.
41. Bullock, P. A., Seo, Y. S., and Hurwitz, J. (1991). *Mol. Cell. Biol.*, **11**, 2350.
42. Denis, D. and Bullock, P. A. (1993). *Mol. Cell. Biol.*, **13**, 2882.
43. Bullock, P. A., Tevosian, S., Jones, C., and Denis, D. (1994). *Mol. Cell. Biol.*, **14**, 5043.
44. Nethanel, T. and Kaufmann, G. (1990). *J. Virol.*, **64**, 5912.
45. Takasaki, Y., Fishwild, D., and Tan, E. M. (1984). *J. Exp. Med.*, **159**, 981.
46. Nethanel, T., Reisfeld, S., Dinter-Gottlieb, G., and Kaufmann, G. (1988). *J. Virol.*, **62**, 2867.
47. Orrù, D., DelGrosso, N., Angeletti, B., and D'Ambrosio, E. (1993). *Bio-Techniques*, **14**, 905.
48. Sanger, F., Nicklen, S., and Coulson, A. R. (1977). *Proc. Natl. Acad. Sci. USA*, **74**, 5463.
49. Maxam, A. M. and Gilbert, W. (1980). In *Methods in enzymology* (ed. L. Grossman and K. Holdave), Vol. 65, p. 499. Academic Press.
50. Bullock, P. A. and Denis, D. (1995). *Mol. Cell. Biol.*, **15**, 173.
51. Bullock, P. A., Joo, W. S., Sreekumar, K. R., and Mello, C. (1997). *Virology*, **227**, 460.

11

Analysis of DNA replication complexes by DNA probing

JAMES A. BOROWIEC, THOMAS G. GILLETTE, NATALIA V. SMELKOVA, and CRISTINA IFTODE

1. Introduction

This chapter describes techniques used to characterize complexes formed in solution between replication proteins and DNA. Although general DNA probing or 'footprinting' techniques used to probe standard protein:DNA complexes are provided, special emphasis will be placed on those methods most useful to investigators studying the interaction of DNA replication proteins with DNA replication substrates. Thus, the reader will find a variety of DNA probing techniques that can reveal the presence and location of distorted DNA within complexes. We discuss chemical and enzymatic probes useful for examining the disposition of single-stranded DNA (ssDNA) often found in replication intermediates. We also describe the use of interference assays which allow the determination of DNA contacts important for protein binding, and for the characterization of transient complexes formed between replication proteins and DNA.

2. Applications of DNA footprinting techniques

Characterization of protein:DNA complexes by DNA footprinting in one manifestation employs a DNA cleavage or modifying agent whose pattern of modification along the DNA is altered by the binding of a protein to the DNA. Alternatively, DNA containing a binding site for the protein of interest can first be modified, and the sites where modification interferes with protein binding determined subsequently. In either case, characterization of differences in the pattern of modification between those DNA molecules bound by protein and those remaining unbound can be used to achieve various types of useful information. The DNA footprinting techniques that we discuss can be used to:

(a) Reveal the general placement of replication proteins on DNA.
(b) Determine the base and phosphate residues closely apposed by the bound

protein, or which are critical for complex formation between the DNA binding protein and DNA.

(c) Detect regions of structurally distorted DNA within replication protein: DNA complexes, and identify the type of distortion (e.g. melting, bending).

(d) Identify bases and phosphate residues critical for DNA-dependent enzymatic reactions mediated by the replication protein.

These techniques have been successfully used to examine complexes formed between origin binding proteins and origin DNA such as the viral large T-antigen with the simian virus 40 (SV40) origin (1), the viral UL9 with the herpes simplex virus origin (2), and the origin recognition complex (ORC) with *Saccharomyces cerevisiae* ARS elements (3). Moreover, DNA footprinting techniques have also been productively used to examine the binding of proteins to specific DNA structures rather than a certain DNA sequence. Among the enzymes and DNA targets examined have been human DNA polymerase δ and the DNA polymerase accessory proteins PCNA and RF-C with primer templates (4), and the SV40 T-antigen with synthetic DNA replication forks (5). A subset of these methods also have the potential for probing DNA *in vivo* as discussed by others (6).

3. DNA probing techniques—direct probing versus interference analysis

Two DNA footprinting techniques, which we term 'direct probing' and 'interference analysis', are most often used to analyse protein:DNA complexes. Each technique by itself gives rise to an important subset of key information, but when used in tandem the techniques can yield an extensive amount of information on how the protein interacts with DNA.

3.1 Direct probing

In this approach, the protein:DNA complexes are formed and then probed with an appropriate reagent. After the modification pattern is determined, it is compared to the pattern found in the absence of added replication protein to identify key DNA contacts within the complex (*Figure 1*). The approach is relatively simple and allows one to directly probe complexes in solution with a suitable enzymatic or chemical reagent. Importantly, the technique allows regions of structurally distorted DNA to be identified.

The DNA substrate is normally a DNA molecule radioactively labelled on one strand, generally a $^{32}PO_4$ tag on either the 5′ or 3′ terminus. If the chosen probe does not directly cleave the DNA (e.g. dimethyl sulfate), a specific DNA cleavage reaction is subsequently performed to break the DNA at the sites of modification. These sites are then identified by subjecting the broken

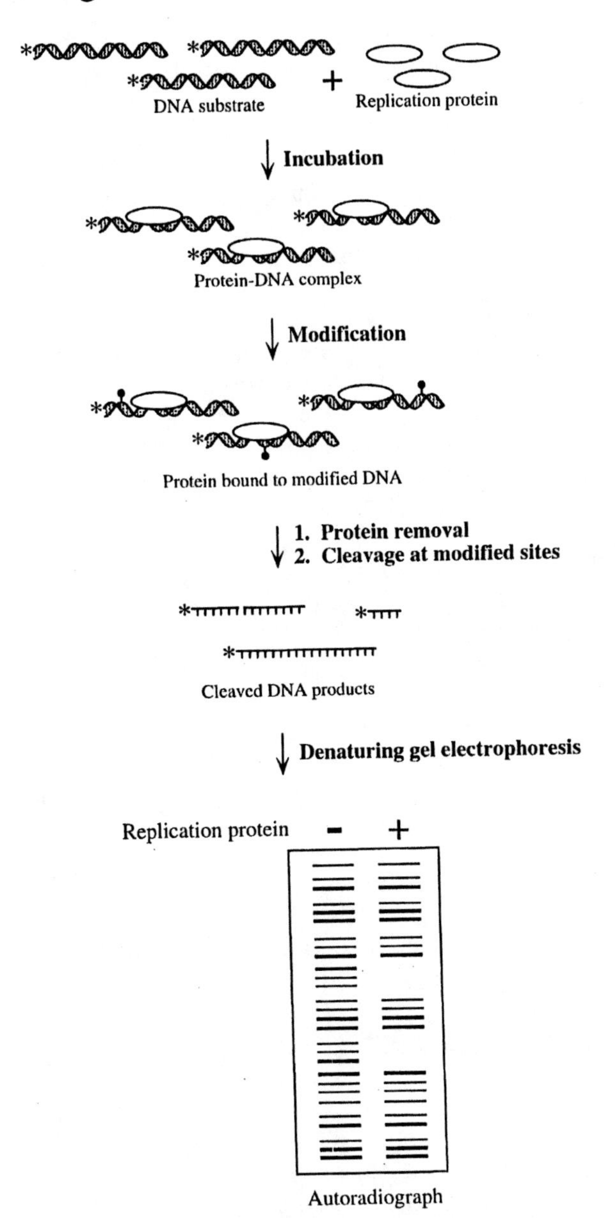

Figure 1. Schematic of direct probing assay. A scheme is shown in which a DNA probe modifies, but does not cleave the DNA substrate (e.g. dimethyl sulfate). For reagents that directly cleave the DNA (e.g. the 1,10-phenanthroline–copper complex), the DNA would be deproteinized after cleavage and then directly subjected to denaturing gel electrophoresis.

DNA to denaturing gel electrophoresis and autoradiography, using standard protocols (7).

On occasion, it is preferable to utilize a DNA substrate that is present on a circular plasmid DNA or a lengthy linear DNA molecule. For example, one may wish to examine a protein-mediated interaction between two distal DNA sequences. Although the protocol will not be covered in this chapter, the modification pattern can be determined after the DNA cleavage by extension of a 5′-[^{32}P]oligonucleotide primer through the target site with a DNA polymerase such as the Klenow fragment of *E. coli* DNA polymerase I (8). The extension products are then visualized by denaturing gel electrophoresis and autoradiography.

Although the direct probing method is powerful, there are two inherent limitations. First, if the replication protein is able to form multiple complexes with the DNA, the modification pattern will represent contributions from each complex rather than being a result of a single species. Secondly, short-lived intermediates will not be observed unless they give rise to enhanced modification at sites that are visible above the background modification.

3.2 Interference analysis

In contrast to direct probing, interference analysis utilizes a pre-modified DNA molecule as substrate (*Figure 2*). This allows the use of reagents that modify DNA under conditions incompatible with protein binding (e.g. *N*-nitroso-*N*-ethylurea or formic acid). The modified substrate is then incubated with the protein of interest, after which the bound DNA molecules are separated from protein-free DNA.

The separation of bound from free DNA molecules is usually accomplished by use of a gel retardation assay. In this approach, protein:DNA reaction mixtures are subjected to non-denaturing polyacrylamide gel electrophoresis. The greater mass of the protein:DNA complex relative to unbound DNA generally causes a reduction in the electrophoretic mobility of the complex. After autoradiography to localize the bound and free DNA, gel slices containing each DNA pool are excised and the DNA isolated. DNA molecules that contain modifications interfering with complex formation will be under-represented in the bound DNA fraction and over-represented in the free DNA pool.

Cleavage of the isolated free and bound DNA pools at the sites of modification will therefore reveal, upon denaturing gel electrophoresis and autoradiography, base and phosphate residues important for protein:DNA complex formation. Because it is difficult to use large DNA substrate molecules in this method, interference assays preclude the use of a primer extension reaction to visualize the modification pattern. The method also requires that the complexes have sufficient stability to allow separation from the free DNA.

If the protein forms multiple complexes with DNA, it is generally possible to isolate each complex and individually determine the modification pattern

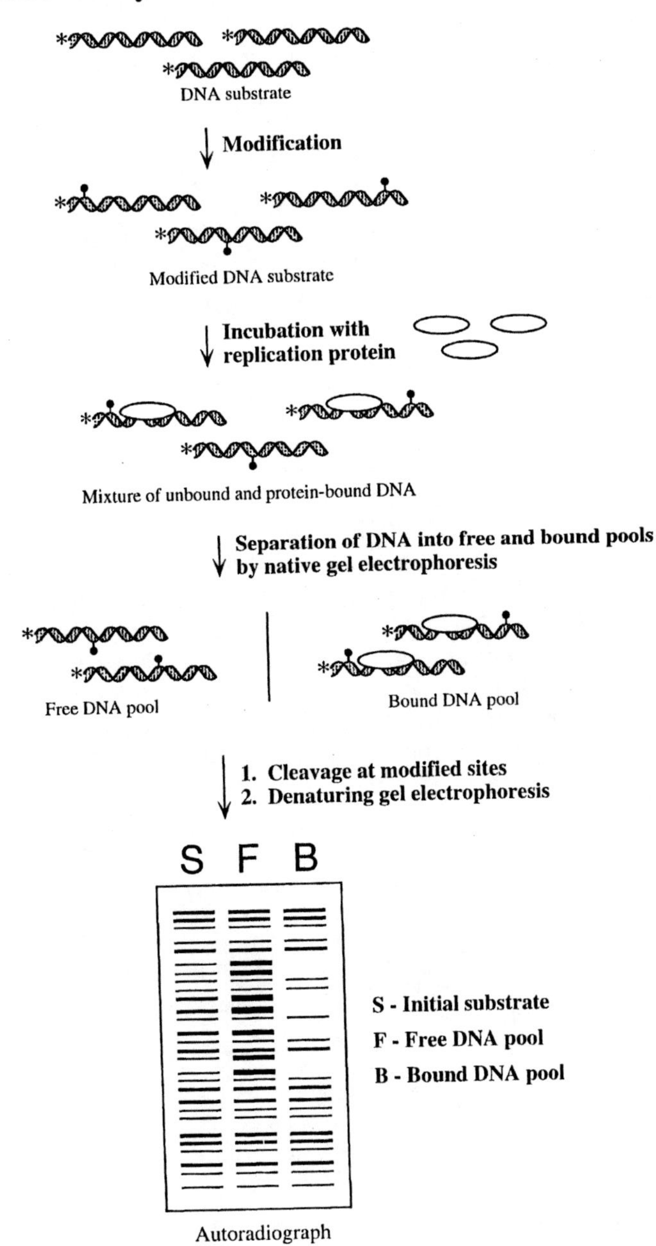

Figure 2. Schematic of interference assay.

of the bound DNA. The approach is also useful in its ability to probe dynamic complexes. For example, the technique has been used to determine bases and phosphates on the SV40 origin required for origin denaturation by T-antigen and a single-stranded DNA binding protein during replication initiation (1).

4. Overview of chemical and enzymatic probes

A large number of chemical probes have been developed that react with various moieties on DNA. The sites modified include phosphate and sugar residues of the backbone and bases at major and minor groove positions and at hydrogen bonding sites. The applicability of these chemical and enzymatic probes for direct probing and interference assays are listed in *Table 1*, and are described in greater detail below.

4.1 Chemical probes

4.1.1 *N*-nitroso-*N*-ethylurea

This reagent ethylates phosphates on the sugar phosphate backbone of DNA to form a phosphotriester (9). Because conditions needed for DNA ethylation (50% ethanol, 50°C) are incompatible with protein binding, use of this reagent is restricted to interference assays. Modification of the DNA substrate at phosphates that are juxtaposed to or contacted by the DNA binding protein can result in the destabilization of the protein:DNA complex. Sites at which phosphate ethylation causes a reduction of complex formation are thus a reliable indicator of phosphate residues important for protein binding. The pattern of phosphate contacts, when plotted on a helix map of protein

Table 1. Applications for DNA probes

	Direct probing of protein:DNA complexes		
DNA probe	**Characterization of protein:DNA contacts?**	**Detection of DNA distortion?**	**Interference assays?**
N-nitroso-*N*-ethylurea	No	No	Yes
Dimethyl sulfate	Yes	Yes	Yes
Diethyl sulfate	No	No	Yes
Potassium permanganate	No	Yes	No
1,10-phananthroline–copper complexes	Yes	Yes	No
Hydrazine	No	No	Yes
Formic acid	No	No	Yes
DNase I (bovine pancreatic)	Yes	No	No
Exonuclease III (*E. coli*)	Yes	No	No
Nuclease P1 (*Pennicillium citrinum*)	Yes (on ssDNA)	Yes (detection of ssDNA)	No

contacts on DNA, can also be useful in determining the helical face of DNA that a protein binds (1).

In addition to modifying phosphates, the reagent also efficiently ethylates nucleotide bases at various positions (10). To use this reagent for detection of critical phosphates, specific cleavage conditions must be used to selectively cleave the DNA at the modified phosphate residues.

4.1.2 Dimethyl sulfate

As DNA is highly reactive to dimethyl sulfate under normal protein binding conditions, it can be used in direct probing assays to characterize close contacts of protein with DNA (*Protocol 3*; see Section 6.3), as well as to detect melted regions within DNA (*Protocol 8*; see Section 7.4), and for methylation interference (*Protocol 9*; see Section 8.2).

On double-stranded DNA (dsDNA), dimethyl sulfate methylates nucleotide bases primarily on the N7 site of guanosine (a major groove position) and, to a lesser extent, the adenosine N3 (a minor groove position). Modification of guanosines is usually most informative when examining protein binding because recognition of specific DNA sequences generally occurs through the major groove. The N1 of adenines and the N3 of cytosines are normally inaccessible to modification by dimethyl sulfate in duplex DNA because of their involvement in hydrogen bonding interactions with the opposite strand. When present in ssDNA these sites become exposed and can be methylated to a significant extent. The ability of a duplex DNA binding protein to allow methylation of the adenine N1 or cytosine N3 is therefore indicative of DNA melting induced by the protein.

Diethyl sulfate is often used in place of dimethyl sulfate for interference assays (11) (*Protocol 10*; see Section 8.2). Although the reactivity of diethyl sulfate is somewhat less than that of dimethyl sulfate, diethyl sulfate produces a bulkier adduct (i.e. an ethyl group versus a methyl group) and perhaps a greater ability to perturb protein binding.

4.1.3 Potassium permanganate ($KMnO_4$)

Duplex DNA in the normal B-form is poorly reactive to $KMnO_4$. When a DNA binding protein causes the DNA to become melted, sharply bent or significantly untwisted (of altered helical pitch), the distorted region(s) becomes highly reactive to $KMnO_4$ attack. The sites of attack are predominantly at thymine residues, in which the C5=C6 double bond of pyrimidines becomes oxidized (forming a *cis*-diol) (12, 13). A lower level of cytosine modification has also been observed although the oxidation of these residues is apparently dependent on reaction conditions. $KMnO_4$-modified residues can be mapped by chemical cleavage of a ^{32}P-end-labelled DNA fragment (14) (*Protocol 5*). Alternatively, as the oxidized nucleotides can stall nucleotide incorporation by various DNA polymerases (e.g. the Klenow fragment of

DNA polymerase I), the modification pattern can be determined by extension of a ^{32}P-labelled DNA primer across the region of interest (15).

4.1.4 1,10-phenanthroline–copper complexes

The cuprous complexes of 1,10-phenanthroline and its relative 5-phenyl-1,10-phenanthroline can be used in direct probing assays to detect distortions within duplex DNA (16). On dsDNA, the phenanthroline–copper complex binds the minor groove, from which a reactive species is generated that cleaves the sugar phosphate backbone of one of the two strands. Local changes in duplex DNA structure that facilitate binding of the probe cause enhanced cleavage of the DNA at these sites. The probe is sensitive to subtle structural changes, such as alterations in minor groove width, that $KMnO_4$ is unable to recognize (C. I. and J. A. B., unpublished observations). Because this reagent is active on ssDNA, it is possible to employ this reagent to probe the binding of proteins to ssDNA although this application for the copper–phenanthroline probe has not been extensively characterized. In addition to the use of this reagent in direct probing assays, it can also be employed to characterize DNA:protein complexes within an acrylamide gel matrix after separation from the unbound DNA (16).

4.1.5 Hydrazine

Hydrazine is used to prepare partially depyrimidated duplex DNA. The reagent destroys the pyrimidine ring, converting the attached sugar to a hydrazone without cleaving the backbone (17). The reagent can therefore be used in interference assays to determine important pyrimidine contacts used by the replication protein when binding DNA (18).

4.1.6 Formic acid

The complement of hydrazine, formic acid protonates nitrogens on the purine ring, weakening the glycosidic bond and promoting release of the purine base. Formic acid is thus used to make partially depurinated duplex DNA and thereby determine critical purines that support productive interaction of the replication protein with DNA (18).

4.2 Enzymatic probes

4.2.1 DNase I

Bovine pancreatic deoxyribonuclease I is the standard enzymatic reagent used to directly probe protein:DNA complexes. The enzyme prefers dsDNA but has lesser activity on ssDNA. Crystal structures of DNase I indicate that the enzyme utilizes minor groove contacts when binding duplex DNA (19). These observations make understandable the sequence dependence of the enzyme, particularly with regard to the relatively poor digestion of AT-rich tracts which have a reduced minor groove width. The enzyme cleaves to the 5′ side of the phosphate, releasing products with a 3′ OH.

4.2.2 Exonuclease III

This enzyme has a 3′-5′ exonuclease activity, removing mononucleotides from the 3′ OH ends of the dsDNA in a processive manner (20). When a protein is bound to the DNA, the exonuclease activity is blocked and the 3′ boundary of the protein binding site can be determined. Using a DNA substrate labelled on the 5′ end of the top or bottom strand allows determination of the DNA region protected by the protein. Although the enzyme gives rise to less information than DNase I, it is useful for examining protein:DNA complexes in which only a small fraction of the DNA substrate is bound, because the resulting signal is generally visible over a low background.

4.2.3 Nuclease P1

This is a single-strand-specific endonuclease from *Pennicillium citrinum* that can be used to detect ssDNA regions in protein:DNA complexes or to probe the disposition of preformed ssDNA regions. Nuclease P1 can be used at neutral pH, and is active under most DNA binding conditions used for replication proteins (21). The enzyme gives rise to products with a 3′ OH.

5. DNA substrate

5.1 General considerations

The standard DNA substrate is a duplex DNA molecule containing a $^{32}PO_4$ label on the 5′ or 3′ termini of one of the two DNA strands. Choice of the optimal length for DNA footprinting is dependent on a few variables. For example, proteins that employ a 50 bp DNA binding site are better analysed using a 250 bp substrate than a 100 bp substrate. DNA lengths of 150–250 bp are generally appropriate for most DNA probing analyses, with the binding site located ~ 75–100 bp from the labelled DNA end. It is possible to use substrates as small as 25 bp or so but it is more challenging to obtain suitable resolution for both the smallest and largest oligonucleotide products of the footprinting reaction.

Study of DNA replication complexes often entails construction of non-standard DNA substrates that contain unusual structures such as a DNA hairpin or DNA fork (4). These substrates are usually prepared using oligonucleotides designed to give rise to the appropriate structure. The substrate is 5′-^{32}P-labelled with [γ-^{32}P]ATP and bacteriophage T4 polynucleotide kinase (7) on one of the oligonucleotides prior to the annealing reaction. Preparation of these substrates is constrained by size limits for synthetic oligonucleotides (~ 100 nt) and the cost of production. Use of high purity DNA oligonucleotides is essential for reducing the background of the final footprinting autoradiographs. If the purity of the oligonucleotide precursors is in doubt, purify the oligonucleotide using standard procedures (e.g. by denaturing gel electrophoresis) (7) before construction of the DNA binding substrate.

5.2 Standard preparation of a ^{32}P-labelled duplex DNA substrate

Preparation of duplex DNA substrates that are singly ^{32}P-end-labelled is a standard protocol well-covered by others (17). In brief, a plasmid DNA molecule containing a recognition element for the protein of interest is cleaved by a restriction enzyme at a site to one side of the element and at which labelling will subsequently occur. For 5′-^{32}P-labelling, the restricted DNA is incubated with [γ-^{32}P]ATP and bacteriophage T4 polynucleotide kinase. Labelling on the 3′ end entails DNA polymerase-mediated incorporation of an [α-^{32}P]dNTP at a recessed 3′ end. The labelled DNA is then cleaved by a second restriction enzyme at a site located on the other side of the recognition element. The DNA molecule, labelled at only one end, is isolated from other fragments by native gel electrophoresis (7).

5.3 Preparation of the DNA substrate by PCR

Although the standard method of duplex DNA preparation (see Section 5.2) is straightforward, it requires the presence of convenient restriction sites on both sides of the protein binding site. In an alternative approach, we have found that use of the polymerase chain reaction (PCR) facilitates substrate preparation for multiple reasons. Use of PCR methodology allows preparation of DNA fragments of desired length, which can become important when specific fragment lengths are needed for resolving different complexes formed between the replication protein and the DNA substrate. Secondly, the protein binding site can be optimally placed in the centre of the fragment. Thirdly, because only one of the two PCR primers is 5′-^{32}P-labelled, restriction digestion of the labelled DNA is not required, allowing a more rapid preparation of the substrate. Finally, characterization of a protein:DNA complex is often facilitated by examining the binding to mutant DNA substrates. If mutant recognition sequences are present on plasmids with identical flanking DNA, the PCR method facilitates the preparation of the mutant substrates with identical specific activities.

Protocol 1 describes the preparation of a duplex substrate by PCR. The average yield from the primers is ~ 50% (i.e. 5 pmol), usually sufficient for > 100 footprinting reactions. Template DNA levels are kept relatively high (1 ng) to shorten the number of cycles required. Although the protocol suggests the use of *Taq* DNA polymerase, other thermostable DNA polymerases can be effectively substituted (e.g. *Pwo* DNA polymerase; Boehringer Mannheim).

Protocol 1. PCR-mediated preparation of duplex DNA substrates

Equipment and reagents

- Thermal cycler
- Two DNA primers: 10 pmoles each, one of which is 5′-^{32}P-labelled to a specific activity of 1–3 × 10^6 c.p.m./pmol, in TE[a]
- *Taq* DNA polymerase: 5 U/μl (Boehringer Mannheim) or equivalent thermostable DNA polymerase
- 10 × *Taq* DNA polymerase buffer: 100 mM Tris–HCl pH 8.3 at 20°C, 15 mM $MgCl_2$, 500 mM KCl
- dNTP solution: 2 mM dATP, dGTP, dCTP, and dTTP
- Template DNA containing the sequence to be amplified

Method

1. Mix the following reagents:
   - 10 pmol 5′-^{32}P-labelled DNA primer
   - 10 pmol second DNA primer
   - 5 μl 10 × *Taq* DNA polymerase buffer
   - 5 μl dNTP solution
   - 1 ng template DNA
   - Distilled water to 50 μl
   - 1 μl *Taq* DNA polymerase
2. Subject the reaction mixture to 20 cycles of annealing, extension, and denaturation in a thermal cycler.
3. Isolate the PCR product by standard protocols (7).[b]

[a] 10 mM Tris–HCl pH 8, 1 mM EDTA.
[b] Various methods are available to separate the PCR product from the unincorporated primer molecules. As described in detail in Sambrook *et al.* (7), these procedures include agarose gel electrophoresis of the product onto DEAE cellulose membranes, and the use of low melting temperature agarose for gel electrophoresis to allow efficient recovery of the product. Various commercial kits also yield DNA of high purity such as Geneclean (Bio 101, Vista, CA).

5.4 Use of competitor DNA

Replication factors often bind DNA non-specifically as a consequence of their activity in DNA replication. This non-specific binding can effectively obscure the specific signal, especially when testing replication factors that recognize a certain DNA structure rather than a DNA sequence. Competitor DNA is therefore routinely added to DNA probing reactions to minimize background binding yet maintain specific protein:DNA complexes at high levels. The best choice of competitor DNA must be determined for the replication protein tested. A simple and inexpensive option is the use of circular or linear plasmid DNA, the latter if it is known that the protein binds DNA ends non-specifically. One may occasionally need to construct the appropriate competitor using oligonucleotides, but because this can be an exercise in trial-

and-error, it is not recommended for most cases. The addition of competitor DNA is also useful to ensure that a small variability in DNA substrate concentration will have no impact on modification levels, and to act as a carrier for efficient ethanol precipitation of the substrate during manipulations.

6. General characterization of replication protein:DNA complexes

6.1 Determination of modification conditions

A key step for the analysis of replication complexes by DNA footprinting is the determination of reaction conditions necessary to appropriately modify the DNA substrate. For both direct probing and interference analysis, the degree of modification is such that only a small fraction of sites on the DNA substrate are modified. Although this level is sometimes referred to as 'one hit per strand', the optimum level is somewhat less. A good rule of thumb is that 70% of the substrate remains unmodified.

For DNA substrates to be used in interference assays (see Section 8), the ^{32}P-labelled DNA substrate is incubated with various concentrations of the probing reagent using specific reaction conditions. The DNA is subsequently cleaved using standard protocols, and the products visualized by denaturing polyacrylamide gel electrophoresis and autoradiography. The optimum concentration of probing reagent is that which gives rise to a similar average intensity for both the largest and smallest product bands, and leaves a significant fraction of the substrate unreacted. At this point, it may be useful to repeat the assay with a narrower range of reagent concentrations and test different incubation times.

It is often more complicated to choose suitable modification conditions for direct probing assays as they must be compatible with replication protein binding. On occasion, one will find that the conditions for the two events are incompatible and it becomes essential to assay numerous buffer conditions to find the best balance between protein:DNA complex formation and DNA probe reactivity. While in some cases it may be necessary to add the DNA probing reagent along with a required reaction component to 'tweak' probe reactivity (e.g. the addition of DNase I, plus Ca^{2+} to stimulate high nuclease activity), such manipulations have the potential of causing a perturbation or destabilization of the protein:DNA complex and should be avoided if possible.

In all protocols given below, a concentration or range of concentrations of the DNA probing reagent are supplied as examples. However, it is important that appropriate modification conditions are individually determined for each experiment.

It is preferable to use siliconized microcentrifuge tubes for all steps involving modified DNA, since certain modifications tend to increase the 'stickiness' of the DNA and hence its retentiveness to non-treated polypropylene tubes, leading to a potential loss of signal. Siliconized microcentrifuge tubes are available commercially or can be prepared in the laboratory as described in Sambrook *et al.* (7).

6.2 DNase I

Perhaps the most widely used probe to examine protein:DNA complexes is bovine pancreatic DNase I. Although the preferred substrate is duplex DNA, the activity on ssDNA is sufficiently high that probing of substrates that contain both structures is possible in the same reaction tube. Although the enzyme is most active in the presence of Ca^{2+}, one can obtain a similar pattern of digestion using standard Mg^{2+} concentrations (1–7 mM) with a slightly higher concentration of the nuclease. The enzyme is active from 0–37 °C although more enzyme is required at the lower temperature. Enzyme digestion is less uniform at 0 °C compared to 25 °C because DNase I tends to cut the DNA at fewer sites (22), likely because of temperature-dependent changes in DNA structure. Highly AT-rich regions are digested less well than GC-rich regions and only limited information about the binding of the replication protein to the former regions may be obtained (see Section 4.2).

Protocol 2 provides conditions for DNase I digestion. Note that a significant cause of irreproducibility in the level of DNA digestion in otherwise 'identical' lanes is the failure to completely quench the nuclease activity. For this reason, it is important to extensively vortex the reaction mixture with the enzyme stop buffer and phenol:chloroform solution.

Protocol 2. DNase I digestion

Equipment and reagents

- Microcentrifuge
- ^{32}P-end-labelled DNA: ~ 0.01–0.10 pmol, prepared either by PCR (*Protocol 1*) or by end-labelling a suitable restriction fragment
- Enzyme stop buffer: 10% SDS, 0.5 M EDTA
- Phenol:chloroform solution (1:1, v/v)
- Bovine pancreatic DNase I: 10–50 $\times$ 10^3 U/ml (Boehringer Mannheim); the enzyme is diluted to the appropriate concentration prior to use with the reaction buffer
- 3 M sodium acetate pH 5.2
- 100% ethanol

Method

1. Incubate the ^{32}P-labelled DNA in the presence and absence of the replication protein(s) with appropriate buffers, in a reaction volume of 20 µl.
2. Add ~ 0.02–0.1 U bovine pancreatic DNase I and incubate at 37 °C for 1 min.
3. Quench digestion by the addition of 0.1 vol. enzyme stop buffer and vortex for 10 sec at room temperature.

Protocol 2. *Continued*

4. Add an equal volume of phenol:chloroform solution and vortex for 10 sec. Remove the aqueous phase.
5. Precipitate the DNA by the addition of 0.1 vol. 3 M sodium acetate and 3 vol. ethanol. Chill on ice for 10 min. Microcentrifuge the sample for 10–30 min to pellet the DNA. Discard the supernatant.
6. Add 70 μl ethanol to the pellet. Microcentrifuge the sample for 5 sec. Discard the supernatant and air dry the DNA pellet.
7. Proceed to Section 9.2 (gel electrophoresis of cleaved DNA products).

6.3 Dimethyl sulfate

A common chemical method to characterize replication protein:DNA complexes is the methylation protection reaction (reviewed in ref. 23). Based on the standard Maxam and Gilbert 'G-only' sequencing chemistry (17), the method relies on the ability of proximate protein:DNA contacts to prevent the methylation of DNA bases, generally the N7 of guanine (Section 4.1). Methylation is quite vigorous at 25–37 °C and works under most buffer conditions.

The typical concentrations of dimethyl sulfate used to modify the DNA substrate are in the range of 10–30 mM. The solubility of dimethyl sulfate in water at 20 °C is slightly greater than 200 mM and it is therefore common to dilute the dimethyl sulfate stock (neat) to 200 mM with distilled water immediately prior to use. Alternatively, one can use the stock directly but a larger reaction volume must be used (e.g. 0.5 μl dimethyl sulfate in a 200 μl reaction volume).

Protocol 3. Methylation protection

Equipment and reagents

- Microcentrifuge
- ^{32}P-end-labelled DNA: ~ 0.01–0.10 pmol, prepared either by PCR (*Protocol 1*) or by end-labelling a suitable restriction fragment
- Dimethyl sulfate stop buffer: 3 M ammonium acetate, 1 M 2-mercapto-ethanol, 20 mM EDTA, 100 μg/ml tRNA
- 200 mM dimethyl sulfate solution: prepare by diluting 4.7 μl dimethyl sulfate (Sigma) to 250 μl with distilled water. *Dimethyl sulfate is a known carcinogen. Use gloves and, if possible, employ a fume-hood when working with this reagent.*
- 100% ethanol

Method

1. Incubate the ^{32}P-end-labelled DNA in the presence and absence of the replication protein(s) with appropriate buffers, in a reaction volume of 20 μl.
2. Add 3 μl 200 mM dimethyl sulfate, and incubate an additional 3 min.

3. Quench the reaction by the addition of 2 vol. dimethyl sulfate stop buffer. Precipitate the DNA by the addition of 2 vol. (i.e. 140 μl) ethanol. Chill on ice for 10 min. Microcentrifuge the sample for 10–30 min to pellet the DNA. Discard the supernatant.
4. Add 70 μl ethanol to the pellet. Microcentrifuge the sample for 5 sec. Discard the supernatant and air dry the DNA pellet.
5. Proceed to *Protocol 15* (piperidine cleavage of DNA substrates).

6.4 Exonuclease III

Escherichia coli exonuclease III, a 3′-5′ exonuclease, is used to determine the boundaries of the protein binding site on duplex DNA. Because of the symmetrical digestion by two enzymes starting from each 3′ end of the DNA, a fully complementary duplex DNA is digested by excess exonuclease III into two ssDNA molecules, each with a length equal to approximately half of the intact DNA (half-molecules) (20, 24). Due to the pausing of the enzyme at preferred sites, additional digestion products may be detected (exonuclease stops). When a protein is bound to the DNA, the exonuclease activity is blocked, and the boundaries of the binding site can be determined by individually probing each strand.

There are some important caveats to remember when using this enzyme. It has been observed that exonuclease III can displace the bound protein, digesting DNA regions that are in fact covered by the protein (24). As a consequence, the protected regions will appear smaller than their real size. A second complication arises when studying a protein that has multiple binding sites. In this case, distinct stops for exonuclease III will be detected, but it may be difficult in practice to distinguish them from the artefactual exonuclease stops. To minimize this problem, various exonuclease III concentrations should be tested. The optimal exonuclease III concentration are those which give rise to clear half-molecules and a minimal number of erroneous exonuclease stops. The enzyme requires the presence of 1–5 mM $MgCl_2$ for efficient activity.

Protocol 4. Exonuclease III digestion

Equipment and reagents

- Microcentrifuge
- 5′-^{32}P-end-labelled DNA: ~ 0.01–0.10 pmol,[a] prepared either using a PCR reaction (*Protocol 1*) or by end-labelling a suitable restriction fragment
- Enzyme stop buffer: 1% SDS, 20 mM EDTA
- Phenol:chloroform solution (1:1, v/v)
- *E. coli* exonuclease III: 100–200 U/μl (Boehringer Mannheim); the enzyme is diluted to the appropriate concentration prior to use
- 3 M sodium acetate pH 5.2
- 100% ethanol

Protocol 4. *Continued*

Method

1. Incubate the 5′-^{32}P-labelled DNA in the presence and absence of the replication protein(s) with appropriate buffers, at 37 °C in a reaction volume of 20 μl.
2. Digest the DNA with 1–10 U exonuclease III for 10 min at 37 °C.
3. Quench the digestion reaction by the addition of 20 μl enzyme stop buffer and vortex for 10 sec at room temperature.
4. Add an equal volume of phenol:chloroform (1:1, v/v) and vortex for 10 sec. Remove the aqueous phase.
5. Ethanol precipitate the DNA (see *Protocol 2*, steps 5 and 6).
6. Proceed to Section 9.2 (gel electrophoresis of cleaved DNA products).

[a] Because the enzyme proceeds in the 3′-5′ direction, the ^{32}P-label of the DNA substrate must be on the 5′ end.

7. Analysis of DNA structure within DNA replication complexes

7.1 $KMnO_4$ modification

Potassium permanganate ($KMnO_4$) oxidizes the C5=C6 double bond in pyrimidines, with single-stranded and distorted DNA being more sensitive to attack than duplex DNA (25). Thymine residues are more susceptible to the $KMnO_4$ modification than cytosine residues. Because duplex DNA is relatively resistant to $KMnO_4$ oxidation, one of the advantages of this method is that signals are generated above a low background, allowing the detection of minor structural transitions.

Although $KMnO_4$ was first developed as a structural probe for DNA in conjunction with the primer extension protocol (15), a simpler and more reliable version of this technique (described below) combines the ability of $KMnO_4$ to preferentially oxidize the pyrimidine rings with standard cleavage chemistry at modified nucleotides (17) (*Protocol 15*). Regions of altered structure will be visible as dark bands following gel separation and autoradiography of the reaction products. These regions are often seen at similar positions on the top and bottom strands.

While the induction of $KMnO_4$ hyperreactivity is evidence of a protein-mediated structural distortion, it does not directly indicate the type of distortion generated. This must be determined by other means. For example, sensitivity of the same region to nuclease P1 attack (*Protocol 7*) is strong evidence for the presence of single-stranded DNA. Similarly, one may find that a region of $KMnO_4$ hypersensitivity is not reactive to nuclease S1 after dimethyl sulfate modification (*Protocol 8*), suggesting the absence of DNA melting. Coupled with the finding that the sites of $KMnO_4$ modification are on the same face of the DNA helix (1), this data would be indicative of the replication protein severely bending the DNA substrate.

Because an oxidation reaction is employed, reaction components can have severe effects on the efficiency of the reaction. For example, high levels of dithiothreitol (often used in combination with the poorly hydrolysed analogue of ATP, ATP-γ-S) will cause inactivation of the reagent. Destruction of the reagent can be observed visually by a change in reagent colour from dark purple to a brownish/orange.

Protocol 5. $KMnO_4$ oxidation

Equipment and reagents

- Microcentrifuge
- ^{32}P-end-labelled DNA substrate: ~0.01–0.10 pmol, prepared either using a PCR reaction (*Protocol 1*) or by end-labelling a suitable restriction fragment
- 2-mercaptoethanol (Sigma)
- Freshly prepared 60 mM $KMnO_4$ (J. T. Baker)[a]
- 3 M sodium acetate pH 5.2
- 100% ethanol

Method

1. Incubate the ^{32}P-labelled DNA in the presence and absence of the replication protein(s) with appropriate buffers, at 37 °C in a reaction volume of 30 μl.
2. Add 3 μl 60 mM $KMnO_4$ to give a final concentration of 6 mM. Incubate for 4 min at 37 °C.
3. Quench the reaction by the addition of 2.5 μl 2-mercaptoethanol to give a final concentration of ~ 1 M.
4. Ethanol precipitate the DNA (see *Protocol 2*, steps 5 and 6).
5. Proceed to *Protocol 15* (piperidine cleavage of DNA substrates).

[a] The reagent can be prepared by mixing 47 mg $KMnO_4$ with 5 ml distilled water. The reagent takes approx. 1 h to dissolve.

7.2 Copper–phenanthroline cleavage

A group of probes very sensitive to subtle DNA conformational changes are the cuprous complexes of 1,10-phenanthroline and the related 5-phenyl-1,10-phenanthroline (16). As these probes nick both dsDNA and ssDNA substrates, the reaction products can be directly analysed by denaturing gel electrophoresis without need for an additional cleavage reaction. The two phenanthroline probes have somewhat different sensitivities with respect to the induced structural transition. It is therefore useful to compare these two probes for the specific combination of replication protein, DNA, and reaction conditions tested. In addition to their ability to probe DNA distortion, the copper–phenanthroline probes can also be used to characterize close protein contacts with the minor groove. However, as compression of the minor groove can inhibit the interaction of the probe with DNA, these contacts should be verified by other methodology.

In the protocol below, it is important to test various ratios of the phenanthroline and copper solutions as well as the amount of the copper–phenanthroline solution to obtain the best cleavage pattern.

Protocol 6. Copper–phenanthroline probing

Equipment and reagents

- Microcentrifuge
- ^{32}P-end-labelled DNA substrate: ~ 0.01–0.10 pmol, prepared either using a PCR reaction (*Protocol 1*) or by end-labelling a suitable restriction fragment
- 40 mM 1,10-phenanthroline (monohydrate, Aldrich) or 5-phenyl-1,10-phenanthroline (GFS Chemicals, Columbus, OH) in 100% ethanol
- 9 mM $CuSO_4$ (Aldrich) in water
- 58 mM 3-mercaptopropionic acid (Aldrich) in water
- 28 mM 2,9-dimethyl-phenanthroline (neucuproine hydrate, Aldrich) in 100% ethanol
- Phenol:chloroform solution (1:1, v/v)
- 3 M sodium acetate pH 5.2
- 100% ethanol

Method

1. Incubate the ^{32}P-labelled DNA in the presence and absence of the replication protein(s) with appropriate buffers, at 37 °C in a reaction volume of 25 μl.
2. Just prior to use, combine 5 μl 40 mM 1,10-phenanthroline or 5-phenyl-1,10-phenanthroline with 5 μl 9 mM $CuSO_4$. The reagents should be vigorously mixed until a blue colour develops.
3. To the protein:DNA reaction mixture, add 2.5 μl of the copper–phenanthroline solution and 2.5 μl 58 mM 3-mercaptopropionic acid.
4. Incubate the reaction for 4 min at 37 °C.
5. Quench the reaction by the addition of 3 μl 28 mM 2,9-dimethyl-phenanthroline.
6. Add an equal volume of phenol:chloroform (1:1, v/v) and vortex for 10 sec. Remove the aqueous phase.
7. Ethanol precipitate the DNA (see *Protocol 2*, steps 5 and 6).
8. Proceed to Section 9.2 (gel electrophoresis of cleaved DNA products).

7.3 Nuclease P1

A simple method for the detection of melted DNA in protein:DNA complexes is digestion with nuclease P1, an enzyme 200-fold more active on ssDNA compared to dsDNA (26). Although nuclease P1 has a pH optimum between 4.5–6, the enzyme can be productively used under most protein binding conditions (pH ~ 7). The nuclease requires Zn^{2+} for activity with 0.2–0.5 mM commonly used.

One drawback of nuclease P1 is that it may not reveal the presence of ssDNA if the bound replication protein blocks access to the melted region. Therefore, while the detection of nuclease P1 hyperreactive regions is strong evidence for the presence of ssDNA, the lack of hypersensitivity should be re-examined using dimethyl sulfate, as described in *Protocol 8*.

Protocol 7. Nuclease P1 digestion

Equipment and reagents

- Microcentrifuge
- ^{32}P-end-labelled DNA substrate: ~ 0.01–0.10 pmol, prepared either using a PCR reaction (*Protocol 1*) or by end-labelling a suitable restriction fragment
- Nuclease P1 (United States Biochemical)
- Enzyme stop buffer: 10% SDS, 0.5 M EDTA
- Phenol:chloroform solution (1:1, v/v)
- 3 M sodium acetate pH 5.2
- 100% ethanol

Method

1. Incubate the ^{32}P-end-labelled DNA in the presence and absence of the replication protein(s) with appropriate buffers, in a reaction volume of 20 μl.
2. Add nuclease P1 and incubate the reaction mixture at 37 °C for 1 min.
3. Quench digestion by the addition of 2.5 μl of enzyme stop buffer and vortex for 10 sec at room temperature.
4. Add an equal volume of phenol:chloroform (1:1, v/v) and vortex for 10 sec. Remove the aqueous phase.
5. Ethanol precipitate the DNA (see *Protocol 2*, steps 5 and 6).
6. Proceed to Section 9.2 (gel electrophoresis of cleaved DNA products).

7.4 Detection of single-stranded DNA using dimethyl sulfate

Melting of duplex DNA by a replication protein exposes hydrogen bonding sites (i.e. the N1 of adenine and the N3 of cytosine) on nucleotide bases to dimethyl sulfate modification. Two different protocols have been developed that reveal single-stranded DNA regions by virtue of the methylation of these sites.

The first method (*Protocol 8*) relies on the inability of DNA methylated at hydrogen binding sites to completely renature. That is, a single-stranded DNA generated by the binding of a DNA replication protein will, upon methylation by dimethyl sulfate, remain partially melted upon removal of the protein. The deproteinized DNA will therefore become sensitive to cleavage by the ssDNA-specific nuclease S1 (27).

Alternatively, we have found that DNA methylated at internal hydrogen bonding positions on cytosines can be chemically cleaved using the standard piperidine cleavage reaction (25) (*Protocol 15*). Although this reaction has been reported by Kirkegaard *et al.* (28) to require treatment of the methylated DNA with hydrazine prior to piperidine cleavage, this step was not necessary in our hands. Therefore, footprints from the methylation protection reaction (*Protocol 3*) should be examined carefully for cleavage at cytosine residues as evidence of melted DNA.

Protocol 8. Detection of ssDNA by methylation and nuclease S1 cleavage

Equipment and reagents

- Microcentrifuge
- ^{32}P-end-labelled DNA: ~ 0.01–0.10 pmol, prepared either by PCR (*Protocol 1*) or by end-labelling a suitable restriction fragment
- 200 mM dimethyl sulfate solution: prepare by diluting 4.7 μl dimethyl sulfate (Sigma) to 250 μl with distilled water. *Dimethyl sulfate is a known carcinogen. Use gloves and, if possible, employ a fume-hood when working with this reagent.*
- Dimethyl sulfate stop buffer: 3 M ammonium acetate, 1 M 2-mercaptoethanol, 20 mM EDTA, 100 μg/ml tRNA
- Nuclease S1 buffer: 30 mM sodium acetate pH 4.5, 50 mM NaCl, 1 mM $ZnCl_2$
- Nuclease S1 (Boehringer Mannheim)
- 20% SDS
- Phenol:chloroform solution (1:1, v/v)
- 3 M sodium acetate pH 5.2
- 100% ethanol

Method

1. Incubate the ^{32}P-end-labelled DNA in the presence and absence of the replication protein(s) with appropriate buffers, at 37 °C in a reaction volume of 20 μl.
2. Add 3 μl 200 mM dimethyl sulfate, and incubate an additional 3 min.
3. Quench the reaction by the addition of 2 vol. dimethyl sulfate stop buffer. Precipitate the DNA by the addition of 2 vol. (140 μl) ethanol. Chill on ice for 10 min. Microcentrifuge the sample for 10 min to pellet the DNA. Discard the supernatant.
4. Add 70 μl ethanol to the pellet. Microcentrifuge the sample for 5 sec. Discard the supernatant and air dry the DNA pellet.
5. Resuspend the pellet in 50 μl nuclease S1 buffer.
6. Add 1–5 U nuclease S1 (Boehringer Mannheim) and incubate 1 min at 37 °C.
7. Quench the reaction by the addition of 2.5 μl 20% SDS.
8. Add an equal volume of phenol:chloroform (1:1, v/v) and vortex for 10 sec. Remove the aqueous phase.

9. Ethanol precipitate the DNA (see *Protocol 2*, steps 5 and 6).
10. Proceed to Section 9.2 (gel electrophoresis of cleaved DNA products).

8. Interference techniques

8.1 Overview

The interference assay can provide detailed information on DNA contacts required for complex formation with the replication protein. The basic assay consists of incubating the protein with a ^{32}P-labelled DNA substrate that was previously treated with a DNA modification reagent. Following the binding reaction, the protein:DNA complex is separated from the free (unbound) DNA. These two DNA pools and an aliquot of the initial DNA substrate are cleaved at the modified sites (*Figure 1B*). The cleaved fragments are separated by denaturing gel electrophoresis and visualized by autoradiography. Bands corresponding to modifications that interfere with complex formation will be under-represented in the bound DNA lane and over-represented in the free DNA lane, as compared to the starting substrate. The nucleotides corresponding to these sites can be strongly implicated as playing critical roles for protein:DNA complex formation. On occasion, it is also possible to detect the enhancement of bands in the bound species, indicative of the modified residues leading to stimulation of complex formation.

The separation of protein bound DNA from protein-free DNA is generally achieved by a gel retardation assay (see Section 8.3). If the replication protein can form multiple complexes with the DNA, each complex can be isolated and the important contacts for each individually determined. Nitrocellulose filters have also been successfully used to separate the free and bound pools (29). However, this latter method does not allow separation of individual complexes and should be considered a secondary alternative.

In examining the results of an interference assay, it is important to note that while a protein may contact many sites on the DNA, only those sites where modification interferes with complex formation will be revealed. If the protein binding proceeds through multiple steps, it is also possible that the identified phosphates or bases are not required within the final complex per se, but rather are necessary to stabilize an intermediate in the pathway of complex formation.

This latter point can be used to advantage for the examination of critical contacts on DNA used by the replication protein in dynamic reactions. For example, we have previously used interference analysis to determine sites on a model DNA fork used by a DNA helicase, the SV40 T-antigen, during DNA unwinding (5), and to examine the interaction of T-antigen and *E. coli* SSB with the SV40 origin during origin denaturation (1). Such reactions are obviously difficult in practice to examine using direct probing assays. In principle, other types of reactions can be examined and this avenue should be explored if possible.

8.2 DNA modification for interference assays

We provide protocols below for four types of interference reactions. Protein contacts with specific guanine bases are examined by dimethyl sulfate (23) (*Protocol 9*) or diethyl sulfate modification (11) (*Protocol 10*) which lead primarily to the methylation and ethylation, respectively, of guanines. Likewise, the effect of missing bases can be examined by depurination with formic acid (18) (*Protocol 11*) or depyrimidation with hydrazine (18) (*Protocol 12*). In addition, contacts made with the sugar phosphate backbone can be examined by treatment with *N*-nitroso-*N*-ethylurea which ethylates phosphates (29) (*Protocol 13*). The reactions each use 5 pmol of DNA substrate which is generally sufficient for 20–100 interference reactions.

Protocol 9. Base methylation interference

Equipment and reagents

- Microcentrifuge
- ^{32}P-end-labelled DNA: 5 pmol, prepared either by PCR (*Protocol 1*) or by end-labelling a suitable restriction fragment
- 300 ng non-specific DNA
- 2-mercaptoethanol (Sigma)
- 3 M sodium acetate pH 5.2
- 200 mM dimethyl sulfate: prepare by diluting 4.7 μl dimethyl sulfate (Sigma) to 250 μl with distilled water. *Dimethyl sulfate is a known carcinogen. Use gloves and, if possible, employ a fume-hood when working with this reagent.*
- 100% ethanol

Method

1. Mix the ^{32}P-end-labelled DNA substrate and non-specific DNA in 30 μl TE.
2. Add 3 μl 200 mM dimethyl sulfate.
3. Incubate at 37 °C for 5 min.
4. Quench the reaction by the addition of 5 μl 2-mercaptoethanol.
5. Precipitate the DNA by the addition of 0.1 vol. 3 M sodium acetate and 3 vol. ethanol. Chill on ice for 10 min. Microcentrifuge the sample for 10 min to pellet the DNA. Discard the supernatant.
6. Resuspend DNA in 30 μl TE. Reprecipitate the DNA by the addition of 0.1 vol. 3 M sodium acetate and 3 vol. ethanol. Chill on ice for 10 min. Microcentrifuge the sample for 10–30 min to pellet the DNA. Discard the supernatant.
7. Add 70 μl ethanol to the pellet. Microcentrifuge the sample for 5 sec. Discard the supernatant and air dry the DNA pellet.
8. Resuspend the DNA in an appropriate volume of TE.

9. Incubate the replication protein(s) with the modified substrate DNA to allow complex formation.
10. Proceed with protein:DNA complex isolation (Section 8.3).

Protocol 10. Base ethylation interference

Equipment and reagents

- Microcentrifuge
- ^{32}P-end-labelled DNA: 5 pmol, prepared either by PCR (*Protocol 1*) or by end-labelling a suitable restriction fragment in 100 mM sodium cacodylate, 1 mM EDTA pH 8
- 300 ng non-specific DNA
- Diethyl sulfate (Aldrich). *Careful! Diethyl sulfate is highly toxic and a cancer suspect agent. Treat with extreme care.*
- 2-mercaptoethanol (Sigma)
- 3 M sodium acetate pH 5.2
- 100% ethanol

Method

1. Mix the ^{32}P-end-labelled DNA substrate and non-specific DNA in 100 μl 100 mM sodium cacodylate, 1 mM EDTA pH 8.
2. Add 4 μl diethyl sulfate, and incubate at 37 °C for 30 min.
3. Quench the reaction by the addition of 5 μl 2-mercaptoethanol.
4. Continue as described in *Protocol 9*, steps 5–10.

Protocol 11. Depurination

Equipment and reagents

- Microcentrifuge
- ^{32}P-end-labelled DNA: 5 pmol, prepared either by PCR (*Protocol 1*) or by end-labelling a suitable restriction fragment
- 300 ng non-specific DNA
- Formic acid (Sigma)
- 3 M sodium acetate pH 5.2
- 100% ethanol

Method

1. Mix the ^{32}P-end-labelled DNA (5 pmol) and 300 ng non-specific DNA in 30 μl TE.
2. Add 15 μl formic acid and incubate 5 min at room temperature.
3. Continue as described in *Protocol 9*, steps 5–10.

Protocol 12. Depyrimidation

Equipment and reagents

- Microcentrifuge
- ^{32}P-end-labelled DNA, 5 pmol, prepared either by PCR (*Protocol 1*) or by end-labelling a suitable restriction fragment
- 300 ng non-specific DNA
- Hydrazine (Sigma). *Careful! Hydrazine is a strong oxidizing agent and should be treated with extreme care.*
- 3 M sodium acetate pH 5.2
- 100% ethanol

Method

1. Mix the ^{32}P-end-labelled DNA and 300 ng non-specific DNA in 30 μl TE.
2. Add 30 μl hydrazine and incubate for 5 min at 37 °C.
3. Continue as described in *Protocol 9*, steps 5–10.

Protocol 13. Phosphate ethylation

Equipment and reagents

- Microcentrifuge
- ^{32}P-end-labelled DNA: 5 pmol, prepared either by PCR (*Protocol 1*) or by end-labelling a suitable restriction fragment in 50 mM sodium cacodylate
- 300 ng non-specific DNA
- Saturated solution of *N*-nitroso-*N*-ethylurea (Sigma) in ethanol
- 3 M sodium acetate pH 5.2
- 4 M ammonium acetate
- 100% ethanol

Method

1. Mix the ^{32}P-end-labelled DNA substrate and non-specific DNA in 100 μl 50 mM sodium cacodylate.
2. Add 100 μl of a saturated solution of *N*-nitroso-*N*-ethylurea in ethanol.
3. Incubate for 30 min at 50 °C.
4. Precipitate the DNA by the addition of 12 μl 4 M ammonium acetate and 200 μl ethanol. Microcentrifuge the sample for 10–30 min to pellet the DNA. Chill on ice for 10 min. Discard the supernatant.
5. Continue as described in *Protocol 9*, steps 6–10.

8.3 Separation of replication protein:DNA complexes by gel retardation

Standard protocols, such as a gel retardation assay, are used to separate the individual protein:DNA complexes. Because the choice of gel electrophoresis conditions is dependent on the type of complex formed by the replication protein with the DNA substrate, no protocol will be provided here. The reader is advised to consult reviews for detailed description of the methodology (30).

For a number of protein:DNA interactions that we have studied (hRPA: ssDNA) (31), (SV40 T-antigen and the viral origin) (1), it was found necessary to employ a cross-linking agent (glutaraldehyde) to stabilize the complex prior to native gel electrophoresis. After protein:DNA complex formation, cross-linking is performed by the addition of glutaraldehyde (Sigma) to a final concentration of 0.03–0.1%, and further incubation for 5–15 min. This reagent causes the formation of both intramolecular cross-links within the protein as well as protein:DNA cross-links. In our use of glutaraldehyde, we invariably find that a significant fraction of the DNA, isolated after gel separation, is not cross-linked with the protein. Thus, this DNA can be treated similarly to non-cross-linked samples after isolation.

Once the complexes have been separated by native gel electrophoresis, the complexes are visualized by autoradiography. To facilitate DNA isolation, the gels are not dried. Below, we describe a simple technique to isolate the protein bound and unbound DNA from the gel, using a traditional crush–soak method (*Protocol 14*). Use of luminescent autoradiogram markers on the gel permits alignment of the autoradiograph with the gel.

Protocol 14. Isolation of DNA from acrylamide gels by the crush–soak method

Equipment and reagents

- Plastic film (e.g. Saran Wrap)
- Luminescent autoradiograph markers (Stratagene)
- Light box
- Gel elution buffer: 0.5 M ammonium acetate, 0.1% SDS, 1 mM EDTA
- 50 μl 150 mM NaOH (for DNA ethylated on phosphates)
- Razor blade
- Sealed Pasteur pipette. To prepare, hold each end of the pipette and flame the narrow pipette tube with a Bunsen burner. When glass is pliable (yellow hot), pull the ends apart and roll the narrow pipette end over the flame to seal.

Method

1. After gel electrophoresis, separate the two gel plates. Wrap the gel-attached plate with plastic film. Tape the plastic film to the back of the gel plate, keeping the plastic film taut. Place two luminescent autoradiograph markers on opposite corners of the gel front.
2. Autoradiograph the gel.
3. Place the autoradiograph on the light box, and position the gel/plate on top (with the gel front up). Align the gel to the autoradiograph using the autoradiograph markers.
4. With a razor blade, excise the gel slices containing the protein bound

Protocol 14. *Continued*

complex(s) and the free DNA. Place each gel slice (without the plastic film) into individual microcentrifuge tubes.

5. Crush the gel slice with the sealed Pasteur pipette. Add ~ 300 μl gel elution buffer to the crushed gel and soak at room temperature for 2–3 h (or overnight).
6. Remove the supernatant. To determine the efficiency of DNA isolation, monitor the level of radioactivity in the supernatant and remaining gel with a Gieger counter. If the recovery of DNA is inefficient, add ~ 150 μl gel elution buffer and soak for 1–2 h. Remove the supernatant and combine with the original eluted material.
7. Add an equal volume of phenol:chloroform (1:1, v/v) and vortex for 10 sec. Remove the aqueous phase.
8. Precipitate the DNA by the addition of 2 vol. ethanol. Chill on ice for 10 min. Microcentrifuge the sample for 10–30 min to pellet the DNA. Discard the supernatant.
9. Add 70 μl ethanol to the pellet. Microcentrifuge the sample for 5 sec. Discard the supernatant and air dry the DNA pellet.
10. (a) For DNA modified by dimethyl sulfate, diethyl sulfate, formic acid, or hydrazine, the DNA is subsequently cleaved as described in *Protocol 15*.

    (b) For DNA ethylated at phosphate residues (*Protocol 13*), the DNA can be cleaved at the ethylated phosphates by resuspending the DNA pellet in 50 μl 150 mM NaOH and incubating at 90°C for 30 min. The DNA is ethanol precipitated (see *Protocol 2*, steps 5 and 6). The cleaved DNA is then prepared for denaturing gel electrophoresis (Section 9.2).

9. Common techniques used for DNA probing studies

9.1 Piperidine cleavage of DNA substrates

Modification of DNA bases generally results in a change in the resonance structure of the nucleotide base, which weakens the glycosidic bond between base and sugar. Treatment of the modified DNA with piperidine at high temperatures (*Protocol 15*) catalyses the β-elimination of both phosphates surrounding the modified nucleoside and destruction of the sugar (17). The remaining oligonucleotide products contain phosphates at their cleaved termini. That is, the 5′ oligonucleotide contains a 3′ phosphate, while the 3′ oligonucleotide product will have a 5′ phosphate.

Protocol 15. Piperidine cleavage

Equipment and reagents

- 1 M piperidine (Aldrich): prepare by mixing 100 μl of the free base with 900 μl distilled water. *Piperidine is highly toxic! Use a fume-hood when adding this reagent to samples.*
- Microcentrifuge tubes with strong seal or screw-cap[a]
- Centrifugal concentrator

Method

1. Resuspend the DNA pellet in 50 μl 1 M piperidine.
2. Heat the sample at 95 °C for 30 min.
3. Remove all traces of piperidine by evaporation using a centrifugal concentrator (e.g. a SpeedVac).
4. Resuspend the DNA pellet in 10 μl distilled water. Evaporate water using a centrifugal concentrator.
5. Repeat step 4 three times. Note that it is necessary to remove the piperidine completely because trace amounts can cause distortion of bands on the sequencing gel.
6. Proceed to Section 9.2 (gel electrophoresis of cleaved DNA products).

[a] Heat treatment of the piperidine-containing solution is performed in sealed microcentrifuge tubes. It is therefore essential that the tubes used are sufficiently sturdy to withstand the pressure generated during the course of the incubation. If problems arise, a reliable alternative is the use of screw-cap microcentrifuge tubes.

9.2 Denaturing gel electrophoresis of cleaved DNA products

The DNA cleavage products are separated by denaturing gel electrophoresis to allow visualization of the modification pattern. Choice of the acrylamide concentration is dependent on the length of the DNA substrate used. For DNA substrates that are 150–250 bp in length, a standard gel contains 6–8% acrylamide (29:1, acrylamide to *bis*acrylamide), 40% urea (w/v), with 89 mM Tris base, 89 mM boric acid, and 2 mM EDTA as running buffer.

For sequence-independent cleavage reactions (e.g. nuclease P1, *N*-nitroso-*N*-ethylurea), the sites of cleavage can be determined by use of the formic acid (purines; *Protocol 11*) and hydrazine (pyrimidines; *Protocol 12*) cleavage products as electrophoretic markers. For preparation of samples for denaturing gel electrophoresis (*Protocol 16*), it is assumed that the ^{32}P-labelled DNA samples have been ethanol precipitated and are in the form of dried pellets.

Protocol 16. Denaturing gel electrophoresis

Equipment and reagents

- Electrophoresis apparatus
- Denaturing dye mix: mix 100 mg urea, 130 μl deionized formamide, 11 μl 50 mM NaOH/1 mM EDTA, 5.5 μl 1% xylene cyanol, and 5.5 μl 1% bromphenol blue

Method

1. Resuspend the pellet in 4 μl dye mix. For DNA samples that are difficult to resuspend, a useful alternative is to first add 2 μl distilled water to the dried DNA pellet, followed by the addition of 4 μl dye mix.
2. Immediately prior to gel electrophoresis, denature samples by heating at 95 °C for 5 min. Note caution related to microcentrifuge tubes in Section 9.1.
3. Separate reaction products by electrophoresis through the appropriate denaturing polyacrylamide gel. Dry gel and expose to film to obtain autoradiograph.

10. Concluding remarks

The methods described in this chapter can provide a high resolution view of how a protein of interest binds a specific DNA substrate. The binding site of the protein, the DNA structural changes the protein makes upon binding, and the DNA contacts the protein utilizes can all be efficiently addressed by DNA probing techniques. On occasion, it is of value to supplement these methods with other approaches that can provide additional information concerning the protein:DNA complex. One method of high value is electron microscopy, particularly scanning transmission electron microscopy (STEM) (31, 32). STEM can provide detailed molecular images as well as an accurate determination of the molecular masses within a complex, the latter allowing an identification of the oligomeric state of the complex. The method requires that the protein: DNA complex be separated from the unbound protein and DNA. Combining DNA probing analysis with other methods of analysis can yield a satisfying view of how a protein interacts with DNA.

Acknowledgements

Work from the J. A. B. laboratory was supported by National Institutes of Health grants CA-62198 and AI-29963, and Kaplan Cancer Center Developmental Funding and Kaplan Cancer Center Support Core Grant (NCI P30CA-16087).

References

1. SenGupta, D.J. and Borowiec, J.A. (1994). *EMBO J.*, **13**, 982.
2. Koff, A., Schwedes, J.F., and Tegtmeyer, P. (1991). *J. Virol.*, **65**, 3284.
3. Diffley, J.F. and Cocker, J.H. (1992). *Nature*, **357**, 169.
4. Tsurimoto, T. and Stillman, B. (1991). *J. Biol. Chem.*, **266**, 1950.
5. SenGupta, D. and Borowiec, J.A. (1992). *Science*, **256**, 1656.
6. Sasse-Dwight, S. and Gralla, J.D. (1991). In *Methods in enzymology* (ed. R. T. Sauer), Vol. 208, p. 146. Academic Press, San Diego, CA.
7. Sambrook, J., Fritsch, E.F., and Maniatis, T. (ed.) (1989). *Molecular cloning: a laboratory manual* (2nd edn). Cold Spring Harbor Laboratory Press, Cold Spring Harbor, NY.
8. Gralla, J.D. (1985). *Proc. Natl. Acad. Sci. USA*, **82**, 3078.
9. Singer, B. and Fraenkel-Conrat, H. (1975). *Biochemistry*, **14**, 772.
10. Sun, L. and Singer, B. (1975). *Biochemistry*, **14**, 1795.
11. Johnston, B.H. and Rich, A. (1985). *Cell*, **42**, 713.
12. Hayatsu, H. and Ukita, T. (1967). *Biochem. Biophys. Res. Commun.*, **29**, 556.
13. Ide, H., Kow, Y.W., and Wallace, S.S. (1985). *Nucleic Acids Res.*, **13**, 8035.
14. Gillette, T.G., Lusky, M., and Borowiec, J.A. (1994). *Proc. Natl. Acad. Sci. USA*, **91**, 8846.
15. Borowiec, J.A., Zhang, L., Sasse-Dwight, S., and Gralla, J.D. (1987). *J. Mol. Biol.*, **196**, 101.
16. Sigman, D.S., Kuwabara, M.D., Chen, C.-H.B., and Bruice, T.W. (1991). In *Methods in enzymology* (ed. R. T. Sauer), Vol. 208, p. 414. Academic Press, San Diego, CA.
17. Maxam, A.M. and Gilbert, W. (1980). In *Methods in enzymology* (ed. L. Grossman and K. Moldave), Vol. 65, p. 499. Academic Press, NY.
18. Brunelle, A. and Schleif, R.F. (1987). *Proc. Natl. Acad. Sci. USA*, **84**, 6673.
19. Suck, D. and Oefner, D. (1986). *Nature*, **321**, 620.
20. Jost, J.-P. (1991). *BioMethods*, **5**, 35.
21. Kowalski, D. (1984). *Nucleic Acids Res.*, **12**, 7071.
22. Borowiec, J.A. and Hurwitz, J. (1988). *Proc. Natl. Acad. Sci. USA*, **85**, 64.
23. Wissmann, A. and Hillen, W. (1991). In *Methods in enzymology* (ed. R. T. Sauer), Vol. 208, p. 365. Academic Press, San Diego, CA.
24. Metzger, W. and Heumann, H. (1994). In *DNA–protein interactions: principles and protocols* (ed. G.G. Kneale), Vol. 30, p. 11. Humana Press Inc., Totowa, NJ.
25. Borowiec, J.A. and Hurwitz, J. (1988). *EMBO J.*, **7**, 3149.
26. Volbeda, A., Lahm, A., Sakiyama, F., and Suck, D. (1991). *EMBO J.*, **10**, 1607.
27. Siebenlist, U. (1979). *Nature*, **279**, 651.
28. Kirkegaard, K., Buc, H., Spassky, A., and Wang, J.C. (1983). *Proc. Natl. Acad. Sci. USA*, **80**, 2544.
29. Siebenlist, U. and Gilbert, W. (1980). *Proc. Natl. Acad. Sci. USA*, **77**, 122.
30. Carey, J. (1991). In *Methods in enzymology* (ed. R. T. Sauer), Vol. 208, p. 103. Academic Press, San Diego, CA.
31. Blackwell, L.J., Borowiec, J.A., and Mastrangelo, I.A. (1996). *Mol. Cell. Biol.*, **16**, 4798.
32. Mastrangelo, I.A., Hough, P.V.C., Wall, J.S., Dodson, M., Dean, F.B., and Hurwitz, J. (1989). *Nature*, **338**, 658.

A1

List of suppliers

Amersham

Amersham, Little Chalfont, Buckinghamshire HP7 9NA, UK.

Amersham International plc., Lincoln Place, Green End, Aylesbury, Buckinghamshire HP20 2TP, UK.

Amersham Corporation, 2636 South Clearbrook Drive, Arlington Heights, IL 60005, USA.

Anderman and Co. Ltd., 145 London Road, Kingston-Upon-Thames, Surrey KT17 7NH, UK.

Beckman Instruments

Beckman Instruments UK Ltd., Progress Road, Sands Industrial Estate, High Wycombe, Buckinghamshire HP12 4JL, UK.

Beckman Instruments Inc., PO Box 3100, 2500 Harbor Boulevard, Fullerton, CA 92634, USA.

Becton Dickinson

Becton Dickinson and Co., Between Towns Road, Cowley, Oxford OX4 3LY, UK.

Becton Dickinson and Co., 2 Bridgewater Lane, Lincoln Park, NJ 07035, USA.

Becton Dickinson, 2350 Qume Drive San Jose, CA 95131–1807, USA.

Bio

Bio 101 Inc., c/o Stratech Scientific Ltd., 61–63 Dudley Street, Luton, Bedfordshire LU2 0HP, UK.

Bio 101 Inc., PO Box 2284, La Jolla, CA 92038–2284, USA.

Bio-Rad Laboratories

Bio-Rad Laboratories Ltd., Bio-Rad House, Maylands Avenue, Hemel Hempstead HP2 7TD, UK.

Bio-Rad Laboratories, Division Headquarters, 3300 Regatta Boulevard, Richmond, CA 94804, USA.

BioSpec Products

Stratech Scientific Ltd., 61–63 Dudley Street, Luton, Bedfordshire LU2 OHP, UK.

BioSpec Products Inc., PO Box 788, Bartlesville, OK 74005–0788, USA.

Boehringer Mannheim

Boehringer Mannheim UK (Diagnostics and Biochemicals) Ltd., Bell Lane, Lewes, East Sussex BN17 1LG, UK.

Boehringer Mannheim Corporation, Biochemical Products, 9115 Hague Road, PO Box 504, Indianopolis, IN 46250–0414, USA.

Boehringer Mannheim Biochemica, GmbH, Sandhofer Str. 116, Postfach 310120, D-6800 Ma 31, Germany.

British Drug Houses (BDH) Ltd., Poole, Dorset, UK.

Carl Zeiss, Inc., One Zeiss Drive,Thornwood, NY 10594, USA.

Difco Laboratories

Difco Laboratories Ltd., PO Box 14B, Central Avenue, West Molesey, Surrey KT8 2SE, UK.

Difco Laboratories, PO Box 331058, Detroit, MI 48232–7058, USA.

Du Pont

Dupont (UK) Ltd. (Industrial Products Division), Wedgwood Way, Stevenage, Hertfordshire SG1 4Q, UK.

Du Pont Co. (Biotechnology Systems Division), PO Box 80024, Wilmington, DE 19880–002, USA.

European Collection of Animal Cell Culture, Division of Biologics, PHLS Centre for Applied Microbiology and Research, Porton Down, Salisbury, Wiltshire SP4 0JG, UK.

Falcon (Falcon is a registered trademark of Becton Dickinson and Co.)

Fisher Scientific Co., 711 Forbest Avenue, Pittsburgh, PA 15219–4785, USA.

Flow Laboratories, Woodcock Hill, Harefield Road, Rickmansworth, Hertfordshire WD3 1PQ, UK.

Fluka

Fluka-Chemie AG, CH-9470, Buchs, Switzerland.

Fluka Chemicals Ltd., The Old Brickyard, New Road, Gillingham, Dorset SP8 4JL, UK.

Gibco BRL

Gibco BRL (Life Technologies Ltd.), Trident House, Renfrew Road, Paisley PA3 4EF, UK.

Gibco BRL (Life Technologies Inc.), 3175 Staler Road, Grand Island, NY 14072–0068, USA.

Arnold R. Horwell, 73 Maygrove Road, West Hampstead, London NW6 2BP, UK.

Hybaid

Hybaid Ltd., 111–113 Waldegrave Road, Teddington, Middlesex TW11 8LL, UK.

Hybaid, National Labnet Corporation, PO Box 841, Woodbridge, NJ 07095, USA.

HyClone Laboratories, 1725 South HyClone Road, Logan, UT 84321, USA.

International Biotechnologies Inc., 25 Science Park, New Haven, Connecticut 06535, USA.

Invitrogen Corporation

Invitrogen Corporation, 3985 B Sorrenton Valley Building, San Diego, CA 92121, USA.

Invitrogen Corporation, c/o British Biotechnology Products Ltd., 4–10 The Quadrant, Barton Lane, Abingdon, Oxon OX14 3YS, UK.

Jackson ImmunoResearch Labs Inc., PO Box 9, West Grove, PA 19390–0009, USA.

Kodak: Eastman Fine Chemicals, 343 State Street, Rochester, NY, USA.

Life Technologies Inc., 8451 Helgerman Court, Gaithersburg, MN 20877, USA.

Merck

Merck Industries Inc., 5 Skyline Drive, Nawthorne, NY 10532, USA.

Merck, Frankfurter Strasse, 250, Postfach 4119, D-64293, Germany.

Millipore

Millipore (UK) Ltd., The Boulevard, Blackmoor Lane, Watford, Hertfordshire WD1 8YW, UK.

Millipore Corp./Biosearch, PO Box 255, 80 Ashby Road, Bedford, MA 01730, USA.

New England Biolabs (NBL)

New England Biolabs (NBL), 32 Tozer Road, Beverley, MA 01915–5510, USA.

New England Biolabs (NBL), c/o CP Labs Ltd., PO Box 22, Bishops Stortford, Hertfordshire CM23 3DH, UK.

New England Nuclear, DuPont NEN Research Products, 549 Albany Street, Boston, MA 02118, USA.

Nikon Corporation, Fuji Building, 2–3 Marunouchi 3-chome, Chiyoda-ku, Tokyo, Japan.

Oncor Inc., PO Box 870, Gaithersberg, MD 20877, USA.

Pell Freez Biologicals, Rogers, Arkansas, USA.

Perkin-Elmer

Perkin-Elmer Ltd., Maxwell Road, Beaconsfield, Buckinghamshire HP9 1QA, UK.

Perkin Elmer Ltd., Post Office Lane, Beaconsfield, Buckinghamshire HP9 1QA, UK.

Perkin Elmer-Cetus (The Perkin-Elmer Corporation), 761 Main Avenue, Norwalk, CT 0689, USA.

Pharmacia Biotech Europe, Procordia EuroCentre, Rue de la Fuse-e 62, B-1130 Brussels, Belgium.

Pharmacia

Pharmacia, 800 Centennial Avenue, Piscataway, NJ 08855, USA.

Pharmacia, Midsummer Boulevard, Central Milton Keynes, Buckinghamshire MK9 3HP, UK.

Pharmacia Biosystems

Pharmacia Biosystems Ltd. (Biotechnology Division), Davy Avenue, Knowlhill, Milton Keynes MK5 8PH, UK.

Pharmacia LKB Biotechnology AB, Björngatan 30, S-75182 Uppsala, Sweden.

Pierce Chemical Co., PO Box 117, Rockford, IL 61105, USA.

Promega

Promega Ltd., Delta House, Enterprise Road, Chilworth Research Centre, Southampton, UK.

Promega Corporation, 2800 Woods Hollow Road, Madison, WI 53711–5399, USA.

Qiagen

Qiagen Inc., c/o Hybaid, 111–113 Waldegrave Road, Teddington, Middlesex TW11 8LL, UK.

Qiagen Inc., 9259 Eton Avenue, Chatsworth, CA 91311, USA.

Sarstedt Canada, 6373 Des Grandes Prairie, St. Leonard, PQ H1P 1A5, Canada.

Schleicher and Schuell

Schleicher and Schuell Inc., PO Box 2012, Keene, NH 3431, USA.

Schleicher and Schuell Inc., D-3354 Dassel, Germany.

Schleicher and Schuell Inc., c/o Andermann and Co. Ltd., UK.

Shandon Scientific Ltd., Chadwick Road, Astmoor, Runcorn, Cheshire WA7 1PR, UK.

Sigma Chemical Company

Sigma Chemical Company (UK), Fancy Road, Poole, Dorset BH17 7NH, UK.

Sigma Chemical Company, 3050 Spruce Street, PO Box 14508, St. Louis, MO 63178–9916, USA.

Sorvall DuPont Company, Biotechnology Division, PO Box 80022, Wilmington, DE 19880–0022, USA.

Stratagene

Stratagene Ltd., Unit 140, Cambridge Innovation Centre, Milton Road, Cambridge CB4 4FG,UK.

Stratagene Inc., 11011 North Torrey Pines Road, La Jolla, CA 92037, USA.

United States Biochemical, PO Box 22400, Cleveland, OH 44122, USA.

Vector Labs

Vector Labs Ltd., 16 Wulfric Square, Bretton, Peterborough PE3 8RF, UK.

Vector Laboratories Inc., 30 Ingold Road, Burlingame, CA 94010, USA.

Vector Laboratories, 3390 South Service Road, Burlington, ON L7N 3J5, Canada.

Wallac Inc., 9238 Gaither Road, Gaithersberg, MD 20877, USA.

Wellcome Reagents, Langley Court, Beckenham, Kent BR3 3BS, UK.

Whatman

Whatman Inc., 9 Bridewell Place, Clifton NJ 07014, USA.

Whatman LabSales, St Leonards Road, Maidstone, Kent ME16 0LS, UK.

YGSC (Yeast Genetic Stock Centre), Department of Molecular and Cell Biology, 229 Stanley Hall No. 3206, University of California, Berkeley, CA 94720–3206, USA.

Index

Index

Printed in the United States
858000001B